MW00581603

# Marketing Data Science

## Modeling Techniques in Predictive Analytics with R and Python

THOMAS W. MILLER

Publisher: Paul Boger
Editor-in-Chief: Amy Neidlinger
Executive Editor: Jeanne Glasser Levine
Operations Specialist: Jodi Kemper
Cover Designer: Alan Clements
Managing Editor: Kristy Hart
Manufacturing Buyer: Dan Uhrig


Published by Pearson Education, Inc.
Old Tappan New Jersey 07675

For information about buying this title in bulk quantities, or for special sales opportunities (which may include electronic versions; custom cover designs; and content particular to your business, training goals, marketing focus, or branding interests), please contact our corporate sales department at corpsales@pearsoned.com or (800) 382-3419.

For government sales inquiries, please contact governmentsales@pearsoned.com.

For questions about sales outside the U.S., please contact international@pearsoned.com.



Printed in the United States of America

14 2021

ISBN-10: 0-13-388655-7
ISBN-13: 978-0-13-388655-9

Pearson Education LTD.
Pearson Education Australia PTY, Limited.
Pearson Education Singapore, Pte. Ltd.
Pearson Education Asia, Ltd.
Pearson Education Canada, Ltd.
Pearson Educación de Mexico, S.A. de C.V.
Pearson Education—Japan
Pearson Education Malaysia, Pte. Ltd.
Library of Congress Control Number: 2015937911

# Contents

# Preface

"Everybody loses the thing that made them. It's even how it's supposed to be in nature. The brave men stay and watch it happen, they don't run."

—QUVENZHANÉ WALLIS AS HUSHPUPPY
IN *Beasts of the Southern Wild* (2012)

Writers of marketing textbooks of the past would promote "the marketing concept," saying that marketing is not sales or selling. Rather, marketing is a matter of understanding and meeting consumer needs. They would distinguish between "marketing research," a business discipline, and "market research," as in economics. And marketing research would sometimes be described as "marketing science" or "marketing engineering."

Ignore the academic pride and posturing of the past. Forget the linguistic arguments. Marketing and sales, marketing and markets, research and science—they are one. In a world transformed by information technology and instant communication, data rule the day.

Data science is the new statistics, a blending of modeling techniques, information technology, and business savvy. Data science is also the new look of marketing research.

In introducing marketing data science, we choose to present research about consumers, markets, and marketing as it currently exists. Research today means gathering and analyzing data from web surfing, crawling, scraping, online surveys, focus groups, blogs and social media. Research today means finding answers as quickly and cheaply as possible.

Finding answers efficiently does not mean we must abandon notions of scientific research, sampling, or probabilistic inference. We take care while designing marketing measures, fitting models, describing research findings, and recommending actions to management.

There are times, of course, when we must engage in primary research. We construct survey instruments and interview guides. We collect data from consumer samples and focus groups. This is traditional marketing research—custom research, tailored to the needs of each individual client or research question.

The best way to learn about marketing data science is to work through examples. This book provides a ready resource and reference guide for modeling techniques. We show programmers how to build on a foundation of code that works to solve real business problems.

The truth about what we do is in the programs we write. The code is there for everyone to see and for some to debug. To promote student learning, programs include step-by-step comments and suggestions for taking analyses further. Data sets and computer programs are available from the website for the *Modeling Techniques* series at `http://www.ftpress.com/miller/`.

When working on problems in marketing data science, some things are more easily accomplished with Python, others with R. And there are times when it is good to offer solutions in both languages, checking one against the other. Together, Python and R make a strong combination for doing data science.

Most of the data in this book come from public domain sources. Supporting data for many cases come from the University of California–Irvine Machine Learning Repository and the Stanford Large Network Dataset Collection. I am most thankful to those who provide access to rich data sets for research.

I have learned from my consulting work with Research Publishers LLC and its ToutBay division, which promotes what can be called "data science as a service." Academic research and models can take us only so far. Eventually, to make a difference, we need to implement our ideas and models, sharing them with one another.

Many have influenced my intellectual development over the years. There were those good thinkers and good people, teachers and mentors for whom

I will be forever grateful. Sadly, no longer with us are Gerald Hahn Hinkle in philosophy and Allan Lake Rice in languages at Ursinus College, and Herbert Feigl in philosophy at the University of Minnesota. I am also most thankful to David J. Weiss in psychometrics at the University of Minnesota and Kelly Eakin in economics, formerly at the University of Oregon.

Thanks to Michael L. Rothschild, Neal M. Ford, Peter R. Dickson, and Janet Christopher who provided invaluable support during our years together at the University of Wisconsin–Madison. While serving as director of the A. C. Nielsen Center for Marketing Research, I met the captains of the marketing research industry, including Arthur C. Nielsen, Jr. himself. I met and interviewed Jack Honomichl, the industry's historian, and I met with Gil Churchill, first author of what has long been regarded as a key textbook in marketing research. I learned about traditional marketing research at the A. C. Nielsen Center for Marketing Research, and I am most grateful for the experience of working with its students and executive advisory board members. Thanks go as well to Jeff Walkowski and Neli Esipova who worked with me in exploring online surveys and focus groups when those methods were just starting to be used in marketing research.

After my tenure with the University of Wisconsin–Madison, I built a consulting practice. My company, Research Publishers LLC, was co-located with the former Chamberlain Research Consultants. Sharon Chamberlain gave me a home base and place to practice the craft of marketing research. It was there that initial concepts for this book emerged:

> What could be more important to a business than understanding its customers, competitors, and markets? Managers need a coherent view of things. With consumer research, product management, competitive intelligence, customer support, and management information systems housed within separate departments, managers struggle to find the information they need. Integration of research and information functions makes more sense (Miller 2008).

My current home is the Northwestern University School of Professional Studies. I support courses in three graduate programs: Master of Science in Predictive Analytics, Advanced Certificate in Data Science, and Master of Arts in Sports Administration. Courses in marketing analytics, database systems and data preparation, web and network data science, and data visualization provide inspiration for this book.

I expect Northwestern's graduate programs to prosper as they forge into new areas, including analytics entrepreneurship and sports analytics. Thanks to colleagues and staff who administer these exceptional graduate programs, and thanks to the many students and fellow faculty from whom I have learned.

Amy Hendrickson of TEXnology Inc. applied her craft, making words, tables, and figures look beautiful in print—another victory for open source. Lorena Martin reviewed the book and provided much needed feedback. Roy Sanford provided advice on statistical explanations. Candice Bradley served dual roles as a reviewer and copyeditor for all books in the *Modeling Techniques* series. I am grateful for their guidance and encouragement.

Thanks go to my editor, Jeanne Glasser Levine, and publisher, Pearson/FT Press, for making this and other books in the *Modeling Techniques* series possible. Any writing issues, errors, or items of unfinished business, of course, are my responsibility alone.

My good friend Brittney and her daughter Janiya keep me company when time permits. And my son Daniel is there for me in good times and bad, a friend for life. My greatest debt is to them because they believe in me.

Thomas W. Miller
Glendale, California
April 2015

# Figures

# Tables

# Exhibits

# 1 Understanding Markets

"What makes the elephant guard his tusk in the misty mist, or the dusky dusk? What makes a muskrat guard his musk?"

—Bert Lahr as Cowardly Lion in *The Wizard of Oz* (1939)

While working on the first book in the *Modeling Techniques* series, I moved from Madison, Wisconsin to Los Angeles. I had a difficult decision to make about mobile communications. I had been a customer of U.S. Cellular for many years. I had one smartphone and two data modems (a 3G and a 4G) and was quite satisfied with U.S. Cellular services. In May of 2013, the company had no retail presence in Los Angeles and no 4G service in California. Being a data scientist in need of an example of preference and choice, I decided to assess my feelings about mobile phone services in the Los Angeles market.

The attributes in my demonstration study were the mobile provider or brand, startup and monthly costs, if the provider offered 4G services in the area, whether the provider had a retail location nearby, and whether the provider supported Apple, Samsung, or Nexus phones in addition to tablet computers. Product profiles, representing combinations of these attributes, were easily generated by computer. My consideration set included AT&T, T-Mobile, U.S. Cellular, and Verizon. I generated sixteen product profiles and presented them to myself in a random order. Product profiles, their attributes, and my ranks, are shown in table 1.1.

*Table 1.1. Preference Data for Mobile Communication Services*

| brand | startup | monthly | service | retail | apple | samsung | google | ranking |
|---|---|---|---|---|---|---|---|---|
| "AT&T" | "$100" | "$100" | "4G NO" | "Retail NO" | "Apple NO" | "Samsung NO" | "Nexus NO" | 11 |
| "Verizon" | "$300" | "$100" | "4G NO" | "Retail YES" | "Apple YES" | "Samsung YES" | "Nexus NO" | 12 |
| "US Cellular" | "$400" | "$200" | "4G NO" | "Retail NO" | "Apple NO" | "Samsung YES" | "Nexus NO" | 9 |
| "Verizon" | "$400" | "$400" | "4G YES" | "Retail YES" | "Apple NO" | "Samsung NO" | "Nexus NO" | 2 |
| "Verizon" | "$200" | "$300" | "4G NO" | "Retail NO" | "Apple NO" | "Samsung YES" | "Nexus YES" | 8 |
| "Verizon" | "$100" | "$200" | "4G YES" | "Retail NO" | "Apple YES" | "Samsung NO" | "Nexus YES" | 13 |
| "US Cellular" | "$300" | "$300" | "4G YES" | "Retail NO" | "Apple YES" | "Samsung NO" | "Nexus NO" | 7 |
| "AT&T" | "$400" | "$300" | "4G NO" | "Retail YES" | "Apple YES" | "Samsung NO" | "Nexus YES" | 4 |
| "AT&T" | "$200" | "$400" | "4G YES" | "Retail NO" | "Apple YES" | "Samsung YES" | "Nexus NO" | 5 |
| "T-Mobile" | "$400" | "$100" | "4G YES" | "Retail NO" | "Apple YES" | "Samsung YES" | "Nexus YES" | 16 |
| "US Cellular" | "$100" | "$400" | "4G NO" | "Retail YES" | "Apple YES" | "Samsung YES" | "Nexus YES" | 3 |
| "T-Mobile" | "$200" | "$200" | "4G NO" | "Retail YES" | "Apple YES" | "Samsung NO" | "Nexus NO" | 6 |
| "T-Mobile" | "$100" | "$300" | "4G YES" | "Retail YES" | "Apple NO" | "Samsung YES" | "Nexus NO" | 10 |
| "US Cellular" | "$200" | "$100" | "4G YES" | "Retail YES" | "Apple NO" | "Samsung NO" | "Nexus YES" | 15 |
| "T-Mobile" | "$300" | "$400" | "4G NO" | "Retail NO" | "Apple NO" | "Samsung NO" | "Nexus YES" | 1 |
| "AT&T" | "$300" | "$200" | "4G YES" | "Retail YES" | "Apple NO" | "Samsung YES" | "Nexus YES" | 14 |

A linear model fit to preference rankings is an example of *traditional conjoint analysis,* a modeling technique designed to show how product attributes affect purchasing decisions. Conjoint analysis is really *conjoint measurement.* Marketing analysts present product profiles to consumers. Product profiles are defined by their attributes. By ranking, rating, or choosing products, consumers reveal their preferences for products and the corresponding attributes that define products. The computed attribute importance values and part-worths associated with levels of attributes represent measurements that are obtained as a group or jointly—thus the name conjoint analysis. The task—ranking, rating, or choosing—can take many forms.

When doing conjoint analysis, we utilize *sum contrasts,* so that the sum of the fitted regression coefficients across the levels of each attribute is zero. The fitted regression coefficients represent conjoint measures of utility called *part-worths.* Part-worths reflect the strength of individual consumer preferences for each level of each attribute in the study. Positive part-worths add to a product's value in the mind of the consumer. Negative part-worths subtract from that value. When we sum across the part-worths of a product, we obtain a measure of the utility or benefit to the consumer.

To display the results of the conjoint analysis, we use a special type of dot plot called the *spine chart,* shown in figure 1.1. In the spine chart, part-worths can be displayed on a common, standardized scale across attributes. The vertical line in the center, the spine, is anchored at zero.

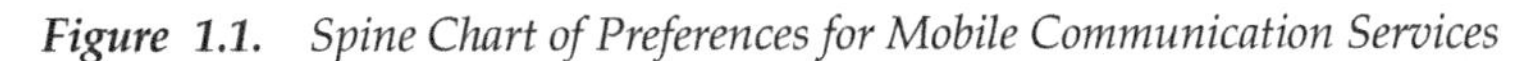

***Figure 1.1.*** *Spine Chart of Preferences for Mobile Communication Services*

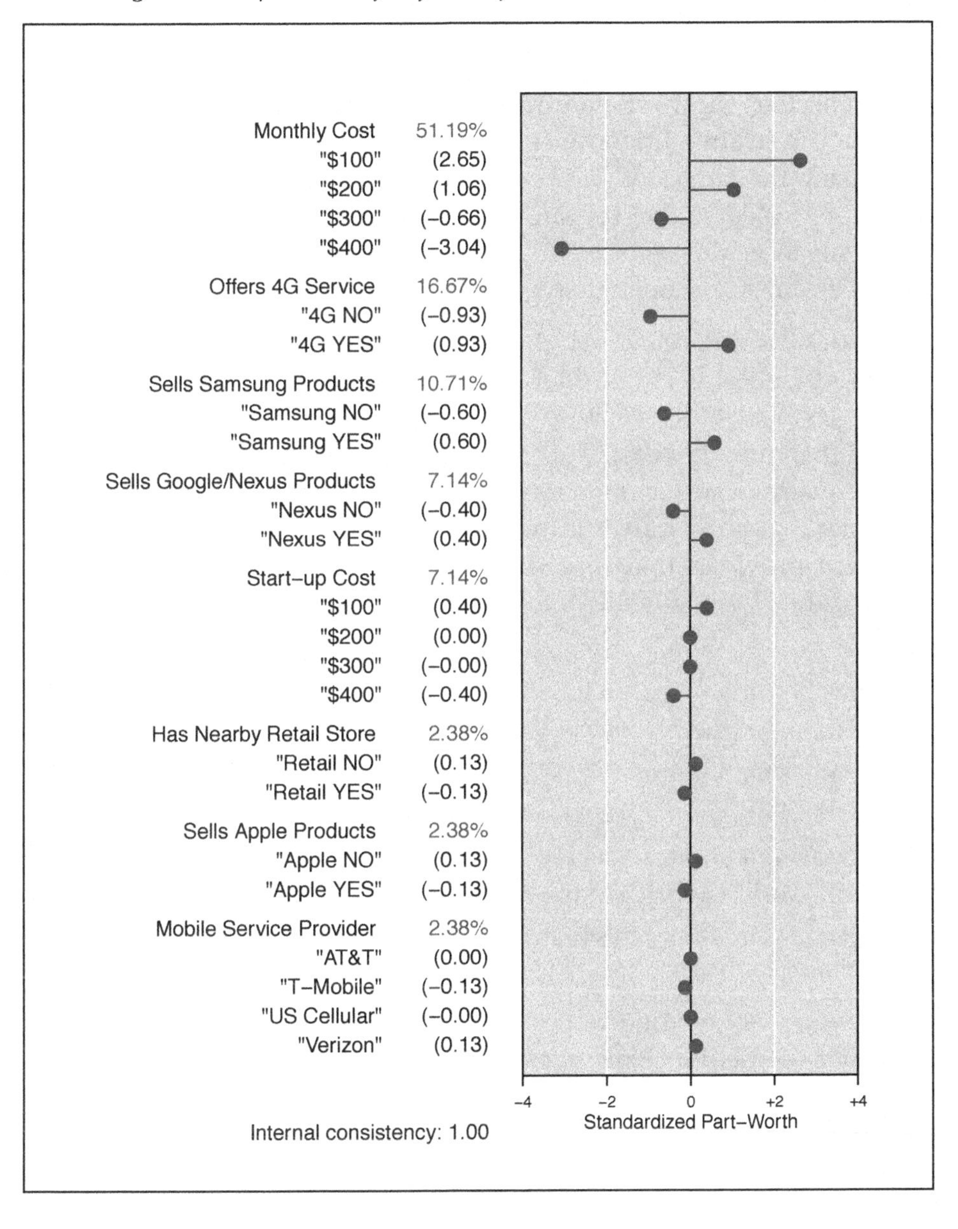

The part-worth of each level of each attribute is displayed as a dot with a connecting horizontal line, extending from the spine. Preferred product or service characteristics have positive part-worths and fall to the right of the spine. Less preferred product or service characteristics fall to the left of the spine.

The spine chart shows standardized part-worths and attribute importance values. The relative importance of attributes in a conjoint analysis is defined using the ranges of part-worths within attributes. These importance values are scaled so that the sum across all attributes is 100 percent. Conjoint analysis is a measurement technology. Part-worths and attribute importance values are conjoint measures.

What does the spine chart say about this consumer's preferences? It shows that monthly cost is of considerable importance. Next in order of importance is 4G availability. Start-up cost, being a one-time cost, is much less important than monthly cost. This consumer ranks the four service providers about equally. And having a nearby retail store is not an advantage. This consumer is probably an Android user because we see higher importance for service providers that offer Samsung phones and tablets first and Nexus second, while the availability of Apple phones and tablets is of little importance.

This simple study reveals a lot about the consumer—it measures consumer preferences. Furthermore, the linear model fit to conjoint rankings can be used to predict what the consumer is likely to do about mobile communications in the future.

Traditional conjoint analysis represents a modeling technique in predictive analytics. Working with groups of consumers, we fit a linear model to each individual's ratings or rankings, thus measuring the utility or part-worth of each level of each attribute, as well as the relative importance of attributes.

The measures we obtain from conjoint studies may be analyzed to identify consumer segments. Conjoint measures can be used to predict each individual's choices in the marketplace. Furthermore, using conjoint measures, we can perform marketplace simulations, exploring alternative product designs and pricing policies. Consumers reveal their preferences in responses to surveys and ultimately in choices they make in the marketplace.

*Figure 1.2. The Market: A Meeting Place for Buyers and Sellers*

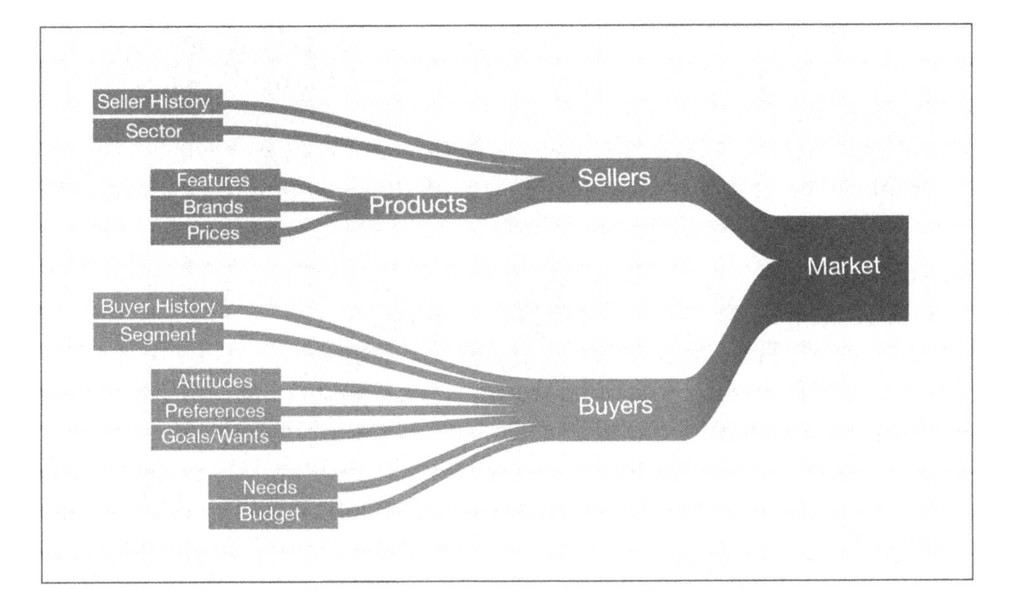


Marketing data science, a specialization of predictive analytics or data science, involves building models of seller and buyer preferences and using those models to make predictions about future marketplace behavior. Most of the examples in this book concern consumers, but the ways we conduct research—data preparation and organization, measurements, and models—are relevant to all markets, business-to-consumer and business-to-business markets alike.

Managers often ask about what drives buyer choice. They want to know what is important to choice or which factors determine choice. To the extent that buyer behavior is affected by product features, brand, and price, managers are able to influence buyer behavior, increasing demand, revenue, and profitability.

Product features, brands, and prices are part of the mobile phone choice problem in this chapter. But there are many other factors affecting buyer behavior—unmeasured factors and factors outside management control. Figure 1.2 provides a framework for understanding marketplace behavior—the choices of buyers and sellers in a market.

A market, as we know from economics, is the location where or channel through which buyers and sellers get together. Buyers represent the demand side, and sellers the supply side. To predict what will happen in a market—products to be sold and purchased, and the market-clearing prices of those products—we assume that sellers are profit-maximizers, and we study the past behavior and characteristics of buyers and sellers. We build models of market response. This is the job of marketing data science as we present it in this book.

Ask buyers what they want, and they may say, *the best of everything.* Ask them what they would like to spend, and they may say, *as little as possible.* There are limitations to assessing buyer willingness to pay and product preferences with direct-response rating scales, or what are sometimes called self-explicative scales. Simple rating scale items arranged as they often are, with separate questions about product attributes, brands, and prices, fail to capture tradeoffs that are fundamental to consumer choice. To learn more from buyer surveys, we provide a context for responding and then gather as much information as we can. This is what conjoint and choice studies do, and many of them do it quite well. In the appendix B (pages 312 to 337) we provide examples of consumer surveys of preference and choice.

Conjoint measurement, a critical tool of marketing data science, focuses on buyers or the demand side of markets. The method was originally developed by Luce and Tukey (1964). A comprehensive review of conjoint methods, including traditional conjoint analysis, choice-based conjoint, best-worst scaling, and menu-based choice, is provided by Bryan Orme (2013). Primary applications of conjoint analysis fall under the headings of new product design and pricing research, which we discuss later in this book.

Exhibits 1.1 and 1.2 show R and Python programs for analyzing ranking or rating data for consumer preferences. The programs perform traditional conjoint analysis. The spine chart is a customized data visualization for conjoint and choice studies. We show the R code for making spine charts in appendix D, exhibit D.1 starting on page 400. Using standard R graphics, we build this chart one point, line, and text string at a time. The precise placement of points, lines, and text is under our control.

**Exhibit 1.1.** *Measuring and Modeling Individual Preferences (R)*

```
# Traditional Conjoint Analysis (R)

# R preliminaries to get the user-defined function for spine chart:
# place the spine chart code file <R_utility_program_1.R>
# in your working directory and execute it by
#     source("R_utility_program_1.R")
# Or if you have the R binary file in your working directory, use
#     load(file="mtpa_spine_chart.Rdata")

# spine chart accommodates up to 45 part-worths on one page
# |part-worth| <= 40 can be plotted directly on the spine chart
# |part-worths| > 40 can be accommodated through standardization

print.digits <- 2  # set number of digits on print and spine chart

library(support.CEs)  # package for survey construction

# generate a balanced set of product profiles for survey
provider.survey <- Lma.design(attribute.names =
  list(brand = c("AT&T","T-Mobile","US Cellular","Verizon"),
  startup = c("$100","$200","$300","$400"),
  monthly = c("$100","$200","$300","$400"),
  service = c("4G NO","4G YES"),
  retail = c("Retail NO","Retail YES"),
  apple = c("Apple NO","Apple YES"),
  samsung = c("Samsung NO","Samsung YES"),
  google = c("Nexus NO","Nexus YES")), nalternatives = 1, nblocks=1, seed=9999)
print(questionnaire(provider.survey))  # print survey design for review

sink("questions_for_survey.txt")  # send survey to external text file
questionnaire(provider.survey)
sink() # send output back to the screen

# user-defined function for plotting descriptive attribute names
effect.name.map <- function(effect.name) {
  if(effect.name=="brand") return("Mobile Service Provider")
  if(effect.name=="startup") return("Start-up Cost")
  if(effect.name=="monthly") return("Monthly Cost")
  if(effect.name=="service") return("Offers 4G Service")
  if(effect.name=="retail") return("Has Nearby Retail Store")
  if(effect.name=="apple") return("Sells Apple Products")
  if(effect.name=="samsung") return("Sells Samsung Products")
  if(effect.name=="google") return("Sells Google/Nexus Products")
  }

# read in conjoint survey profiles with respondent ranks
conjoint.data.frame <- read.csv("mobile_services_ranking.csv")

# set up sum contrasts for effects coding as needed for conjoint analysis
options(contrasts=c("contr.sum","contr.poly"))
```

```
# main effects model specification
main.effects.model <- {ranking ~ brand + startup + monthly + service +
  retail + apple + samsung + google}

# fit linear regression model using main effects only (no interaction terms)
main.effects.model.fit <- lm(main.effects.model, data=conjoint.data.frame)
print(summary(main.effects.model.fit))

# save key list elements of the fitted model as needed for conjoint measures
conjoint.results <-
  main.effects.model.fit[c("contrasts","xlevels","coefficients")]

conjoint.results$attributes <- names(conjoint.results$contrasts)

# compute and store part-worths in the conjoint.results list structure
part.worths <- conjoint.results$xlevels  # list of same structure as xlevels
end.index.for.coefficient <- 1  # intitialize skipping the intercept
part.worth.vector <- NULL # used for accumulation of part worths
for(index.for.attribute in seq(along=conjoint.results$contrasts)) {
  nlevels <- length(unlist(conjoint.results$xlevels[index.for.attribute]))
  begin.index.for.coefficient <- end.index.for.coefficient + 1
  end.index.for.coefficient <- begin.index.for.coefficient + nlevels -2
  last.part.worth <- -sum(conjoint.results$coefficients[
    begin.index.for.coefficient:end.index.for.coefficient])
  part.worths[index.for.attribute] <-
    list(as.numeric(c(conjoint.results$coefficients[
      begin.index.for.coefficient:end.index.for.coefficient],
      last.part.worth)))
  part.worth.vector <-
    c(part.worth.vector,unlist(part.worths[index.for.attribute]))
  }
conjoint.results$part.worths <- part.worths
# compute standardized part-worths
standardize <- function(x) {(x - mean(x)) / sd(x)}
conjoint.results$standardized.part.worths <-
  lapply(conjoint.results$part.worths,standardize)

# compute and store part-worth ranges for each attribute
part.worth.ranges <- conjoint.results$contrasts
for(index.for.attribute in seq(along=conjoint.results$contrasts))
  part.worth.ranges[index.for.attribute] <-
  dist(range(conjoint.results$part.worths[index.for.attribute]))
conjoint.results$part.worth.ranges <- part.worth.ranges

sum.part.worth.ranges <- sum(as.numeric(conjoint.results$part.worth.ranges))
# compute and store importance values for each attribute
attribute.importance <- conjoint.results$contrasts
for(index.for.attribute in seq(along=conjoint.results$contrasts))
  attribute.importance[index.for.attribute] <-
  (dist(range(conjoint.results$part.worths[index.for.attribute]))/
  sum.part.worth.ranges) * 100
conjoint.results$attribute.importance <- attribute.importance
```

```
# data frame for ordering attribute names
attribute.name <- names(conjoint.results$contrasts)
attribute.importance <- as.numeric(attribute.importance)
temp.frame <- data.frame(attribute.name,attribute.importance)
conjoint.results$ordered.attributes <-
  as.character(temp.frame[sort.list(
  temp.frame$attribute.importance,decreasing = TRUE),"attribute.name"])

# respondent internal consistency added to list structure
conjoint.results$internal.consistency <- summary(main.effects.model.fit)$r.squared

# user-defined function for printing conjoint measures
if (print.digits == 2)
  pretty.print <- function(x) {sprintf("%1.2f",round(x,digits = 2))}
if (print.digits == 3)
  pretty.print <- function(x) {sprintf("%1.3f",round(x,digits = 3))}

# report conjoint measures to console
# use pretty.print to provide nicely formated output
for(k in seq(along=conjoint.results$ordered.attributes)) {
  cat("\n","\n")
  cat(conjoint.results$ordered.attributes[k],"Levels: ",
  unlist(conjoint.results$xlevels[conjoint.results$ordered.attributes[k]]))

  cat("\n"," Part-Worths:  ")
  cat(pretty.print(unlist(conjoint.results$part.worths
    [conjoint.results$ordered.attributes[k]])))

  cat("\n"," Standardized Part-Worths:  ")
  cat(pretty.print(unlist(conjoint.results$standardized.part.worths
    [conjoint.results$ordered.attributes[k]])))

  cat("\n"," Attribute Importance:  ")
  cat(pretty.print(unlist(conjoint.results$attribute.importance
    [conjoint.results$ordered.attributes[k]])))
  }

# plotting of spine chart begins here
# all graphical output is routed to external pdf file
pdf(file = "fig_preference_mobile_services_results.pdf", width=8.5, height=11)
spine.chart(conjoint.results)
dev.off()  # close the graphics output device

# Suggestions for the student:
# Enter your own rankings for the product profiles and generate
# conjoint measures of attribute importance and level part-worths.
# Note that the model fit to the data is a linear main-effects model.
# See if you can build a model with interaction effects for service
# provider attributes.
```

***Exhibit 1.2.*** *Measuring and Modeling Individual Preferences (Python)*

```
# Traditional Conjoint Analysis (Python)

# prepare for Python version 3x features and functions
from __future__ import division, print_function

# import packages for analysis and modeling
import pandas as pd  # data frame operations
import numpy as np  # arrays and math functions
import statsmodels.api as sm  # statistical models (including regression)
import statsmodels.formula.api as smf  # R-like model specification
from patsy.contrasts import Sum

# read in conjoint survey profiles with respondent ranks
conjoint_data_frame = pd.read_csv('mobile_services_ranking.csv')

# set up sum contrasts for effects coding as needed for conjoint analysis
# using C(effect, Sum) notation within main effects model specification
main_effects_model = 'ranking ~ C(brand, Sum) + C(startup, Sum) +  \
    C(monthly, Sum) + C(service, Sum) + C(retail, Sum) + C(apple, Sum) + \
    C(samsung, Sum) + C(google, Sum)'

# fit linear regression model using main effects only (no interaction terms)
main_effects_model_fit = \
    smf.ols(main_effects_model, data = conjoint_data_frame).fit()
print(main_effects_model_fit.summary())

conjoint_attributes = ['brand', 'startup', 'monthly', 'service', \
    'retail', 'apple', 'samsung', 'google']

# build part-worth information one attribute at a time
level_name = []
part_worth = []
part_worth_range = []
end = 1  # initialize index for coefficient in params

for item in conjoint_attributes:
    nlevels = len(list(unique(conjoint_data_frame[item])))
    level_name.append(list(unique(conjoint_data_frame[item])))
    begin = end
    end = begin + nlevels - 1
    new_part_worth = list(main_effects_model_fit.params[begin:end])
    new_part_worth.append((-1) * sum(new_part_worth))
    part_worth_range.append(max(new_part_worth) - min(new_part_worth))
    part_worth.append(new_part_worth)
    # end set to begin next iteration

# compute attribute relative importance values from ranges
attribute_importance = []
for item in part_worth_range:
    attribute_importance.append(round(100 * (item / sum(part_worth_range)),2))
```

```
# user-defined dictionary for printing descriptive attribute names
effect_name_dict = {'brand' : 'Mobile Service Provider', \
    'startup' : 'Start-up Cost', 'monthly' : 'Monthly Cost', \
    'service' : 'Offers 4G Service', 'retail' : 'Has Nearby Retail Store', \
    'apple' : 'Sells Apple Products', 'samsung' : 'Sells Samsung Products', \
    'google' : 'Sells Google/Nexus Products'}

# report conjoint measures to console
index = 0  # initialize for use in for-loop
for item in conjoint_attributes:
    print('\nAttribute:', effect_name_dict[item])
    print('    Importance:', attribute_importance[index])
    print('    Level Part-Worths')
    for level in range(len(level_name[index])):
        print('       ',level_name[index][level], part_worth[index][level])
    index = index + 1
```

# 2 Predicting Consumer Choice

"It is not our abilities that show what we truly are. It is our choices."

—RICHARD HARRIS AS PROFESSOR ALBUS DUMBLEDORE
IN *Harry Potter and the Chamber of Secrets* (2002)

I spend much of my life working. This is a choice. When I prepare data for analysis or work on the web, I use Python. For modeling or graphics, I often use R. More choices. And when I am finished programming computers, writing, and teaching, I go to Hermosa Beach—my preference, my choice.

Consumer choice is part of life and fundamental to marketing data science. If we are lucky enough, we choose where we live, whether we rent an apartment or buy a house. We choose jobs, associations, friends, and lovers. Diet and exercise, health and fitness, everything from breakfast cereal to automobiles—these are the vicissitudes of choice. And many of the choices we make are known to others, a record of our lives stored away in corporate databases.

To predict consumer choice, we use explanatory variables from the marketing mix, such as product characteristics, advertising and promotion, or the type of distribution channel. We note consumer characteristics, observable behaviors, survey responses, and demographic data. We build the discrete choice models of economics and generalized linear models of statistics—essential tools of marketing data science.

To demonstrate choice methods, we begin with the Sydney Transportation Study from appendix C (page 375). Commuters in Sydney can choose to go into the city by car or train. The response is binary, so we can use logistic regression, a generalized linear model with a logit (pronounced "low jit") link. The logit is the natural logarithm of the odds ratio.[1]

In the Sydney Transportation Study, we know the time and cost of travel by car and by train. These are the explanatory variables in the case. The scatter plot matrix in figure 2.1 and the correlation heat map in figure 2.2 show pairwise relationships among these explanatory variables.

Time and cost by car are related. Time and cost by train are related. Longer times by the train are associated with longer times by car. These time-of-commute variables depend on where a person lives and may be thought of as proxies or substitutes for distance from Sydney, a variable not in the data set.

We use a linear combination of the four explanatory variables to predict consumer choice. The fitted logistic regression model is shown in table 2.1, with the corresponding analysis of deviance in table 2.2. From this model, we can obtain the predicted probability that a Sydney commuter will take the car or the train.

How well does the model work on the training data? A density lattice conditioned on actual commuter car-or-train choices shows the degree to which these predictions are correct. See figure 2.3.

---

[1] The odds of choosing the train over the car are given by the probability that a commuter chooses the train $p(TRAIN)$ divided by the probability that the commuter chooses car $p(CAR)$. We assume that both probabilities are positive, on the open interval between zero and one. Then the odds ratio will be positive, on the open interval between zero and plus infinity.

$$0 < p(TRAIN) < 1$$
$$0 < p(CAR) < 1$$
$$0 < \frac{p(TRAIN)}{p(CAR)} < +\infty$$

The logit or log of the odds ratio is a logarithm, mapping the set of positive numbers onto the set of all real numbers. This is what logarithms do.

$$-\infty < log\left(\frac{p(TRAIN)}{p(CAR)}\right) < +\infty$$

Using the logit, we can write equations linking choices (or more precisely, probabilities of choices) with linear combinations of explanatory variables. Such is the logic of the logit (or shall we say, the magic of the logit). In generalized linear models we call the logit a *link function*. See appendix A (page 267) for additional discussion of logistic regression.

***Figure 2.1.*** *Scatter Plot Matrix for Explanatory Variables in the Sydney Transportation Study*

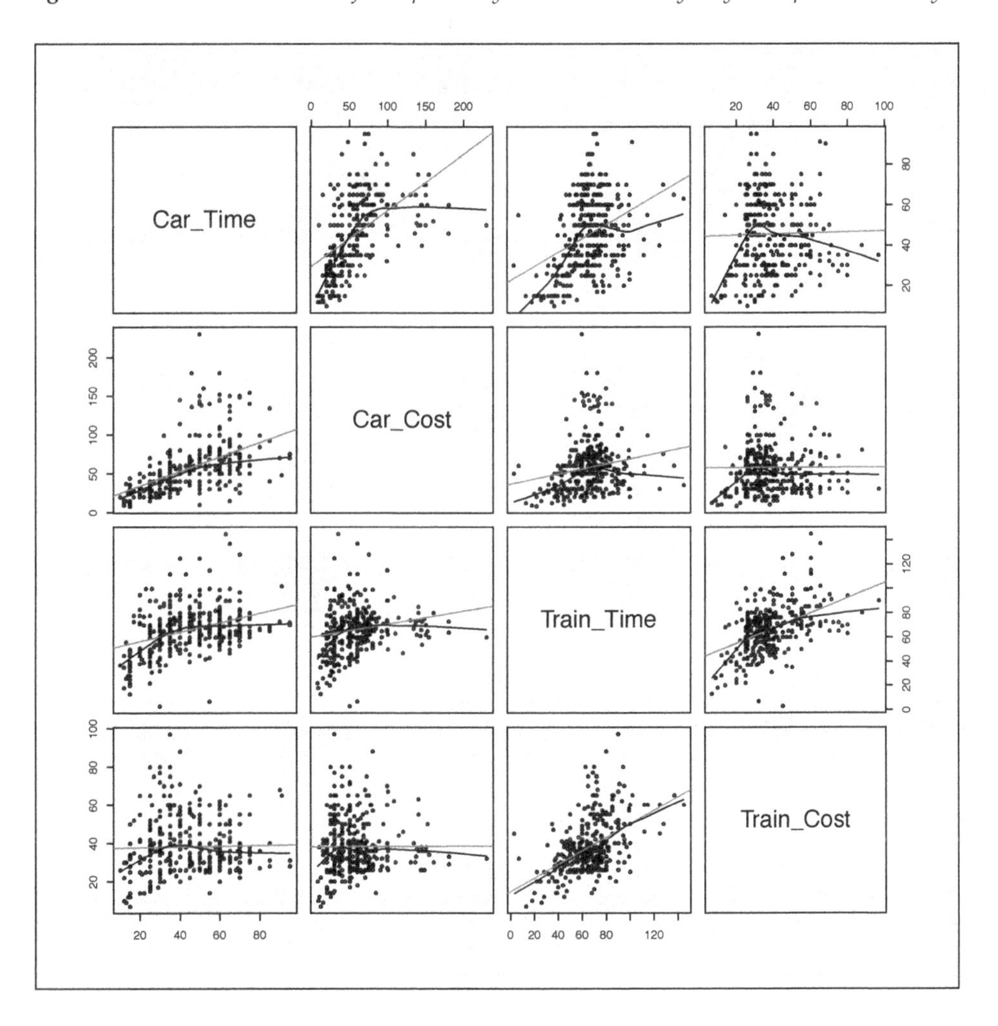

*Figure* 2.2. *Correlation Heat Map for Explanatory Variables in the Sydney Transportation Study*

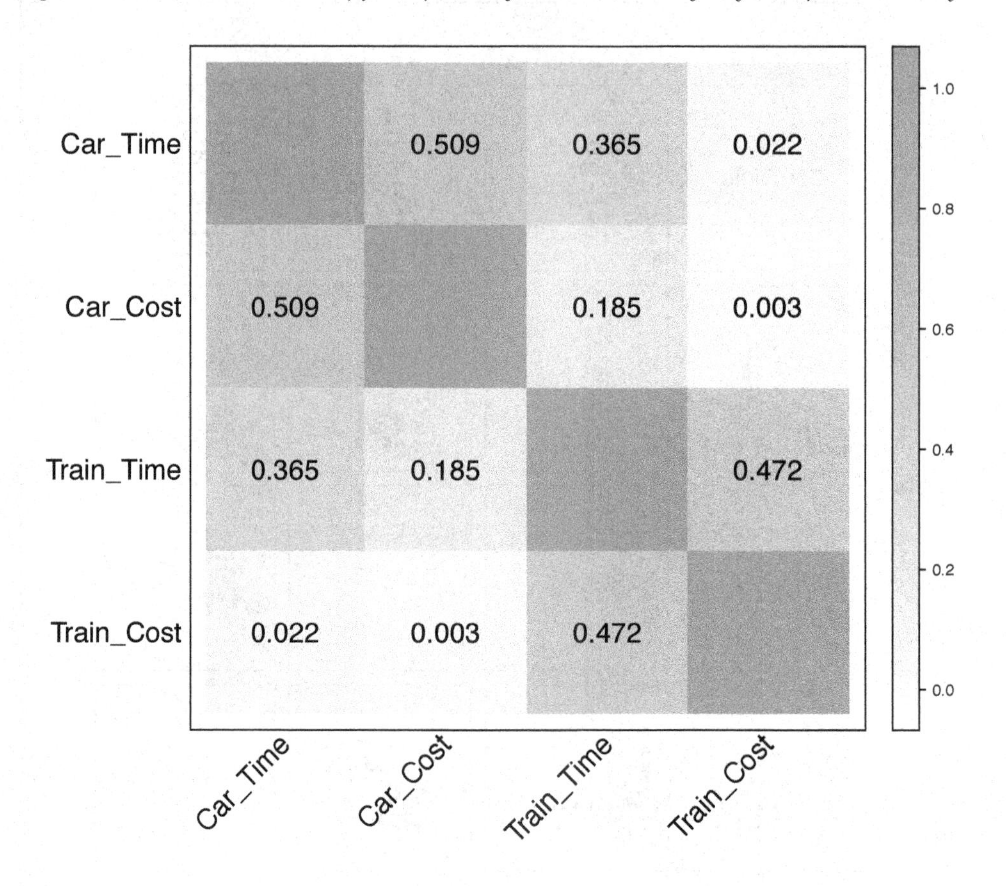

***Table 2.1.*** *Logistic Regression Model for the Sydney Transportation Study*

| Response: Choice (Train) | |
|---|---|
| Car Time | 0.056507 |
| Car Cost | 0.029842 |
| Train Time | 0.014918 |
| Train Cost | -0.111293 |
| Constant | -1.444027 |
| Null Deviance | $458.36 (df = 332)$ |
| Residual Deviance | $272.63 (df = 328)$ |
| AIC | 282.63 |

***Table 2.2.*** *Logistic Regression Model Analysis of Deviance*

| | *df* | *Deviance* | *Residual df* | *Residual Deviance* | *P-Value* |
|---|---|---|---|---|---|
| NULL | | | 332 | 458.36 | |
| Car Time | 1 | 90.402 | 331 | 367.96 | 2.2e-16 |
| Car Cost | 1 | 22.822 | 330 | 345.14 | 1.778e-06 |
| Train Time | 1 | 7.563 | 329 | 337.57 | 0.005956 |
| Train Cost | 1 | 64.938 | 328 | 272.63 | 7.727e-16 |

*Figure 2.3.* *Logistic Regression Density Lattice*

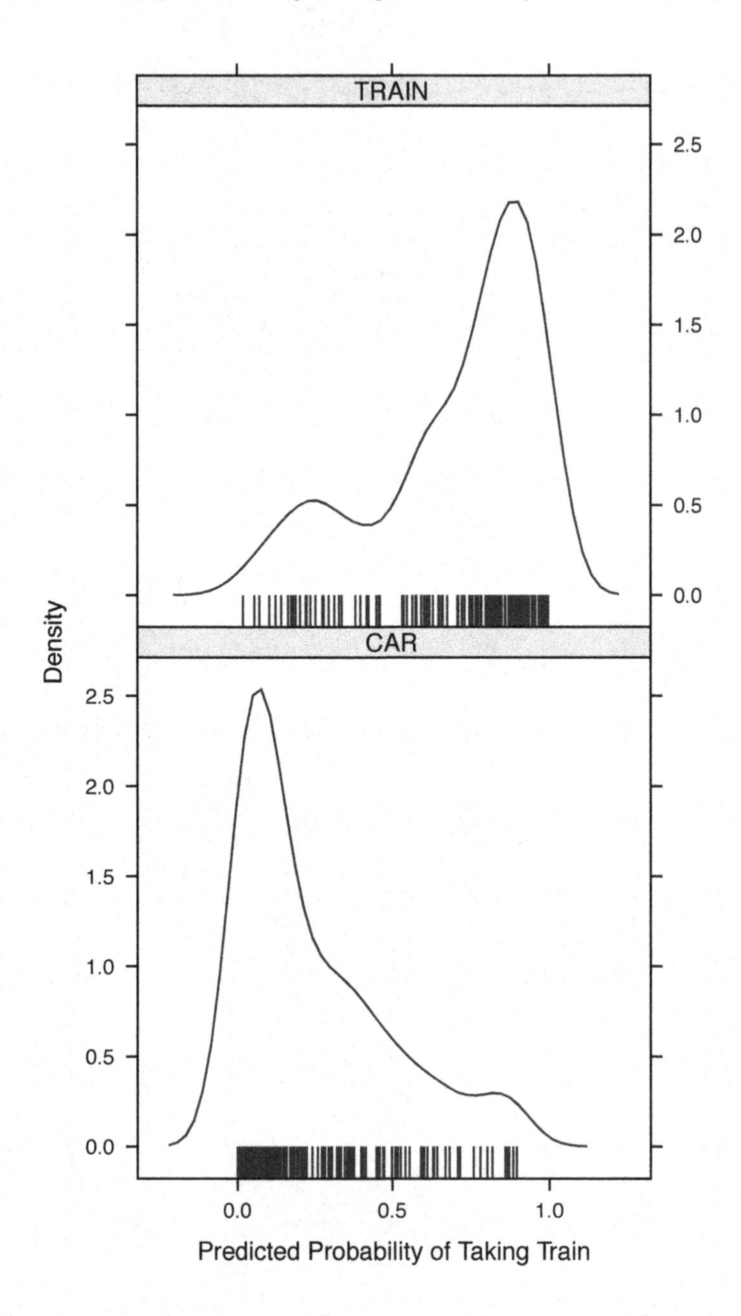

To obtain a car-or-train prediction for each commuter, we set a predicted probability cut-off. Suppose we classify commuters with a 0.50 cut-off. That is, if the predicted probability of taking the train is greater than 0.50, then we predict that the commuter will take the train. Otherwise, we predict the commuter will take the car. The resulting four-fold table or confusion matrix would show that we have correctly predicted transportation choice 82.6 percent of the time. There are many ways to evaluate the predictive accuracy of a classifier such as logistic regression. These are reviewed in appendix A (page 271).

Good data science is more than a matter of constructing equations, more than just mathematics and statistics. Good data science depends on an understanding of business problems.

Time and location variables are important explanatory variables in models of transportation choice, but public administrators have little control over time and location variables. Time and location variables represent control variables rather than decision variables. Cost variables have the potential of being decision variables because to some extent they may be manipulated.

Although public administrators have little to say about the gasoline commodity market, they can raise taxes on gasoline, affecting the cost of transportation by car. More importantly, administrators control ticket prices on public transportation, affecting the cost of transportation by train.

In the Sydney Transportation Study, 150 out of 333 commuters (45 percent) use the train. Suppose public administrators set a goal to increase public transportation usage by 10 percent. How much lower would train ticket prices have to be to achieve this goal, keeping all other variables constant? We can use the fitted logistic regression model to answer this question.

Figure 2.4 provides a convenient summary for administrators. To make this graph, we control car time, car cost and train time variables by setting them to their average values. Then we let train cost vary across a range of values and observe its effect on the estimated probability of taking the train. Explicit calculations from the model suggest that 183 (55 percent) of Sydney commuters would take the train if ticket prices were lowered by 5 cents.

*Figure 2.4.* *Using Logistic Regression to Evaluate the Effect of Price Changes*

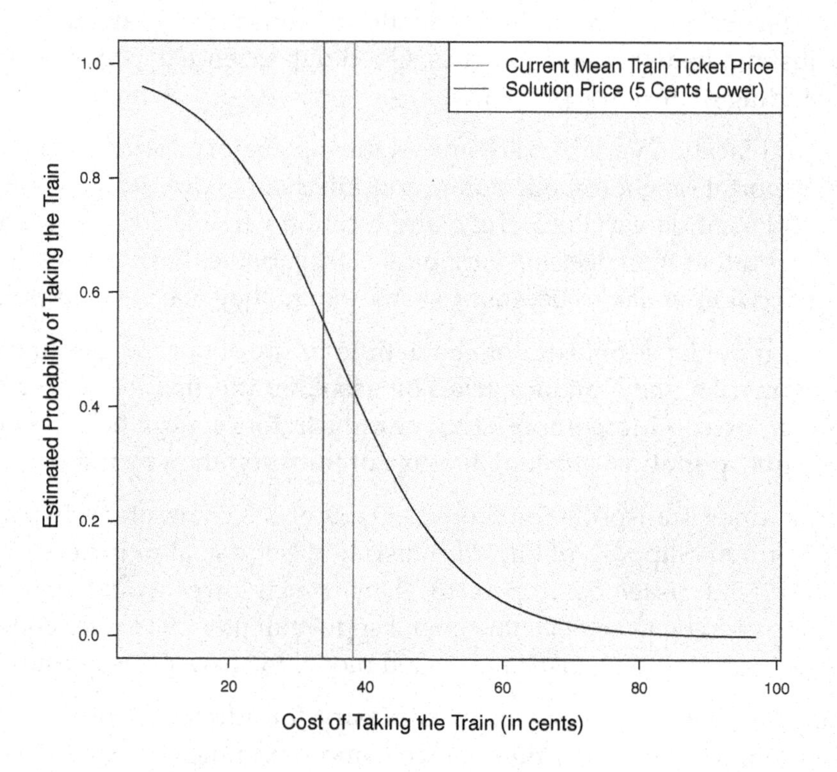

Logistic regression is a generalized linear model. Generalized linear models, as their name would imply, are generalizations of the classical linear regression model. A standard reference for generalized linear models is McCullagh and Nelder (1989). Firth (1991) provides additional review of the underlying theory. Hastie (1992) and Venables and Ripley (2002) give modeling examples relevant to R. Lindsey (1997) discusses a wide range of application examples. See appendix A (pages 266 through 270) for additional discussion of logistic regression and generalized linear models.

There are a number of good resources for understanding discrete choice modeling in economics and market research. Introductory material may be found in econometrics textbooks, such as Pindyck and Rubinfeld (2012) and Greene (2012). More advanced discussion is provided by Ben-Akiva and Lerman (1985). Louviere, Hensher, and Swait (2000) give examples in transportation and market research. Train (2003) provides a thorough review of discrete choice modeling and methods of estimation.

Wassertheil-Smoller (1990) provides an elementary introduction to logistic regression procedures and the evaluation of binary classifiers. For a more advanced treatment, see Hand (1997). Burnham and Anderson (2002) review model selection methods, particularly those using the Akaike information criterion or AIC (Akaike 1973).

As we will see through worked examples in this book, we can answer many management questions by analyzing the choices that consumers make—choices in the marketplace, choices in response to marketing action, and choices in response to consumer surveys such as conjoint surveys. We often use logistic regression and multinomial logit models to analyze choice data.

Exhibit 2.1 shows an R program for analyzing data from the Sydney Transportation Study, drawing on lattice plotting tools from Sarkar (2008, 2014). The corresponding Python program is in exhibit 2.2.

**Exhibit 2.1.** *Predicting Commuter Transportation Choices (R)*

```
# Predicting Commuter Transportation Choices (R)

library(lattice)  # multivariate data visualization

load("correlation_heat_map.RData")  # from R utility programs

# read data from comma-delimited text file... create data frame object
sydney <- read.csv("sydney.csv")
names(sydney) <-
    c("Car_Time", "Car_Cost", "Train_Time", "Train_Cost", "Choice")
plotting_data_frame <- sydney[, 1:4]

# scatter plot matrix with simple linear regression
# models and lowess smooth fits for variable pairs
pdf(file = "fig_predicting_choice_scatter_plot_matrix.pdf",
    width = 8.5, height = 8.5)
pairs(plotting_data_frame,
    panel = function(x, y) {
        points(x, y, cex = 0.5)
        abline(lm(y ~ x), lty = "solid", col = "red")
        lines(lowess(x, y))
        }
    )
dev.off()

# correlation heat map for the explanatory variables
pdf(file = "fig_predicting_choice_correlation_heat_map.pdf",
    width = 8.5, height = 8.5)
sydney_cormat <-
    cor(sydney[, c("Car_Time", "Car_Cost", "Train_Time", "Train_Cost")])
correlation_heat_map(sydney_cormat)
dev.off()

# specify and fit logistic regression model
sydney_model <- {Choice ~ Car_Time + Car_Cost + Train_Time + Train_Cost}
sydney_fit <- glm(sydney_model, family=binomial, data=sydney)
print(summary(sydney_fit))
print(anova(sydney_fit, test="Chisq"))

# compute predicted probability of taking the train
sydney$Predict_Prob_TRAIN <- predict.glm(sydney_fit, type = "response")
pdf(file = "fig_predicting_choice_density_evaluation.pdf",
    width = 8.5, height = 8.5)
plotting_object <- densityplot( ~ Predict_Prob_TRAIN | Choice,
               data = sydney,
               layout = c(1,2), aspect=1, col = "darkblue",
               plot.points = "rug",
               strip=function(...) strip.default(..., style=1),
               xlab="Predicted Probability of Taking Train")
print(plotting_object)
dev.off()
```

```
# predicted car-or-train choice using 0.5 cut-off
sydney$Predict_Choice <- ifelse((sydney$Predict_Prob_TRAIN > 0.5), 2, 1)
sydney$Predict_Choice <- factor(sydney$Predict_Choice,
    levels = c(1, 2), labels = c("CAR", "TRAIN"))
confusion_matrix <- table(sydney$Predict_Choice, sydney$Choice)
cat("\nConfusion Matrix (rows = Predicted Choice, columns = Actual Choice\n")
print(confusion_matrix)
predictive_accuracy <- (confusion_matrix[1,1] + confusion_matrix[2,2])/
                        sum(confusion_matrix)
cat("\nPercent Accuracy: ", round(predictive_accuracy * 100, digits = 1))

# How much lower would train ticket prices have to be to increase
# public transportation usage (TRAIN) by 10 percent?
train_cost_vector <-
     seq(min(sydney$Train_Cost), max(sydney$Train_Cost), length=1000)
beta.vector <- sydney_fit$coefficients
train_probability_vector <- numeric(1000)
for (i in 1:1000) {
       X.vector <- c(1, mean(sydney$Car_Time), mean(sydney$Train_Time),
           mean(sydney$Car_Cost), train_cost_vector[i])
       train_probability_vector[i] <-
           exp(X.vector %*% beta.vector)/
               (1 + exp(X.vector %*% beta.vector))
       }
# currently 150 out of 333 commuters (45 percent) use the train
# determine price required for 55 percent of commuters to take the train
# this is the desired quota set by public administrators
index <- 1  # beginning index for search
while (train_probability_vector[index] > 0.55) index <- index + 1
Solution_Price <- train_cost_vector[index]
cat("\nSolution Price: ", Solution_Price)
Current_Mean_Price <- mean(sydney$Train_Cost)
# how much do administrators need to lower prices?
# use greatest integer function to ensure quota is exceeded
Cents_Lower <- ceiling(Current_Mean_Price - Solution_Price)
cat("\nLower prices by ", Cents_Lower, "cents\n")

pdf(file = "fig_predicting_choice_ticket_price_solution.pdf",
    width = 8.5, height = 8.5)
plot(train_cost_vector, train_probability_vector,
     type="l",ylim=c(0,1.0), las = 1,
     xlab="Cost of Taking the Train (in cents)",
     ylab="Estimated Probability of Taking the Train")

# plot current average train ticket price as vertical line
abline(v = Current_Mean_Price, col = "red", lty = "solid", lwd = 2)
abline(v = Solution_Price, col = "blue", lty = "dashed", lwd = 2)

legend("topright", legend = c("Current Mean Train Ticket Price",
       paste("Solution Price (", Cents_Lower, " Cents Lower)", sep = "")),
    col = c("red", "blue"), pch = c(NA, NA), lwd = c(2, 2),
    border = "black", lty = c("solid", "dashed"), cex = 1.25)
dev.off()
```

```
# Suggestions for the student:
# How much lower must train fares be to encourage more than 60 percent
# of Sydney commuters to take the train? What about car costs? How much
# of a tax would public administrators have to impose in order to have
# a comparable effect to train ticket prices?
# Evaluate the logistic regression model in terms of its out-of-sample
# predictive accuracy (using multi-fold cross-validation, for example).
# Try alternative classification methods such as tree-structured
# classification and support vector machines. Compare their predictive
# performance to that of logistic regression in terms of percentage
# of accurate prediction and other measures of classification performance.
```

**Exhibit 2.2.** *Predicting Commuter Transportation Choices (Python)*

```
# Predicting Commuter Transportation Choices (Python)
# import packages into the workspace for this program
from __future__ import division, print_function
import numpy as np
import pandas as pd
import statsmodels.api as sm
# read data from comma-delimited text file... create DataFrame object
sydney = pd.read_csv("sydney.csv")
# check input DataFrame
print(sydney)

# dictionary object to convert string to binary integer
response_to_binary = {'TRAIN':1, 'CAR':0}
y = sydney['choice'].map(response_to_binary)
cartime = sydney['cartime']
carcost = sydney['carcost']
traintime = sydney['traintime']
traincost = sydney['traincost']
# define design matrix for the linear predictor
Intercept = np.array([1] * len(y))
x = np.array([Intercept, cartime, carcost, traintime, traincost]).T

# generalized linear model for logistic regression
logistic_regression = sm.GLM(y, x, family=sm.families.Binomial())
sydney_fit = logistic_regression.fit()
print(sydney_fit.summary())
sydney['train_prob'] = sydney_fit.predict(linear = False)

# function to convert probability to choice prediction
def prob_to_response(response_prob, cutoff):
    if(response_prob > cutoff):
        return('TRAIN')
    else:
        return('CAR')
# add binary predictions to DataFrame sydney using cutoff value for the case
sydney['choice_pred'] = \
    sydney['train_prob'].apply(lambda d: prob_to_response(d, cutoff = 0.50))
# evaluate performance of logistic regression model
# obtain confusion matrix and proportion of observations correctly predicted
cmat = pd.crosstab(sydney['choice_pred'], sydney['choice'])
a = float(cmat.ix[0,0])
b = float(cmat.ix[0,1])
c = float(cmat.ix[1,0])
d = float(cmat.ix[1,1])
n = a + b + c + d
predictive_accuracy = (a + d)/n

print(cmat)
print('\n Percentage Correctly Predicted',\
     round(predictive_accuracy, 3), "\n")
```

# 3 Targeting Current Customers

"Listen, I—I appreciate this whole seduction scene you've got going, but let me give you a tip: I'm a sure thing. OK?"

—JULIA ROBERTS AS VIVIAN WARD IN *Pretty Woman* (1990)

Mass marketing treats all customers as one group. One-to-one marketing focuses on one customer at a time. Target marketing to selected groups of customers or market segments lies between mass marketing and one-to-one marketing. Target marketing involves directing marketing activities to those customers who are most likely to buy.

Targeting implies selection. Some customers are identified as more valuable than others and these more highly valued customers are given special attention. By becoming skilled at targeting, a company can improve its profitability, increasing revenues and decreasing costs.

Targeting is best executed by companies that keep detailed records for individuals. These are companies that offer loyalty programs or use a customer relationship management system. Sales transactions for individual customers need to be associated with the specific customer and stored in a customer database. Where revenues (cash inflows) and costs (cash outflows) are understood, we can carry out discounted cash-flow analysis and compute the return on investment for each customer.

A target is a customer who is worth pursuing. A target is a profitable customer—sales revenues from the target exceed costs of sales and support. Another way to say this is that a target is a customer with positive lifetime value. Over the course of a company's relationship with the customer, more money comes into the business than goes out of the business.

Managers want to predict responses to promotions and pricing changes. They want to anticipate when and where consumers will be purchasing products. They want to identify good customers for whom sales revenues are higher than the cost of sales and support.

For companies engaging in direct marketing, costs may also be associated with individual customers. These costs include mailings, telephone calls, and other direct marketing activities. For companies that do not engage in direct marketing or lack cost records for individual customers, general cost estimates are used in estimating customer lifetime value.

In target marketing, we need to identify factors that are useful and determine how to use those factors in modeling techniques. A response variable is something we want to predict, such as sales dollars, volume, or whether a consumer will buy a product. Customer lifetime value is a composite response variable, computed from many transactions with each customer, and these transactions include observations of sales and costs.

Explanatory variables are used to predict response variables. Explanatory variables can be continuous (having meaningful magnitude) or categorical (without meaningful magnitude). Statistical models show the relationship between explanatory variables and response variables.

Common explanatory variables in business-to-consumer target marketing include demographics, behavioral, and lifestyle variables. Common explanatory variables in business-to-business marketing include the size of the business, industry sector, and geographic location. In target marketing, whether business-to-consumer or business-to-business, explanatory variables can come from anything that we know about customers, including the past sales and support history with customers.

Regression and classification are two types of predictive models used in target marketing. When the response variable (the variable to be predicted) is continuous or has meaningful magnitude, we use regression to make the prediction. Examples of response variables with meaningful magnitude are sales dollars, sales volume, cost of sales, cost of support, and customer lifetime value.

When the response variable is categorical (a variable without meaningful magnitude), we use classification. Examples of response variables without meaningful magnitude are whether a customer buys, whether a customer stays with the company or leaves to buy from another company, and whether the customer recommends a company's products to another customer.

To realize the benefits of target marketing, we need to know how to target effectively. There are many techniques from which to choose, and we want to find the technique that works best for the company and for the marketing problem we are trying to solve.

All other things being equal, the customers with the highest predicted sales should be the ones the sales team will approach first. Alternatively, we could set a cutoff for predicted sales. Customers above the cutoff are the customers who get sales calls—these are the targets. Customers below the cutoff are not given calls.

When evaluating a regression model using data from the previous year, we can determine how close the predicted sales are to the actual/observed sales. We can find out the sum of the absolute values of the residuals (observed minus predicted sales) or the sum of the squared residuals.

Another way to evaluate a regression model is to correlate the observed and predicted response values. Or, better still, we can compute the squared correlation of the observed and predicted response values. This last measure is called the coefficient of determination, and it shows the proportion of response variance accounted for by the linear regression model. This is a number that varies between zero and one, with one being perfect prediction.

If we plotted observed sales on the horizontal axis and predicted sales on the vertical axis, then the higher the squared correlation between observed sales and predicted sales, the closer the points in the plot will fall along a straight line. When the points fall along a straight line exactly, the squared correlation is equal to one, and the regression model is providing a perfect prediction of sales, which is to say that 100 percent of sales response is accounted for by the model. When we build a regression model, we try to obtain a high value for the proportion of response variance accounted for. All other things being equal, higher squared correlations are preferred.

The focus can be on predicting sales or on predicting cost of sales, cost of support, profitability, or overall customer lifetime value. There are many possible regression models to use in target marketing with regression methods.

To develop a classification model for targeting, we proceed in much the same way as with a regression, except the response variable is now a category or class. For each customer, a logistic regression model, for example, would provide a predicted probability of response. We employ a cut-off value for the probability of response and classify responses accordingly. If the cut-off were set at 0.50, for example, then we would target the customer if the predicted probability of response is greater than 0.50, and not target otherwise. Or we could target all customers who have a predicted probability of response of 0.40, or 0.30, and so on. The value of the cut-off will vary from one problem to the next.

To illustrate the targeting process we consider the Bank Marketing Study from appendix C (page 356). The bank wants its clients to invest in term deposits. A term deposit is an investment such as a certificate of deposit. The interest rate and duration of the deposit are set in advance. A term deposit is distinct from a demand deposit, which has no set rate or duration.

The bank is interested in identifying factors that affect client responses to new term deposit offerings, which are the focus of the marketing campaigns. What kinds of clients are most likely to subscribe to new term deposits? What marketing approaches are most effective in encouraging clients to subscribe?

We begin by looking at the subset of bank clients who are approached with a call for the first time. Part of the challenge in target marketing is dealing with low rates in response to sales and promotional efforts. In this problem only 71 of 3,705 bank clients responded affirmatively by subscribing to the term deposit being offered by the bank.

We examine the relationships between each demographic variable and response to the bank's offer. The demographic variables include age, job type, marital status, and level of education. We also examine relationships between banking experience variables and response to the bank's offer. These variables include the client's average yearly balance, and whether or not the client defaulted on a loan, has a housing loan, or has a personal loan.

Figures 3.1 through 3.5 provide mosaic and lattice plots for selected relationships. Bank clients who subscribe to term deposit offers are older, more highly educated, more likely to be white collar workers than blue collar workers, and more likely to be single, divorced, or widowed than married. They are also less likely to have a housing loan with the bank.

We define a linear predictor using the eight explanatory variables and fit a logistic regression model to the training data. With logistic regression, the left-hand side of the model is a mathematical rendering of the probability of ordering a system upgrade—the logarithm of an odds ratio.

Even though the name of the method is called "logistic regression," the method involves classification, not regression—the response in this problem is categorical, whether or not the client accepts the bank's offer.

We can judge model performance by statistical criteria. After selecting a modeling technique—logistic regression in this case—we employ a probability cut-off to identify target customers. The model provides a predicted probability of response, and we use the cut-off to convert the probability of response into a choice prediction.

When observed binary responses or choices are about equally split between *yes* and *no*, for example, we would use a cut-off probability of 0.50. That is, when the predicted probability of responding *yes* is greater than 0.50, we predict *yes*. Otherwise, we predict *no*.

Logistic regression provides a means for estimating the probability of a favorable (yes) response to the offer. The density lattice in figure 3.6 provides a pictorial representation of the model and a glimpse at model performance.

To evaluate the performance of this targeting model, we look at a two-by-two contingency table or confusion matrix showing the predicted and observed response values. A 50 percent cut-off does not work in the Bank Marketing Study, given the low base rate of responses to the offer.

*Figure 3.1.* *Age and Response to Bank Offer*

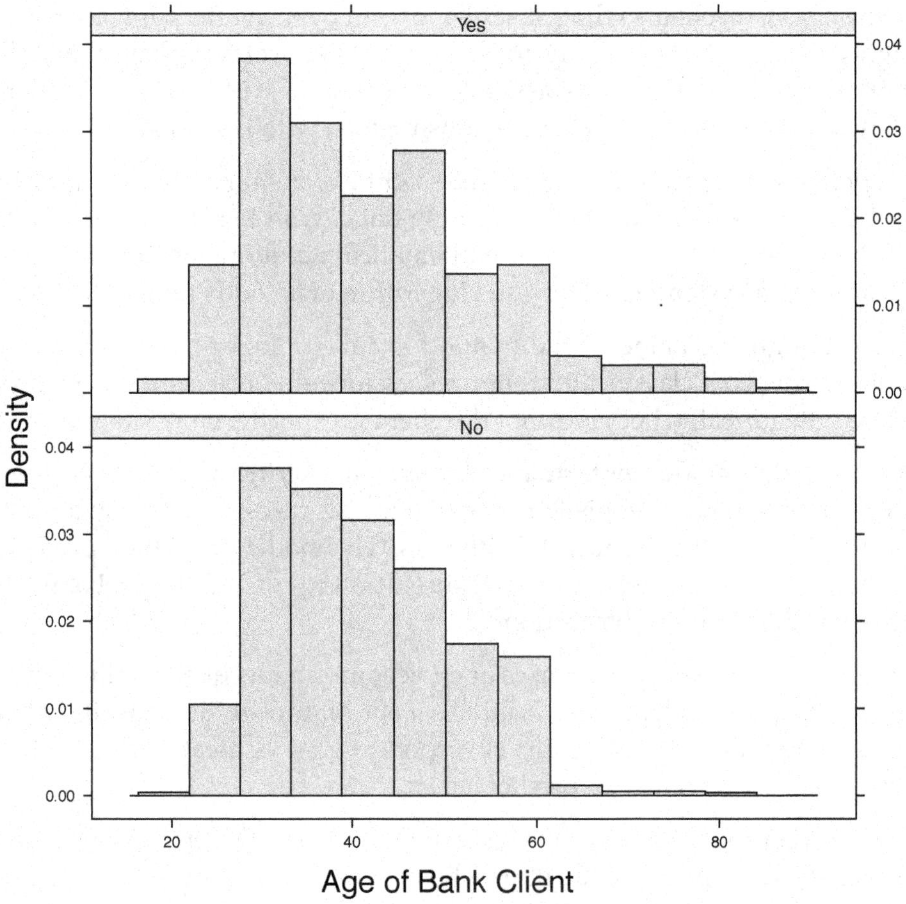

*Figure 3.2.* Education Level and Response to Bank Offer

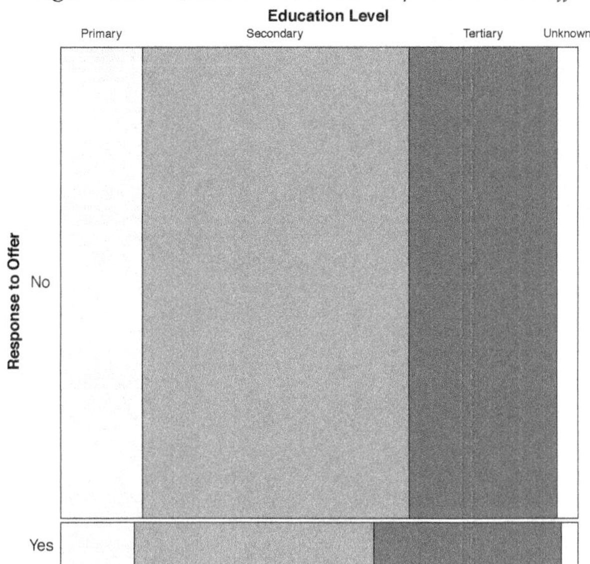

*Figure 3.3.* Job Type and Response to Bank Offer

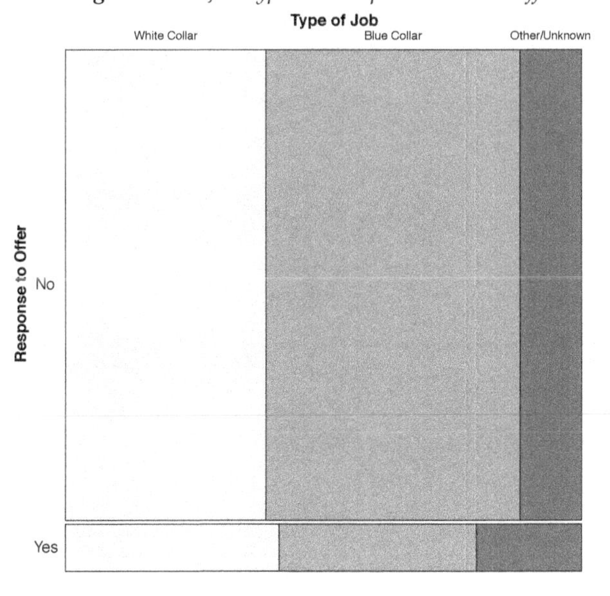

*Figure 3.4.* *Marital Status and Response to Bank Offer*

Marital Status

Divorced Married Single

Response to Offer

No

Yes

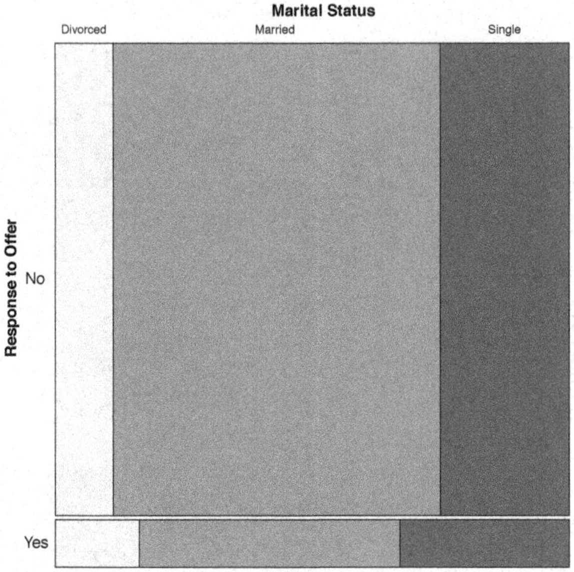

*Figure 3.5.* *Housing Loans and Response to Bank Offer*

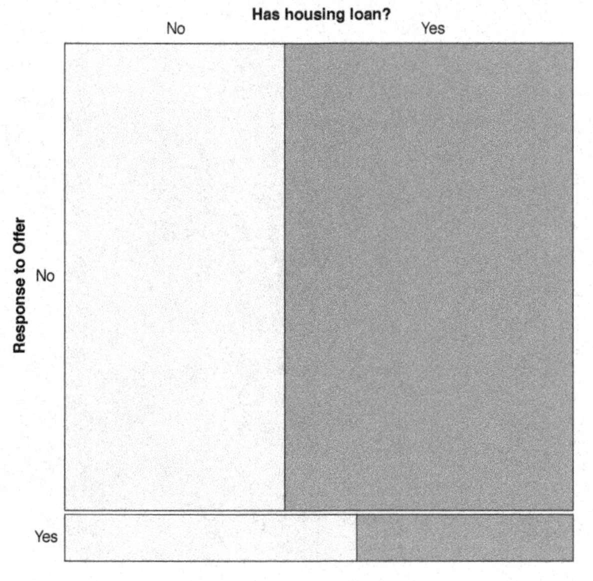

***Figure 3.6.*** *Logistic Regression for Target Marketing (Density Lattice)*

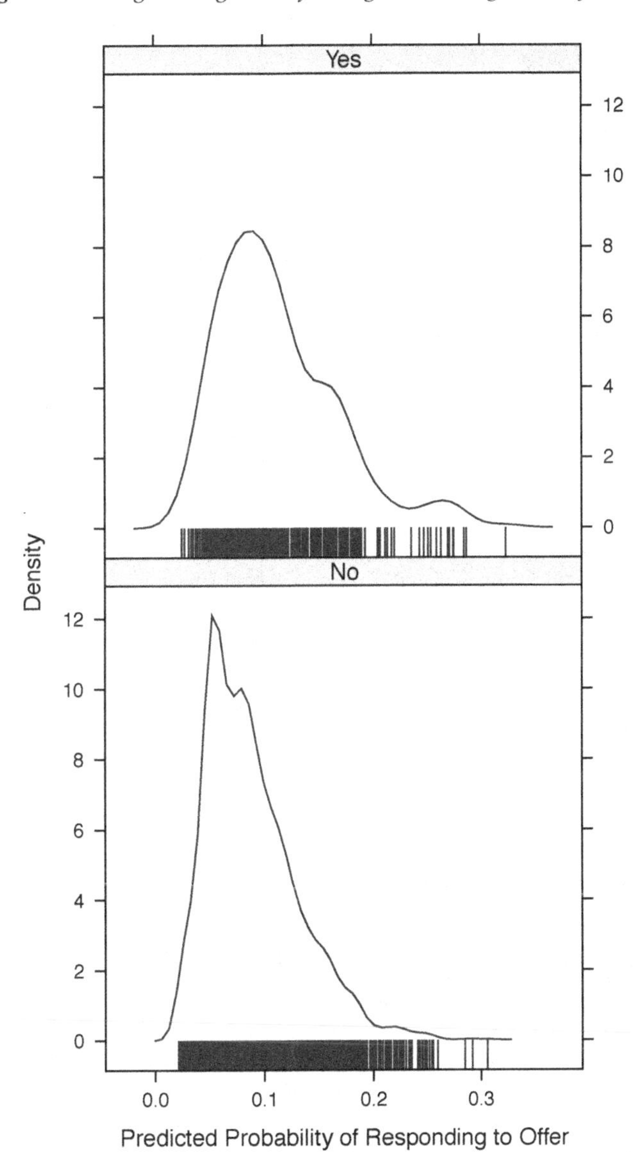

*Figure 3.7. Logistic Regression for Target Marketing (Confusion Mosaic)*

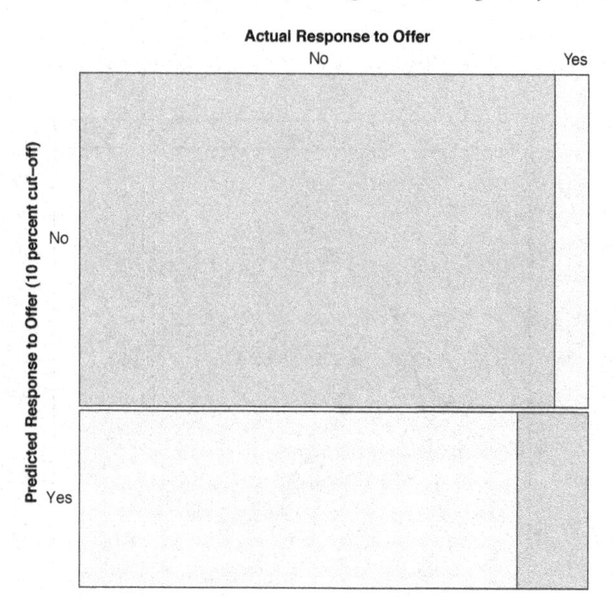


A 50 percent cut-off will not work for the bank, but using a 10 percent cut-off for the response variable (accepting the term deposit offer or not), yields 65.9 percent accuracy in classification. The confusion matrix for the logistic regression and 10 percent cut-off is shown as a mosaic in figure 3.7.

The Bank Marketing Study is typical of target marketing problems. Response rates are low, much lower than 0.50, so a 50 percent cut-off performs poorly. In fact, if bank analysts were to use a 50 percent cut-off, they would predict that every client would respond *no*, and the bank would target no one. Too high a cut-off means the bank will miss out on many potential sales.

Too low a cut-off presents problems as well. Too low a cut-off means the bank will pursue sales with large numbers of clients, many of whom will never subscribe to the term deposit offer. It is wise to pick a cut-off that maximizes profit, given the unit revenues and costs associated with each cell of the confusion matrix. Target marketing, employed in the right situations and with the right cut-offs, yields higher profits for a company.

*Figure 3.8. Lift Chart for Targeting with Logistic Regression*

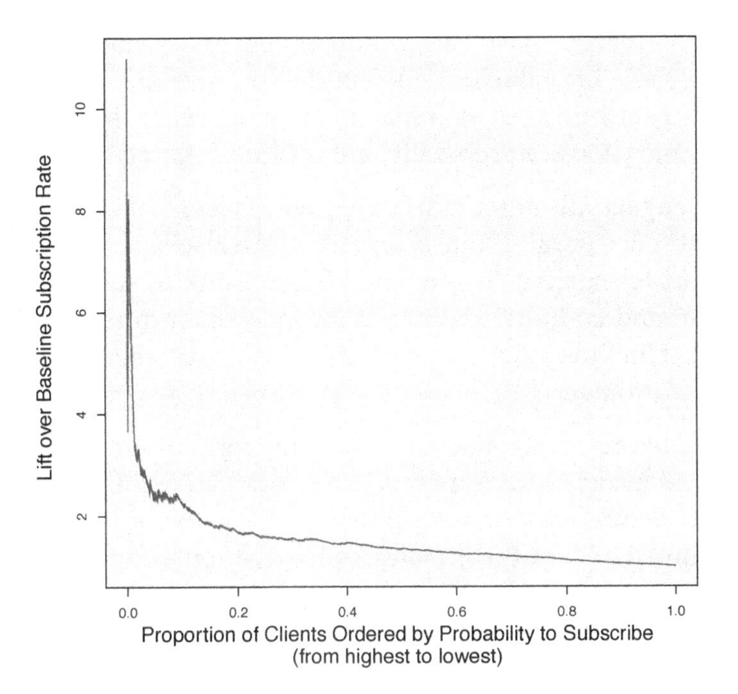


The analyst or data scientist sets the cut-off probability, and the cut-off affects the financial performance of the targeting model. One approach to picking the right cut-off is to compute *lift* or the response rate that the predictive model provides over the response rate observed in the entire customer base. We order customers by their predicted probability of responding to an offer and then note how much this predicted probability is greater than the base rate of responding to the offer. Lift is a ratio of these probabilities or rates of responding.

Figure 3.8 displays the lift chart for the Bank Marketing Study. The horizontal axis shows the proportion of clients ordered by their probability to subscribe, from highest to lowest, and the vertical axis shows the associated values of lift.

To set a cut-off value for the probability of responding, we might determine that we want to contact clients who are at least twice as likely to subscribe than clients at large. Then we would choose the probability cut-off that corresponds to a lift value of two. Lift does not directly translate into revenues and costs, however, so it may make more sense to perform financial calculations to choose a probability cut-off for targeting.

When we engage in target marketing, we review data from current customers, particularly sales transaction data. We also assess the costs of sales and support for current customers. We can think of each customer as an investment, and compute the return on investment for each customer. If the expected lifetime value of a customer is positive, then it makes sense to retain that customer.

Customer lifetime value analysis draws on concepts from financial management. We evaluate investments in terms of cash in-flows and out-flows over time. Before we pursue a prospective customer, we want to know that discounted cash in-flows (sales) will exceed discounted cash out-flows (costs). Similarly, it makes sense to retain a current customer when discounted future cash flows are positive. To review discounted cash flow and investment analysis, see financial management references such as Higgins (2011) and Brealey, Myers, and Allen (2013).

Customer lifetime value is computed from our experience with each customer. For the cash in-flows, we note a customer's purchasing history as recorded in sales transactions. For the cash out-flows, we note past sales and support costs as recorded in customer relationship management systems. Customer lifetime value analysis is best executed when detailed records are maintained for customers.

Data for valuing customers may be organized as panel or longitudinal data. Rows correspond to customers and columns correspond to time periods. Data from past transactions may be incomplete, and future cash-flows are unknown. So, we use predictive models to estimate cash-flows. We draw on available data to impute missing observations from the past, and we use observations from the past to forecast observations in the future.

Direct marketers are the quintessential target marketers. Their work involves contacting prospective and former customers directly through telephone, mail, e-mail, and online channels. Direct marketers collect and maintain information about past contacts, mailings, in-coming and outgoing communications, and business transactions. And they do this on a customer-by-customer basis. These data are used to guide sales promotions and direct marketing programs. Direct mailings and outgoing communications include product brochures and announcements, as well as coupons and information about product prices, bundles, and promotions.

Each direct marketing promotion can be evaluated in terms of its contribution to the profit of the firm. There are costs associated with mailings and online activities. There are revenues coming from people who order. The hit-rate or proportion of people who respond to a direct mail or online offer is a critical number to watch because it determines the success or failure of the promotion.

Let us consider revenues and costs in the Bank Marketing Study. Bank term deposits provide money for loans. A bank generates revenue by charging a higher rate of interest on loans than it pays on deposits. Suppose the difference in interest rates across average deposit and loan amounts is 100 Euros—this corresponds to the revenue from one client term deposit. Suppose the associated sales and marketing costs for each client contact (including mailings and telephone calls) is 5 Euros. Furthermore, suppose that post-deposit/post-sale support costs are 25 Euros. Then a financial analysis of target marketing using logistic regression with a 10 percent cut-off would play out as shown in figure 3.9. We can see that target marketing is financially beneficial to the bank. The money it saves in sales and marketing costs exceeds the money lost as a result of having fewer term deposit subscriptions.

Direct marketing promotions, properly constructed, represent field experiments. Rarely is it wise to mail to an entire list at once. Better to divide the list into sections and vary the direct mail offer or advertising copy across sections. The conditions that yield the highest profit on the test mailing set the stage for subsequent mailings. Numerous treatment conditions can be examined for each direct marketing promotion in a phased rollout of the marketing effort. Methods that have been practiced for many years across physical mail channels are now being employed across online channels.

*Figure 3.9. Financial Analysis of Target Marketing*

**Logistic Regression Model**
**Confusion Matrix (10 percent cut-off)**

| Predicted Response | Actual Response No | Actual Response Yes | |
|---|---|---|---|
| No | 2,262 | 159 | 2,421 |
| Yes | 1,106 | 178 | 1,284 |
| | 3,368 | 337 | 3,705 |

**Pursue All Clients**

| Number of Subscriptions | Revenue per Subscription | |
|---|---|---|
| 337 | 100 | 33,700 Revenue |
| **Supported Clients** | **Unit Cost of Support** | |
| 337 | 25 | 8,425 Expense |
| **Clients Pursued by Sales and Marketing** | **Unit Cost of Sales and Marketing** | |
| 3,705 | 5 | 18,525 Expense |
| | | 6,750 Profit |

**Pursue Only Targeted Clients**

| Number of Subscriptions | Revenue per Subscription | |
|---|---|---|
| 178 | 100 | 17,800 Revenue |
| **Supported Clients** | **Unit Cost of Support** | |
| 178 | 25 | 4,450 Expense |
| **Clients Pursued by Sales and Marketing** | **Unit Cost of Sales and Marketing** | |
| 1,284 | 5 | 6,420 Expense |
| | | 6,930 Profit |

Direct and database marketers build models for predicting who will buy in response to marketing promotions. Traditional models, or what are known as *RFM models*, consider the recency (date of most recent purchase), frequency (number of purchases), and monetary value (sales revenue) of previous purchases. More complicated models utilize a variety of explanatory variables relating to recency, frequency, monetary value, and customer demographics.

Useful reviews of traditional direct marketing are provided by Wunderman (1996), and Nash (2000, 1995). Hughes (2000) discusses strategies associated with database marketing and online direct marketing. Direct and database marketing is a rich area of application in marketing data science. Anand and Büchner (2002) discuss applications in cross-selling, finding prospects for additional products from an existing customer list. Blattberg, Kim, and Neslin (2008) provide a comprehensive review modeling methods in direct and database marketing, including extensive discussion of RFM models, lift charts, and alternative methods for setting probability cut-offs.

Lift charts and ROC curves are common tools in direct and database marketing. Area under the ROC curve is a good way to evaluate the statistical accuracy of classifiers, especially when working on a low-base-rate problem as observed in the Bank Marketing Study. There are many other ways to evaluate the statistical accuracy of a classifier. See appendix A (page 271).

Target marketing (in particular, one-to-one target marketing) has been bolstered by the emergence of hierarchical Bayesian methods. Bayesians use the term *consumer heterogeneity* to refer to individual differences across customers. The thinking is that describing consumers in terms of their positions along underlying attribute parameters is more informative than describing them as being members of segments. Bayesian methods in marketing, reviewed by Rossi, Allenby, and McCulloch (2005), have been implemented in R packages by Rossi (2014) and Sermas (2014).

Exhibit 3.1 shows the R program for identifying target customers in the Bank Marketing Study. The program draws on R packages provided by Meyer, Zeileis, Hornik, and Friendly (2014), Sarkar (2008, 2014), and Sing et al. (2015). The corresponding Python program is on the website for the book.

***Exhibit 3.1.*** *Identifying Customer Targets (R)*

```
# Identifying Customer Targets (R)

# call in R packages for use in this study
library(lattice)  # multivariate data visualization
library(vcd)  # data visualization for categorical variables
library(ROCR)  # evaluation of binary classifiers

# read bank data into R, creating data frame bank
# note that this is a semicolon-delimited file
bank <- read.csv("bank.csv", sep = ";", stringsAsFactors = FALSE)
# examine the structure of the bank data frame
print(str(bank))

# look at the first few rows of the bank data frame
print(head(bank))

# look at the list of column names for the variables
print(names(bank))

# look at class and attributes of one of the variables
print(class(bank$age))
print(attributes(bank$age))  # NULL means no special attributes defined
# plot a histogram for this variable
with(bank, hist(age))

# examine the frequency tables for categorical/factor variables
# showing the number of observations with missing data (if any)

print(table(bank$job , useNA = c("always")))
print(table(bank$marital , useNA = c("always")))
print(table(bank$education , useNA = c("always")))
print(table(bank$default , useNA = c("always")))
print(table(bank$housing , useNA = c("always")))
print(table(bank$loan , useNA = c("always")))

# Type of job (admin., unknown, unemployed, management,
# housemaid, entrepreneur, student, blue-collar, self-employed,
# retired, technician, services)
# put job into three major categories defining the factor variable jobtype
# the "unknown" category is how missing data were coded for job...
# include these in "Other/Unknown" category/level
white_collar_list <- c("admin.","entrepreneur","management","self-employed")
blue_collar_list <- c("blue-collar","services","technician")
bank$jobtype <- rep(3, length = nrow(bank))
bank$jobtype <- ifelse((bank$job %in% white_collar_list), 1, bank$jobtype)
bank$jobtype <- ifelse((bank$job %in% blue_collar_list), 2, bank$jobtype)
bank$jobtype <- factor(bank$jobtype, levels = c(1, 2, 3),
    labels = c("White Collar", "Blue Collar", "Other/Unknown"))
with(bank, table(job, jobtype, useNA = c("always")))  # check definition

# define factor variables with labels for plotting
```

```
bank$marital <- factor(bank$marital,
    labels = c("Divorced", "Married", "Single"))
bank$education <- factor(bank$education,
    labels = c("Primary", "Secondary", "Tertiary", "Unknown"))
bank$default <- factor(bank$default, labels = c("No", "Yes"))
bank$housing <- factor(bank$housing, labels = c("No", "Yes"))
bank$loan <- factor(bank$loan, labels = c("No", "Yes"))
bank$response <- factor(bank$response, labels = c("No", "Yes"))

# select subset of cases never perviously contacted by sales
# keeping variables needed for modeling
bankwork <- subset(bank, subset = (previous == 0),
    select = c("response", "age", "jobtype", "marital", "education",
               "default", "balance", "housing", "loan"))

# examine the structure of the bank data frame
print(str(bankwork))
# look at the first few rows of the bank data frame
print(head(bankwork))
# compute summary statistics for initial variables in the bank data frame
print(summary(bankwork))

# -----------------
# age  Age in years
# -----------------
# examine relationship between age and response to promotion
pdf(file = "fig_targeting_customers_age_lattice.pdf",
    width = 8.5, height = 8.5)
lattice_plot_object <- histogram(~age | response, data = bankwork,
    type = "density", xlab = "Age of Bank Client", layout = c(1,2))
print(lattice_plot_object)  # responders tend to be older
dev.off()
# ----------------------------------------------------------
# education
# Level of education (unknown, secondary, primary, tertiary)
# ----------------------------------------------------------
# examine the frequency table for education
# the "unknown" category is how missing data were coded
with(bankwork, print(table(education, response, useNA = c("always"))))
# create a mosaic plot in using vcd package
pdf(file = "fig_targeting_customers_education_mosaic.pdf",
    width = 8.5, height = 8.5)
mosaic( ~ response + education, data = bankwork,
  labeling_args = list(set_varnames = c(response = "Response to Offer",
  education = "Education Level")),
  highlighting = "education",
  highlighting_fill = c("cornsilk","violet","purple","white",
     "cornsilk","violet","purple","white"),
  rot_labels = c(left = 0, top = 0),
  pos_labels = c("center","center"),
  offset_labels = c(0.0,0.6))
dev.off()
```

```
# ----------------------------------------------------------------
# job status using jobtype
# White Collar: admin., entrepreneur, management, self-employed
# Blue Collar: blue-collar, services, technician
# Other/Unknown
# ----------------------------------------------------------------
# review the frequency table for job types
with(bankwork, print(table(jobtype, response, useNA = c("always"))))
pdf(file = "fig_targeting_customers_jobtype_mosaic.pdf",
    width = 8.5, height = 8.5)
mosaic( ~ response + jobtype, data = bankwork,
  labeling_args = list(set_varnames = c(response = "Response to Offer",
  jobtype = "Type of Job")),
  highlighting = "jobtype",
  highlighting_fill = c("cornsilk","violet","purple",
      "cornsilk","violet","purple"),
  rot_labels = c(left = 0, top = 0),
  pos_labels = c("center","center"), offset_labels = c(0.0,0.6))
dev.off()
# ---------------------------------------------
# Marital status (married, divorced, single)
# [Note: ‘‘divorced’’ means divorced or widowed]
# ---------------------------------------------
# examine the frequency table for marital status
# anyone not single or married was classified as "divorced"
with(bankwork, print(table(marital, response, useNA = c("always"))))
pdf(file = "fig_targeting_customers_marital_mosaic.pdf",
    width = 8.5, height = 8.5)
mosaic( ~ response + marital, data = bankwork,
  labeling_args = list(set_varnames = c(response = "Response to Offer",
  marital = "Marital Status")),
  highlighting = "marital",
  highlighting_fill = c("cornsilk","violet","purple",
      "cornsilk","violet","purple"),
  rot_labels = c(left = 0, top = 0),
  pos_labels = c("center","center"),
  offset_labels = c(0.0,0.6))
dev.off()
# ----------------------------------------
# default  Has credit in default? (yes, no)
# ----------------------------------------
with(bankwork, print(table(default, response, useNA = c("always"))))
pdf(file = "fig_targeting_customers_default_mosaic.pdf",
    width = 8.5, height = 8.5)
mosaic( ~ response + default, data = bankwork,
  labeling_args = list(set_varnames = c(response = "Response to Offer",
  default = "Has credit in default?")),
  highlighting = "default",
  highlighting_fill = c("cornsilk","violet"),
  rot_labels = c(left = 0, top = 0),
  pos_labels = c("center","center"),
  offset_labels = c(0.0,0.6))
dev.off()
```

```
# ------------------------------------------
# balance  Average yearly balance (in Euros)
# ------------------------------------------
# examine relationship between age and response to promotion
pdf(file = "fig_targeting_customers_balance_lattice.pdf",
    width = 8.5, height = 8.5)
lattice_plot_object <- histogram(~balance | response, data = bankwork,
    type = "density",
    xlab = "Bank Client Average Yearly Balance (in dollars)",
    layout = c(1,2))
print(lattice_plot_object)  # responders tend to be older
dev.off()
# ------------------------------------
# housing  Has housing loan? (yes, no)
# ------------------------------------
with(bankwork, print(table(housing, response, useNA = c("always"))))
pdf(file = "fig_targeting_customers_housing_mosaic.pdf",
    width = 8.5, height = 8.5)
mosaic( ~ response + housing, data = bankwork,
  labeling_args = list(set_varnames = c(response = "Response to Offer",
  housing = "Has housing loan?")),
  highlighting = "housing",
  highlighting_fill = c("cornsilk","violet"),
  rot_labels = c(left = 0, top = 0),
  pos_labels = c("center","center"),
  offset_labels = c(0.0,0.6))
dev.off()
# ----------------------------------
# loan  Has personal loan? (yes, no)
# ----------------------------------
with(bankwork, print(table(loan, response, useNA = c("always"))))
pdf(file = "fig_targeting_customers_loan_mosaic.pdf",
    width = 8.5, height = 8.5)
mosaic( ~ response + loan, data = bankwork,
  labeling_args = list(set_varnames = c(response = "Response to Offer",
  loan = "Has personal loan?")),
  highlighting = "loan",
  highlighting_fill = c("cornsilk","violet"),
  rot_labels = c(left = 0, top = 0),
  pos_labels = c("center","center"),
  offset_labels = c(0.0,0.6))
dev.off()
# ----------------------------------
# specify predictive model
# ----------------------------------
bank_spec <- {response ~ age + jobtype + education + marital +
    default + balance + housing + loan}
# ----------------------------------
# fit logistic regression model
# ----------------------------------
bank_fit <- glm(bank_spec, family=binomial, data=bankwork)
print(summary(bank_fit))
print(anova(bank_fit, test="Chisq"))
```

```
# compute predicted probability of taking the train
bankwork$Predict_Prob_Response <- predict.glm(bank_fit, type = "response")

pdf(file = "fig_targeting_customer_log_reg_density_evaluation.pdf",
    width = 8.5, height = 8.5)
plotting_object <- densityplot( ~ Predict_Prob_Response | response,
               data = bankwork,
               layout = c(1,2), aspect=1, col = "darkblue",
               plot.points = "rug",
               strip=function(...) strip.default(..., style=1),
               xlab="Predicted Probability of Responding to Offer")
print(plotting_object)
dev.off()

# predicted response to offer using using 0.5 cut-off
# notice that this does not work due to low base rate
# we get more than 90 percent correct with no model
# (predicting all NO responses)
# the 0.50 cutoff yields all NO predictions
bankwork$Predict_Response <-
    ifelse((bankwork$Predict_Prob_Response > 0.5), 2, 1)

bankwork$Predict_Response <- factor(bankwork$Predict_Response,
    levels = c(1, 2), labels = c("NO", "YES"))

confusion_matrix <- table(bankwork$Predict_Response, bankwork$response)
cat("\nConfusion Matrix (rows=Predicted Response, columns=Actual Choice\n")
print(confusion_matrix)
predictive_accuracy <- (confusion_matrix[1,1] + confusion_matrix[2,2])/
                        sum(confusion_matrix)
cat("\nPercent Accuracy: ", round(predictive_accuracy * 100, digits = 1))

# this problem requires either a much lower cut-off
# or other criteria for evaluation... let's try 0.10 (10 percent cut-off)
bankwork$Predict_Response <-
    ifelse((bankwork$Predict_Prob_Response > 0.1), 2, 1)
bankwork$Predict_Response <- factor(bankwork$Predict_Response,
    levels = c(1, 2), labels = c("NO", "YES"))
confusion_matrix <- table(bankwork$Predict_Response, bankwork$response)
cat("\nConfusion Matrix (rows=Predicted Response, columns=Actual Choice\n")
print(confusion_matrix)
predictive_accuracy <- (confusion_matrix[1,1] + confusion_matrix[2,2])/
                        sum(confusion_matrix)
cat("\nPercent Accuracy: ", round(predictive_accuracy * 100, digits = 1))
# mosaic rendering of the classifier with 0.10 cutoff
with(bankwork, print(table(Predict_Response, response, useNA = c("always"))))
pdf(file = "fig_targeting_customers_confusion_mosaic_10_percent.pdf",
    width = 8.5, height = 8.5)
mosaic( ~ Predict_Response + response, data = bankwork,
  labeling_args = list(set_varnames =
  c(Predict_Response =
      "Predicted Response to Offer (10 percent cut-off)",
      response = "Actual Response to Offer")),
```

```
  highlighting = c("Predict_Response", "response"),
  highlighting_fill = c("green","cornsilk","cornsilk","green"),
  rot_labels = c(left = 0, top = 0),
  pos_labels = c("center","center"),
  offset_labels = c(0.0,0.6))
dev.off()

# compute lift using prediction() from ROCR and plot lift chart
bankwork_prediction <-
    prediction(bankwork$Predict_Prob_Response, bankwork$response)
bankwork_lift <- performance(bankwork_prediction , "lift", "rpp")
pdf(file = "fig_targeting_customers_lift_chart.pdf",
    width = 8.5, height = 8.5)
plot(bankwork_lift,
col = "blue", lty = "solid", main = "", lwd = 2,
    xlab = paste("Proportion of Clients Ordered by Probability",
    " to Subscribe\n(from highest to lowest)", sep = ""),
    ylab = "Lift over Baseline Subscription Rate")
dev.off()

# direct calculation of lift
baseline_response_rate <-
    as.numeric(table(bankwork$response)[2])/nrow(bankwork)
prediction_deciles <- quantile(bankwork$Predict_Prob_Response,
    probs = seq(0, 1, 0.10), na.rm = FALSE)
# reverse the deciles from highest to lowest
reordered_probability_deciles <- rev(as.numeric(prediction_deciles))
lift_values <- reordered_probability_deciles / baseline_response_rate
cat("\nLift Chart Values by Decile:", lift_values, "\n")

# Suggestions for the student:
# Try alternative methods of classification, such as neural networks,
# support vector machines, and random forests. Compare the performance
# of these methods against logistic regression. Use alternative methods
# of comparison, including area under the ROC curve.
# Ensure that the evaluation is carried out using a training-and-test
# regimen, perhaps utilizing multifold cross-validation.
# Check out the R package cvTools for doing this work.
# Examine the importance of individual explanatory variables
# in identifying targets. This may be done by looking at tests of
# statistical significance, classification trees, or random-forests-
# based importance assessment.
```

# 4 Finding New Customers

"Fox, where have you been for the last three hours? If I were you (and I thank my personal god I am not), I wouldn't be sitting around chin wagging. Plenty of six-figure names in that zip code file to cold-call."

—James Karen as Lynch in *Wall Street* (1987)

No business can long survive without finding new customers. The process of finding new customers begins by learning as much as we can about current customers and groups of customers.

We identify the types of customers who are most likely to buy as well as sets of variables that may be used to find those customers. Ranges of values on those sets of variables define market segments, which in turn can be used to identify new customers.

A market segment is a group of consumers that is different from other groups of consumers in ways that matter to marketing managers. For target marketing, a customer's segment membership can be among the most useful of explanatory variables.

We can identify segments by geographic, demographic, psychographic, and behavioral characteristics. Most useful for target marketing are characteristics that are accessible and easily measured.

The process of identifying market segments is called market segmentation. Market segmentation, when executed properly, contributes to marketing strategy and tactics. What consumers like, what they buy, where they buy, and how much they buy may differ across segments. Products, marketing messages, advertising, and promotions may be tailored to specific segments.

Marketing managers use market segments in product development, advertising, promotion, pricing, and target marketing. Some consumers are responsive to advertisements and promotions. Some are more price-conscious than others. By using knowledge of a consumer's market segment, marketing managers can make informed decisions about marketing actions.

To be useful to marketing management, market segments must be discoverable and reachable. We must be able to identify a consumers segment from available data, find our way to that consumer, and take appropriate marketing action. This is the core of sales and marketing: converting a consumer with little knowledge of a product into a customer of that product.

Variables that go into the segmentation should be easily available or accessible. We avoid variables that are difficult to measure. And we prefer publicly available data. Common segmentation variables reflect geographical location, age, income, and behavioral or lifestyle variables. We look for data that we can gather without survey sampling.

Consider a company that sells audio equipment. On the one hand, there are casual listeners—any speakers, headphones, or ear buds would satisfy their needs. Music lovers, on the other hand, may demand the highest quality in audio equipment. Casual listeners and music lovers are market segments for a manufacturer or seller of audio equipment.

Marketing managers sometimes identify segmentation variables before doing research. Knowing the nature of their products, they feel confident describing potential buyers or groups of consumers. Identifying segments prior to data analysis and modeling is called *a priori* segmentation. Identifying segments that are data-based using modeling techniques is called *post hoc* segmentation.

Consumer information is critical to market segmentation regardless of your approach. If an analyst takes an *a priori* approach, defining segments prior to looking at data, then he or she would likely look at actual data to determine how these predefined segments differed from one another.

To be useful in finding new customers, identified market segments should apply to the larger group of consumers, both current customers and potential customers. This means that we must select input variables carefully.

Sales transaction data may be useful in identifying variables to be used for segmentation. But these data themselves should not be used as input to segmentation. We do not use sales transaction data directly because information about sales is available only for current customers or consumers who have had some contact with the company.

Varieties of data for market segmentation are many and varied. Geographic data refer to where people live, such as the region of the country, state, city, street location, or census area. Behavioral data include occupational information, how consumers travel, where they shop, what they read, and to which community groups they belong. Demographic data include variables such as gender, age, education, occupation, and level of income. Psychographic data include psychological factors, attitudes, interests, personality characteristics, and lifestyle factors.

Age is a common variable in *a priori* and data-based (post hoc) segmentation. Many marketing managers think in terms of generations by date of birth. It is common to distinguish among members of the silent generation (birth range 1925–1945), baby boomers (1946–1964), Generation Xers (1965–1978) , and Millennials (also called Generation Y, 1979–1994)). Consumers in generational segments are understood to have diverse needs, attitudes, and behaviors. Generational differences can affect consumer response to marketing messages, products, promotions, and pricing. Millennials, for example, can be expected to be less price-sensitive and more technologically savvy (Kotler and Keller 2012).

Marketing managers often design products and tailor marketing messages for individual segments or combinations of segments. Knowing that Millennials are technology-savvy, for example, may prompt managers to advertise products on computers and mobile devices to reach the segment. Or they may emphasize online sales promotion as opposed to brick-and-mortar promotions.

Using existing customer information for segmentation, we would exclude buyer status and other sales response variables from the input variable set. But we can utilize buyer status and sales response variables in the input variable selection process itself. For example, we may choose demographic variables that are related to sales response. We can also use buyer status and sales response variables to evaluate segments after those segments have been identified.

Traditional methods of cluster analysis are widely used in market segmentation. They represent multivariate techniques for grouping consumers based on their similarity to one another. Distance metrics or measures of agreement between consumers guide the segmentation process.

To illustrate the clustering process we again refer the the Bank Marketing Study (page 356). We begin by considering bank clients who are being approached with a sales call for the first time. Many variables to which we have access—variables that are entirely appropriate for target marketing—are not good candidates for market segmentation. Client information relating to personal finances and banking transactions would not be useful as inputs to segmentation analysis because these variables depend on the client's history with the bank. Non-customers have no such history. So for segmentation work with the Bank Marketing Study, we focus on demographic variables.

Exploratory data analysis can provide an initial picture of demographic variables used in market segmentation. Some variables may have unusual distributions or little variability, suggesting that they would be of little use in identifying market segments. From earlier work with the Bank Marketing Study, we learned that there were low frequencies for some job types, so we formed larger categories of white collar, blue collar, and other jobs. From additional analysis of the case (figures 3.1 through 3.4, pages 32–34), we learned that bank clients who subscribed to term deposit offers were older, more highly educated, more likely to be white collar workers than blue collar workers, and more likely to be single, divorced, or widowed than married.

We often use cluster analysis to identify market segments. There are two major types of cluster analysis: hierarchical cluster analysis and partitioning. For the Bank Marketing Study, we will use a partitioning method that requires that input variables have meaningful magnitude or be binary (two-level) categorical variables. Except for age, demographic variables in the Bank Marketing Study are multi-category demographic variables. So to prepare the bank's data for analysis, we convert multi-category demographic variables into binary categorical variables. It is appropriate, as well, to standardize all input variables prior to clustering.

Cluster analysis involves finding groups in data. Cluster analysis takes many consumer-related variables and uses them to represent differences between consumers. When two consumers have similar values for variables like age, marital status, job type, and education, they are seen as being similar. When two consumers have very different values on these variables, they are seen as dissimilar. Cluster analysis looks at the differences among consumers—their distances from one another—to identify groups of consumers or market segments.

In most cases, we would perform many cluster analyses, providing a set of clustering solutions. We can begin by seeing what happens when we divide the entire set of consumers into just two groups or clusters. Then we can try three clusters, then four, and so on. The number of clusters is an important feature of a clustering solution. If the number of clusters is too large, the clustering solution may be difficult to utilize in product marketing.

Specialized statistics like the silhouette coefficient reflect the extent to which clusters are distinct from one another (Rousseeuw 1987). This information is helpful in selecting a clustering solution. We look for segments that are well defined and few in number.

It is common to look for solutions involving ten or fewer clusters. We also seek solutions for which all clusters or segments are of sufficient size to warrant marketing attention. We note the proportion of consumers in each cluster when evaluating solutions. Clusters that are very small may be of little interest to management. The fewer consumers in a cluster, the lower the potential sales revenue.

Selecting a clustering solution among the many possible solutions we obtain from cluster analysis algorithms is as much art as science. There are numerous factors to consider in the context of market segmentation. We look for segments that are easy to interpret in terms of descriptive statistics. After segments have been identified, we can return to the full database of customer or client information to see if there are relationships between segment membership and sales order history and/or responsiveness to sales and marketing activities.

If we are successful in identifying differences across segments, then managers will have an easier time understanding the meaning of segments. And if managers understand the meaning of segments (how segments are defined, how they differ from one another, and how they relate to the products being offered by the firm), then managers will be more likely to use the results of segmentation to guide business decisions.

For the Bank Marketing Study, we searched across clustering solutions with between two and twenty clusters. We determined that a seven-cluster solution made the most sense, with the first five of seven clusters or segments being well defined.

The lattice plot in figure 4.1 shows age across the five clusters, and the mosaic plot in figure 4.2 shows the response to term deposit offers across the five clusters. Table 4.3 provides a summary of the resulting market segments using demographic and bank client history and response-to-term-deposit factors. For the Bank Marketing Study, we can see that segments identified as B and E would constitute likely targets.

Market segmentation and target marketing often go hand-in-hand. But there are people in marketing who are philosophically opposed to using segments for targeting. Instead of targeting market segments, these researchers promote a one-to-one marketing approach that targets each individual as an individual, much as we did in the previous chapter on targeting customers.

The process described here is an example of data-based, demographic segmentation. Information about age, education, job type, and income is more easily obtained than information about needs, attitudes, and interests. We want identified segments to be discoverable.

*Figure 4.1.* Age of Bank Client by Market Segment

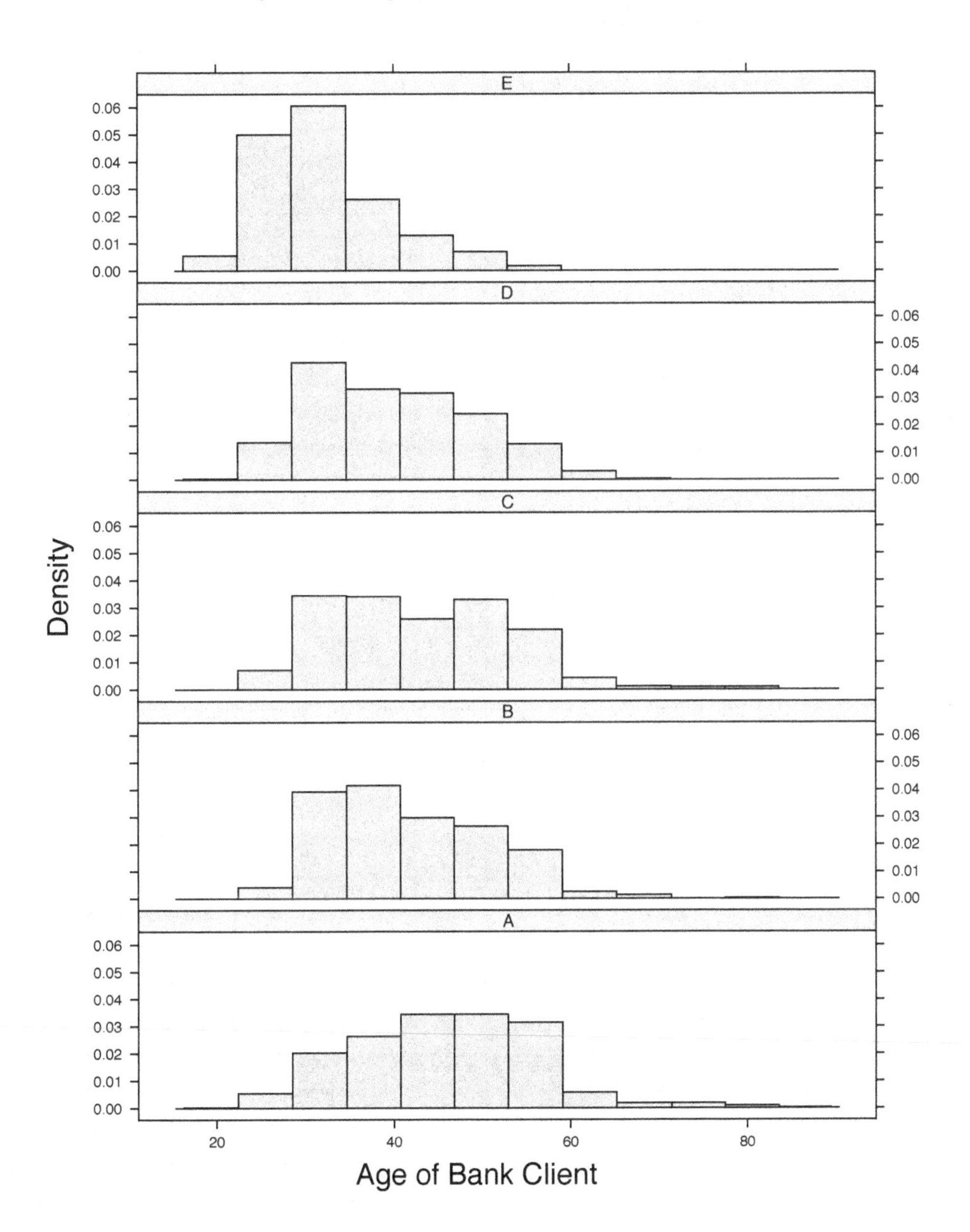

*Figure 4.2. Response to Term Deposit Offers by Market Segment*

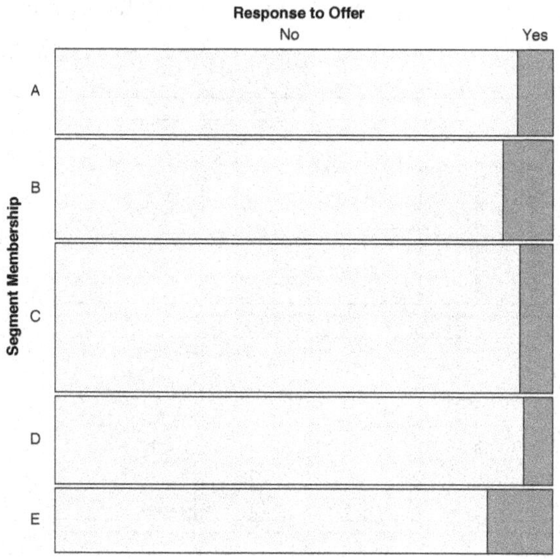

Response to Offer
No Yes

Segment Membership
A
B
C
D
E

*Figure 4.3. Describing Market Segments in the Bank Marketing Study*

| | *Segment Membership* | | | | |
|---|---|---|---|---|---|
| | A | B | C | D | E |
| Percentage of Customers in Segment | 13.7 | 16.0 | 23.8 | 14.0 | 10.3 |
| *Average Age* | 45.9 | 41.7 | 43.0 | 40.0 | 32.3 |
| *Education* | Primary | Tertiary | Secondary | Secondary | Secondary |
| *Job Type* | Mixed | Mostly White Collar | Mostly Blue Collar | White Collar | Blue Collar |
| *Marital Status* | Mostly Married | Married | Married | Mostly Married | Single |
| *Percentage Responding to Term Deposit Offer* | 7.1 | 9.9 | 6.6 | 5.8 | 13.1 |

The benefits of market segmentation are described in marketing management references, such as Dickson (1997) and Kotler and Keller (2012). Sternthal and Tybout (2001) and Cespedes, Dougherty, and Skinner (2013) review management issues in segmentation and targeting. Frank, Massey, and Wind (1972), Neal (2000), and Wedel and Kamakura (2000) discuss the objectives and methods of market segmentation.

Exhibit 4.1 shows the R program for identifying market segments in the Bank Marketing Study. It brings in the client data, recodes selected explanatory variables, generates clustering solutions, selects an appropriate solution for working with bank managers, and provides a statistical description of the resulting market segments. We use a partitioning method developed by Kaufman and Rousseeuw (1990) that is similar in concept to K-means cluster analysis. Partitioning methods are good for working with large data sets. The R program draws on packages for cluster analysis and data visualization provided by Maechler (2014a), Meyer, Zeileis, Hornik, and Friendly (2014), and Sarkar (2008, 2014).

Exhibit 4.2 shows a Python program that uses K-means clustering to identify client segments for the Bank Marketing Study. The selected clustering solution in this example suggests that two segments are best.

***Exhibit 4.1.*** *Identifying Consumer Segments (R)*

```
# Identifying Consumer Segments (R)

# call in R packages for use in this study
library(lattice)  # multivariate data visualization
library(vcd)  # data visualization for categorical variables
library(cluster)  # cluster analysis methods

# read bank data into R, creating data frame bank
# note that this is a semicolon-delimited file
bank <- read.csv("bank.csv", sep = ";", stringsAsFactors = FALSE)

# examine the structure of the bank data frame
print(str(bank))
print(head(bank))

print(table(bank$job , useNA = c("always")))
print(table(bank$marital , useNA = c("always")))
print(table(bank$education , useNA = c("always")))
print(table(bank$default , useNA = c("always")))
print(table(bank$housing , useNA = c("always")))
print(table(bank$loan , useNA = c("always")))

# Type of job (admin., unknown, unemployed, management,
# housemaid, entrepreneur, student, blue-collar, self-employed,
# retired, technician, services)
# put job into three major categories defining the factor variable jobtype
# the "unknown" category is how missing data were coded for job...
# include these in "Other/Unknown" category/level
white_collar_list <- c("admin.","entrepreneur","management","self-employed")
blue_collar_list <- c("blue-collar","services","technician")
bank$jobtype <- rep(3, length = nrow(bank))
bank$jobtype <- ifelse((bank$job %in% white_collar_list), 1, bank$jobtype)
bank$jobtype <- ifelse((bank$job %in% blue_collar_list), 2, bank$jobtype)
bank$jobtype <- factor(bank$jobtype, levels = c(1, 2, 3),
    labels = c("White Collar", "Blue Collar", "Other/Unknown"))
with(bank, table(job, jobtype, useNA = c("always")))  # check definition
# define binary indicator variables as numeric 0/1 variables
bank$whitecollar <- ifelse((bank$jobtype == "White Collar"), 1, 0)
bank$bluecollar <- ifelse((bank$jobtype == "Blue Collar"), 1, 0)
with(bank, print(table(whitecollar, bluecollar)))  # check definition
with(bank, print(table(jobtype)))  # check definition

# define factor variables with labels for plotting and binary factors
bank$marital <- factor(bank$marital,
    labels = c("Divorced", "Married", "Single"))

# define binary indicator variables as numeric 0/1 variables
bank$divorced <- ifelse((bank$marital == "Divorced"), 1, 0)
bank$married <- ifelse((bank$marital == "Married"), 1, 0)
with(bank, print(table(divorced, married)))  # check definition
with(bank, print(table(marital)))  # check definition
```

```
bank$education <- factor(bank$education,
    labels = c("Primary", "Secondary", "Tertiary", "Unknown"))
# define binary indicator variables as numeric 0/1 variables
bank$primary <- ifelse((bank$education == "Primary"), 1, 0)
bank$secondary <- ifelse((bank$education == "Secondary"), 1, 0)
bank$tertiary <- ifelse((bank$education == "Tertiary"), 1, 0)
with(bank, print(table(primary, secondary, tertiary)))  # check definition
with(bank, print(table(education)))  # check definition

# client experience variables will not be useful for segmentation
# but can be referred to after segments have been defined
bank$default <- factor(bank$default, labels = c("No", "Yes"))
bank$housing <- factor(bank$housing, labels = c("No", "Yes"))
bank$loan <- factor(bank$loan, labels = c("No", "Yes"))
bank$response <- factor(bank$response, labels = c("No", "Yes"))

# select subset of cases never perviously contacted by sales
# keeping variables needed for cluster analysis and post-analysis
bankfull <- subset(bank, subset = (previous == 0),
    select = c("response", "age", "jobtype", "marital", "education",
               "default", "balance", "housing", "loan",
               "whitecollar", "bluecollar", "divorced", "married",
               "primary", "secondary", "tertiary"))

# examine the structure of the full bank data frame
print(str(bankfull))
print(head(bankfull))

# select subset of variables for input to cluster analysis
data_for_clustering <- subset(bankfull,
    select = c("age",
               "whitecollar", "bluecollar",
               "divorced", "married",
               "primary", "secondary", "tertiary"))

# -----------------------------------------------------
# clustering solutions (min_clusters to max_clusters)
# this step may take 10 minutes or more to complete
# -----------------------------------------------------
# set file for graphical output from the clustering solutions
pdf(file = "fig_finding_new_customers_cluster_search.pdf",
        width = 8.5, height = 11)
min_clusters <- 2
max_clusters <- 20
# evaluate alternative numbers of clusters/segments
# we use the average silhouette width as a statistical criterion
evaluation_vector <- NULL  # initialize evaluation vector
# selected algorithm is pam (partitioning around medoids)
# with so many binary variables, manhattan distances seemed
# to work better than Euclidean distances
for (number_of_clusters in min_clusters:max_clusters) {
    try_clustering <- pam(data_for_clustering, k = number_of_clusters,
        metric = "manhattan", stand = TRUE)
```

```
    evaluation_vector <- rbind(evaluation_vector,
        data.frame(number_of_clusters,
            average_silhouette_width =
                try_clustering$silinfo$avg.width))
    # show plot for this clustering solution
    plot(try_clustering)  # add this clustering solution to results file
    }
dev.off()  # close the pdf results file for the clustering solution
# examine the cluster solution results,
# look for average silhouette width > 0.5
# look for last big jump in average silhoutte width

print(evaluation_vector)

# provide a single summary plot for the clustering solutions
pdf(file = "fig_finding_new_customers_cluster_summary.pdf",
        width = 8.5, height = 8.5)
with(evaluation_vector, plot(number_of_clusters,
    average_silhouette_width))
dev.off()  # close summary results file

# -----------------------------------------------------
# select clustering solution and examine it
# -----------------------------------------------------
# examine the seven-cluster solution in more detail
seven_cluster_solution <- pam(data_for_clustering, k = 8,
        metric = "manhattan", stand = TRUE)
pdf(file = "fig_finding_new_customers_seven_cluster_solution.pdf",
    width = 8.5, height = 8.5)
plot(seven_cluster_solution)
dev.off()
# from the silhouette plot, the first five of the seven
# clusters appear to be large and well-defined

# add the cluster membership information and select first five
bankfull$cluster <- seven_cluster_solution$clustering
bankpart <- subset(bankfull, subset = (cluster < 6))
bankpart$cluster <- factor(bankpart$cluster,
    labels = c("A", "B", "C", "D", "E"))

# look at demographics across the clusters/segments
# -----------------
# age  Age in years
# -----------------
# examine relationship between age and response to promotion
with(bankpart, print(by(age, cluster, mean)))
pdf(file = "fig_finding_new_customers_age_lattice.pdf",
    width = 8.5, height = 11)
lattice_plot_object <- histogram(~age | cluster, data = bankpart,
    type = "density",
    xlab = "Age of Bank Client", layout = c(1,5))
print(lattice_plot_object)  # responders tend to be older
dev.off()
```

```
# ----------------------------------------------------------
# education
# Level of education (unknown, secondary, primary, tertiary)
# ----------------------------------------------------------
with(bankpart, print(table(cluster, education)))

# --------------------------------------------------------------
# job status using jobtype
# White Collar: admin., entrepreneur, management, self-employed
# Blue Collar: blue-collar, services, technician
# Other/Unknown
# --------------------------------------------------------------
with(bankpart, print(table(cluster, jobtype)))

# ---------------------------------------------
# marital status
# Marital status (married, divorced, single)
# [Note: ''divorced'' means divorced or widowed]
# ---------------------------------------------
with(bankpart, print(table(cluster, marital)))

# look at bank client history across the clusters/segments

# ----------------------------------------
# default  Has credit in default? (yes, no)
# ----------------------------------------
with(bankpart, print(table(cluster, default)))

# -----------------------------------------
# balance  Average yearly balance (in Euros)
# -----------------------------------------
with(bankpart, print(by(balance, cluster, mean)))
pdf(file = "fig_finding_new_customers_blance_lattice.pdf",
    width = 8.5, height = 11)
lattice_plot_object <- histogram(~balance | cluster, data = bankpart,
    type = "density", xlab = "Age of Bank Client",
    layout = c(1,5))
print(lattice_plot_object)  # responders tend to be older
dev.off()

# ------------------------------------
# housing  Has housing loan? (yes, no)
# ------------------------------------
with(bankpart, print(table(cluster, housing)))

# ----------------------------------
# loan  Has personal loan? (yes, no)
# ----------------------------------
with(bankpart, print(table(cluster, loan)))

# ---------------------------------------------------
# response  Response to term deposit offer (yes, no)
# ---------------------------------------------------
```

```
with(bankpart, print(table(cluster, response)))
pdf(file = "fig_finding_new_customers_response_mosaic.pdf",
    width = 8.5, height = 8.5)
mosaic( ~ response + cluster, data = bankpart,
  labeling_args = list(set_varnames = c(response = "Response to Term Deposit Offer",
  cluster = "Segment Membership")),
  highlighting = "response",
  highlighting_fill = c("cornsilk","violet"),
  rot_labels = c(left = 0, top = 0),
  pos_labels = c("center","center"),
  offset_labels = c(0.0,0.6))
dev.off()

# compute percentage of yes responses to term deposit offer
response_table <- table(bankpart$cluster, bankpart$response)
cat("\nPercentage Responses\n")
for (i in 1:5)
     cat("\n", toupper(letters[i]),
         round(100 * response_table[i,2] /
             sum(response_table[i,]), digits = 1))

# note the percentage of the customers receiving offers
# for the first time falling into each of the clusters/segments
# A = 1 ... E = 5 ...
print(round(100 * table(bankfull$cluster) / nrow(bankfull), digits = 1))

# Suggestions for the student:
# Try alternative clustering methods, using various
# distance measures and clustering algorithms.
# Try your hand at slecting a best clustering solution
# and interpreting the solution for bank managers.
# Could your clustering solution be useful in
# target marketing?
```

**Exhibit 4.2.** *Identifying Consumer Segments (Python)*

```
# Identifying Consumer Segments (Python)

# prepare for Python version 3x features and functions
from __future__ import division, print_function

# import packages for data manipulation and multivariate analysis
import pandas as pd  # DataFrame structure and operations
import numpy as np  # arrays and numerical processing
from sklearn.cluster import KMeans  # cluster analysis by partitioning
from sklearn.metrics import silhouette_score as silhouette_score

# read data from comma-delimited text file... create DataFrame object
bank = pd.read_csv('bank.csv', sep = ';')
print(bank.head)  # check the structure of the data frame
print(bank.shape)
# look at the list of column names
list(bank.columns.values)

# examine the demographic variable age
print(bank['age'].unique())
print(bank['age'].value_counts(sort = True))
print(bank['age'].describe())

# examine the demographic variable job
print(bank['job'].unique())
print(bank['job'].value_counts(sort = True))
print(bank['job'].describe())
# define job indicator variables
job_indicators = pd.get_dummies(bank['job'], prefix = 'job')
print(job_indicators.head())
bank = bank.join(job_indicators)
bank['whitecollar'] = bank['job_admin.'] + bank['job_management'] + \
    bank['job_entrepreneur'] + bank['job_self-employed']
bank['bluecollar'] = bank['job_blue-collar'] + bank['job_services'] + \
    bank['job_technician'] + bank['job_housemaid']

# examine the demographic variable marital
print(bank['marital'].unique())
print(bank['marital'].value_counts(sort = True))
print(bank['marital'].describe())
# define marital indicator variables
marital_indicators = pd.get_dummies(bank['marital'], prefix = 'marital')
print(marital_indicators.head())
bank = bank.join(marital_indicators)
bank['divorced'] = bank['marital_divorced']
bank['married'] = bank['marital_married']

# examine the demographic variable education
print(bank['education'].unique())
print(bank['education'].value_counts(sort = True))
print(bank['education'].describe())
# define education indicator variables
```

```
education_indicators = pd.get_dummies(bank['education'], prefix = 'education')
print(education_indicators.head())
bank = bank.join(education_indicators)
bank['primary'] = bank['education_primary']
bank['secondary'] = bank['education_secondary']
bank['tertiary'] = bank['education_tertiary']

# select/filter for cases never previously contacted by sales
bank_selected = bank[bank['previous'] == 0]
print(bank_selected.shape)
# select subset of variables needed for cluster analysis and post-analysis
bankfull = pd.DataFrame(bank_selected, \
    columns = ['response', 'age', 'whitecollar', 'bluecollar',
               'divorced', 'married',
               'primary', 'secondary', 'tertiary'])
# examine the structure of the full bank DataFrame
print(bankfull.head)  # check the structure of the data frame
print(bankfull.shape)
# look at the list of column names
list(bankfull.columns.values)

# select subset of variables for input to cluster analysis
data_for_clustering = pd.DataFrame(bank_selected, \
    columns = ['age', 'whitecollar', 'bluecollar',
               'divorced', 'married',
               'primary', 'secondary', 'tertiary'])
# convert to matrix/numpy array for input to cluster analysis
data_for_clustering_matrix = data_for_clustering.as_matrix()

# investigate alternative numbers of clusters using silhouette score
silhouette_value = []
k = range(2,21)  # look at solutions between 2 and 20 clusters
for i in k:
    clustering_method = KMeans(n_clusters = i, random_state = 9999)
    clustering_method.fit(data_for_clustering_matrix)
    labels = clustering_method.predict(data_for_clustering_matrix)
    silhouette_average = silhouette_score(data_for_clustering_matrix, labels)
    silhouette_value.append(silhouette_average)

# highest silhouette score is for two clusters... use that solution here
clustering_method = KMeans(n_clusters = 2, random_state = 9999)
clustering_method.fit(data_for_clustering_matrix)
labels = clustering_method.predict(data_for_clustering_matrix)

# add cluster labels to bankfull DataFrame and review the solution
bankfull['cluster'] = labels
# pivot table and cross-tabulation examples
bankfull.pivot_table(rows = ['cluster'])
pd.crosstab(bankfull.cluster, bankfull.bluecollar, margins = True)
# groupby example
segments = bankfull.groupby('cluster')
segments.describe()
```

# 5 Retaining Customers

"Italian men think that 'Fidelity' is the name
of the woman that lives across the hall."

—STÉPHANE DEBAC AS FRANÇOIS IN *Girl on a Bicycle* (2013)

I first saw someone wearing Vibram FiveFingers® a couple years ago. These are shoes with separate compartments for each toe. I call them toe-shoes.

Recently I noticed a man wearing toe-shoes at the grocery store. I asked him how they felt, and he said, "It is like walking in your bare feet."

Feeling adventurous, I ordered a pair of toe-shoes from Zappos. I tried men's $8\frac{1}{2}$ medium. They were much too small, so I returned them. Then I tried $9\frac{1}{2}$ to 10—they were tight and uncomfortable. For tennis and walking shoes, I wear men's size 9 medium. The next larger size toe-shoes would have been men's size 11. Reluctant to be seen as "mister big foot" or the man with big toes, I decided toe-shoes were not for me.

I used to shop for shoes and clothing in brick-and-mortar stores, thinking I had to try things on before buying. That has changed. The selection online is large, and I no longer worry about returns. Curiously, my horrible experience with toe-shoes turned into a good experience with Zappos. This company knows what it takes to retain customers.

Following the breakup of the firm in 1986, AT&T customers had a choice: they could stay with AT&T or move to some other common carrier. In the data set for the AT&T Choice Study (appendix C, page 353), the response variable is binary, taking values *ATT* or *OCC*. What are the odds that a customer will switch telephone services and choose some other common carrier? How can AT&T managers predict which customers will switch to another service provider? And what can AT&T do, if anything, to reduce the probability of a customer switching?

Markets for communication services are highly competitive. The word *churn* is used to describe the activities of service providers known for their attempts at stealing another's customers. The odds of a customer going to another service provider is a ratio of probabilities. In the AT&T Choice Study, we have the odds ratio $\frac{p(OCC)}{p(ATT)}$ that shows the odds of an AT&T customer moving to some other common carrier. AT&T wants this odds ratio to be small.[1]

Using odds and log odds transformations, we can write equations linking choices (or more precisely, the probability of choices) to linear predictors. This is the logic of the logit as employed in logistic regression. Explanatory variables comprise the linear predictor.

There are nine explanatory variables in the AT&T Choice Study, including information about household income and mobility, and demographic information about survey respondents (age level, education level, and employment status). There is also information about telephone usage and participation in AT&T marketing programs. There are nearly complete data for telephone usage (average minutes per month) and two AT&T marketing programs.

---

[1] $\frac{p(OCC)}{p(ATT)}$ gives AT&T's odds of losing a customer (seeing the customer switch from AT&T to some other common carrier). Assuming that $p(OCC)$ and $p(ATT)$ are numbers between zero and one, their odds ratio will be positive. And if the probability of switching is greater than the probability of staying with AT&T, then the odds ratio will be greater than 1, and the corresponding logit or log of the odds ratio will be positive:

$$p(OCC) > p(ATT)$$

$$\frac{p(OCC)}{p(ATT)} > 1$$

$$log\left(\frac{p(OCC)}{p(ATT)}\right) > 0$$

We start our analysis by looking at customers having complete data on the response variable common carrier choice (pick) and on the two AT&T marketing programs: AT&T Reach Out America (reachout) and the AT&T Calling Card (card).

Figure 5.1 shows the density histogram lattice for telephone usage by service provider choice. Another view of these data is provided by fitting a smooth curve through the pick-by-usage scatter plot, where the pick is set to the value zero for customers choosing AT&T and one for customers choosing some other common carrier. The resulting visualization is shown in figure 5.2.[2] We see that customers with higher levels of usage are less likely to switch from AT&T to some other common carrier.

Mosaic plots for the marketing programs show a similar pattern for service provider choice. Customers who are more heavily involved with AT&T are much less likely to switch service providers. Figure 5.3 shows the relationship for the AT&T Reach Out America plan and service provider choice. Figure 5.4 shows the relationship for the AT&T Calling Card and service provider choice.

Suppose we set the linear predictor in the AT&T Choice Study to involve the three explanatory variables reviewed above. Tables 5.1 and 5.2 show the fitted model to the data set with complete information for these explanatory variables.

Logistic regression provides a means for estimating the probability of switching service providers. The density lattice in figure 5.5 provides a pictorial representation of the model and a glimpse at model performance.

To evaluate the performance of this targeting model, we look at a two-by-two contingency table or confusion matrix showing the predicted and observed response values. A 50 percent cut-off works for this problem, yielding a 58.1 percent accuracy in classification. The confusion matrix for the logistic regression and 50 percent cut-off is shown as a mosaic plot in figure 5.6.

---

[2] I affectionately call this plot a *probability smooth*. A similar scatter plot smoothing method was employed by Chambers and Hastie (1992) in their analysis of the AT&T Choice Study. Those authors used a generalized additive model from the S to fit a smooth curve through the points.

*Figure 5.1.* Telephone Usage and Service Provider Choice (Density Lattice)

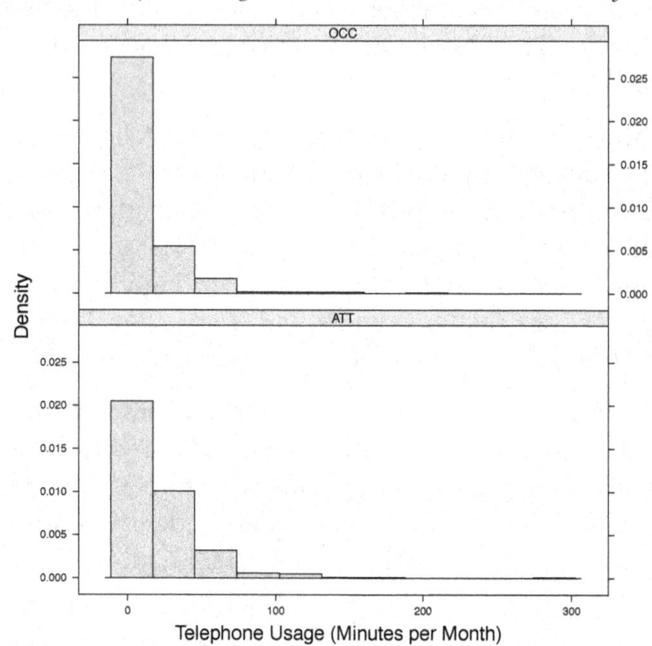


*Figure 5.2.* Telephone Usage and the Probability of Switching (Probability Smooth)

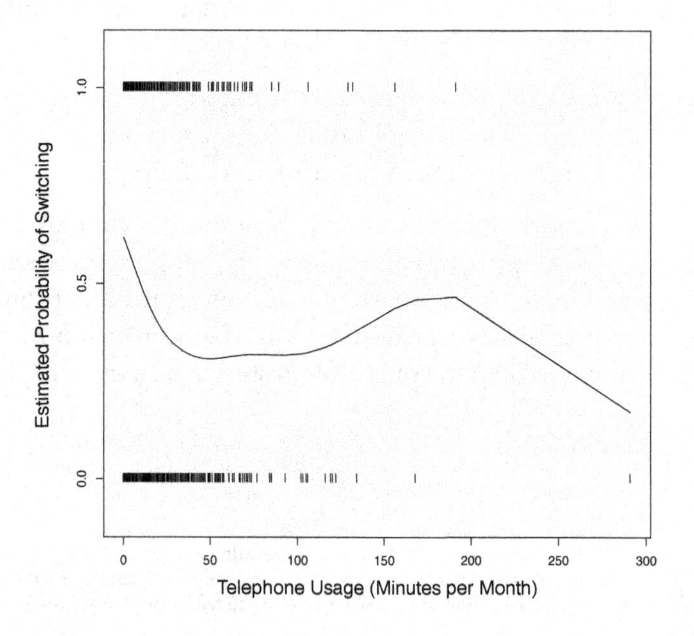

*Figure 5.3. AT&T Reach Out America Plan and Service Provider Choice*

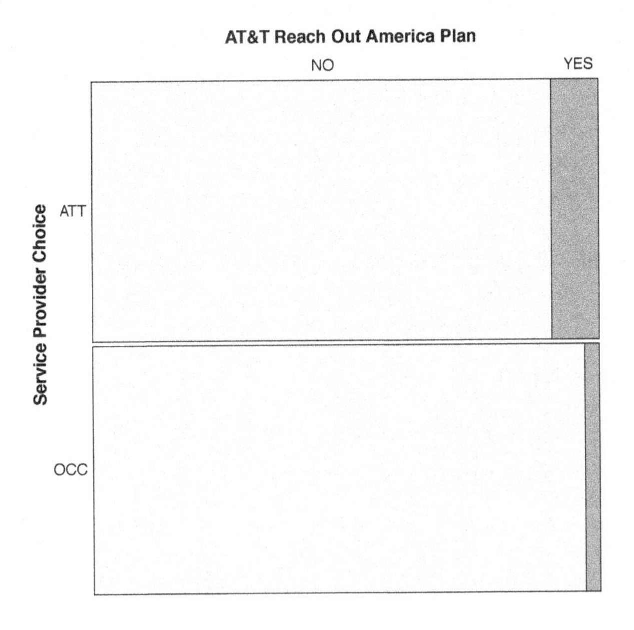


*Figure 5.4. AT&T Calling Card and Service Provider Choice*

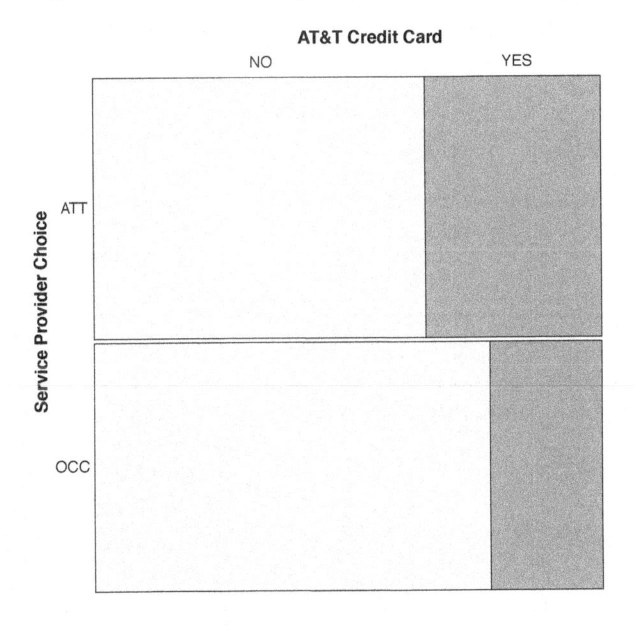

**Table 5.1.** *Logistic Regression Model for the AT&T Choice Study*

| Response: Switch to Other Common Carrier | |
|---|---|
| Telephone Usage | -0.013080 |
| AT&T Reach Out America Plan | -0.869531 |
| AT&T Credit Card | -0.475578 |
| Constant | 0.338302 |
| Null Deviance | 1359.4($df = 980$) |
| Residual Deviance | 1307.0($df = 977$) |
| AIC | 1315 |

**Table 5.2.** *Logistic Regression Model Analysis of Deviance*

| | *df* | *Deviance* | *Residual df* | Residual Deviance | *P-Value* |
|---|---|---|---|---|---|
| NULL | | | 980 | 1359.4 | |
| Telephone Usage | 1 | 33.303 | 979 | 1326.1 | 7.886e-09 |
| AT&T Reach Out America | 1 | 8.872 | 978 | 1317.2 | 0.002895 |
| AT&T Credit Card | 1 | 10.227 | 977 | 1307.0 | 0.001384 |

*Figure 5.5.* Logistic Regression for the Probability of Switching (Density Lattice)

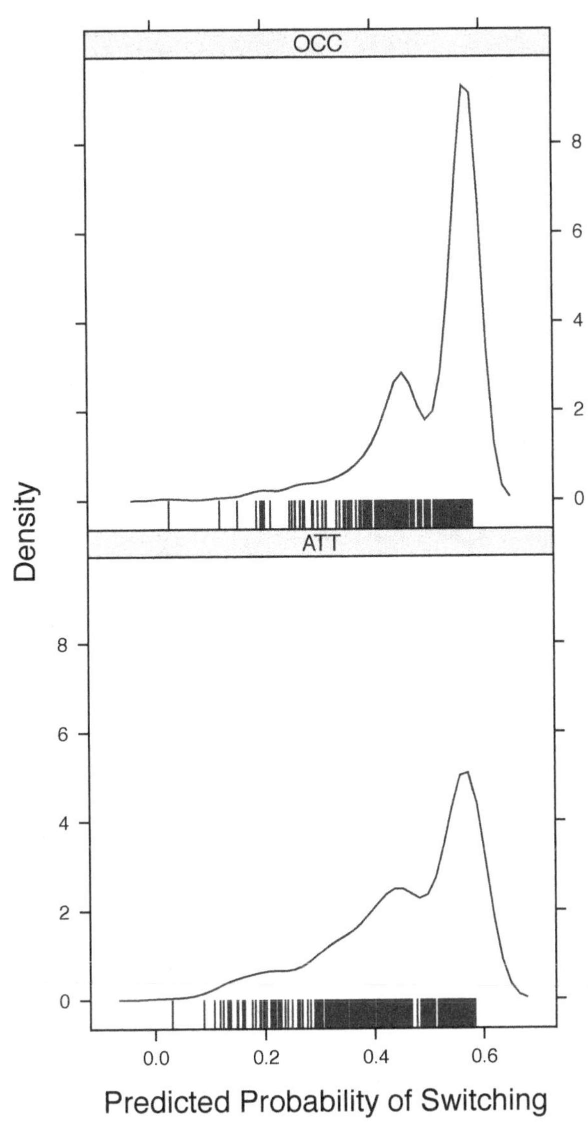

*Figure 5.6.* *Logistic Regression for the Probability of Switching (Confusion Mosaic)*

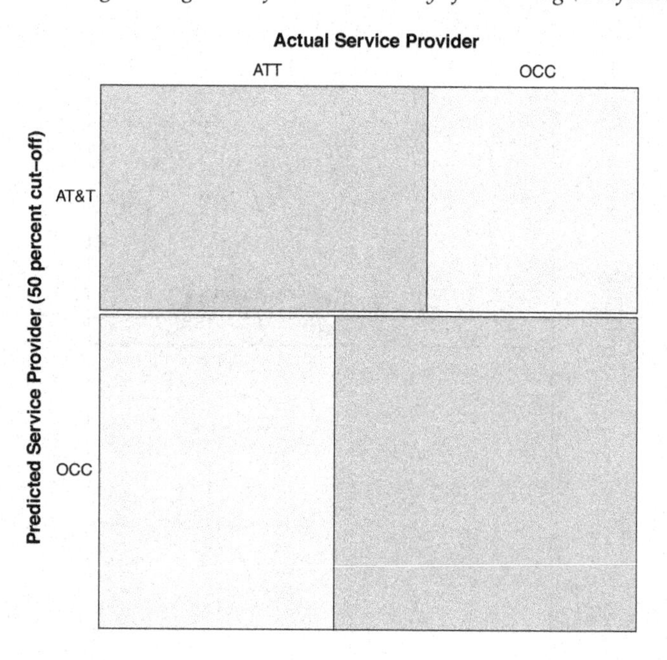

*Figure 5.7.* *A Classification Tree for Predicting Consumer Choices about Service Providers*

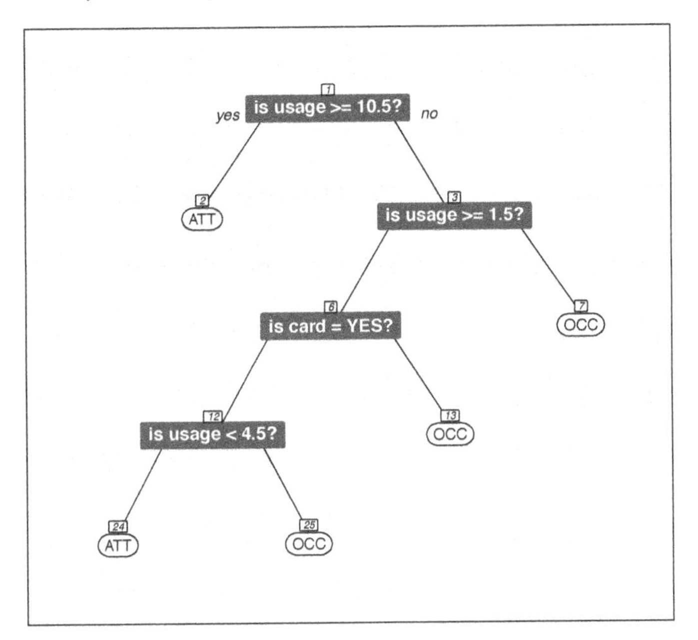


One way of demonstrating the importance of explanatory variables in predictive models, in a way that is easy for most managers to understand, is to show a classification tree for the data, like that shown in figure 5.7. Telephone usage is the key variable in predicting the probability of switching. Random forests, which are collections of trees, can also be used to screen for the most important variables in a set of explanatory variables.

Logistic regression is one of many classification methods. Part of the job of marketing data science is to see which methods work best for each particular situation. Careful evaluation requires a training-and-test regimen, as described in appendix A (page 259). We can use the AT&T Choice Study to illustrate techniques for evaluating models.

Machine learning methods of classification include naïve Bayes classification, neural networks, support vector machines, and random forests. To conduct a fair evaluation of these methods versus logistic regression, we split the AT&T data into training and test sets. Then for each of the modeling methods, we fit a model to the training set and evaluate its performance on the training set.

Then using the model fit to the training set, we evaluate it on the hold-out test set. For both training and test sets, we use the area under the receiver operation characteristic (ROC) curve as the measure of predictive performance. The further to the upper left-hand side of the plotting region the ROC curve falls (away from the diagonal line), the better the classifier. Perfect classification across all cut-off values has an area under the ROC curve equal to one. Results for a few classification models are shown in figures 5.8 through 5.10. Table 5.3 provides a summary of the results from tests across five alternative models. Looking at performance in the hold-out test set, logistic regression and naïve Bayes methods compare favorably to other methods for this case.

After selecting a trustworthy modeling method, one that performs well in out-of-sample tests, we can return to management questions about customer retention. We note those variables under management control. Our studies show that telephone usage is key to customer retention. If AT&T management can encourage additional usage through pricing, promotions, or product bundles, then management may be able to affect the proportion of customers who stay with AT&T rather than switch to some other common carrier.

Working with any of the models at our disposal, it would be possible to estimate the number of additional customers retained for each additional minute of telephone usage. Similar analyses would be possible for the AT&T marketing plans in this study.

There is overlap between customer retention and satisfaction research. Customer satisfaction is a large area of application for which many providers offer specialized services. Businesses want to know how to keep the customers they already have. Satisfaction translates into loyalty, and loyalty translates into customer retention. Survey research methods with attitude scales are common in this area of inquiry.

There is overlap between customer retention and risk analysis in selected industries. An auto insurance company wants to identify drivers who are most likely to have accidents. A financial services officer needs to identify consumers most likely to pay off their loans. A manufacturer is concerned about failure rates of new products. A retailer asks about the risk of losing a customer to another store.

***Figure 5.8.*** *Logistic Regression for Predicting Customer Retention (ROC Curve)*

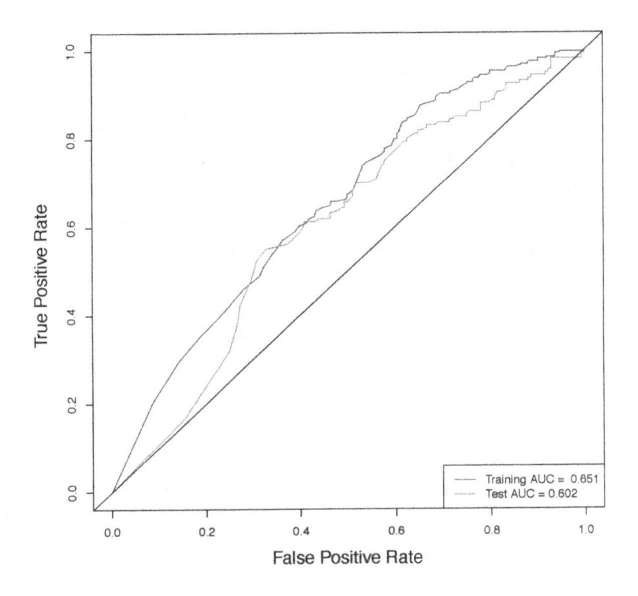


***Figure 5.9.*** *Naïve Bayes Classification for Predicting Customer Retention (ROC Curve)*

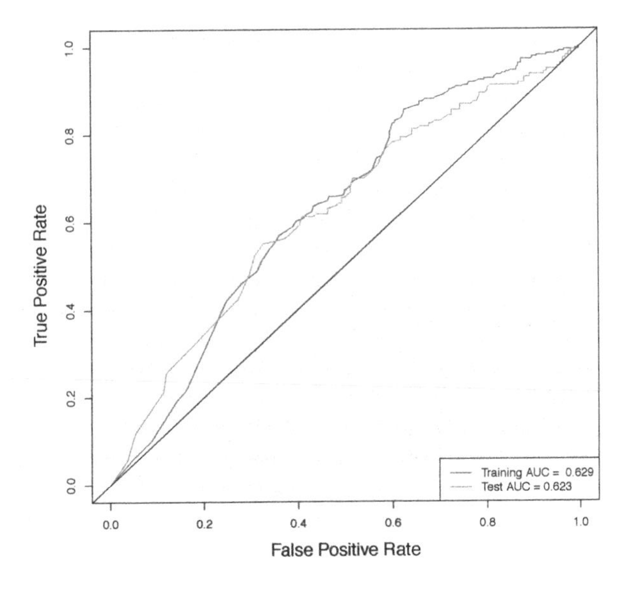

*Figure 5.10. Support Vector Machines for Predicting Customer Retention (ROC Curve)*

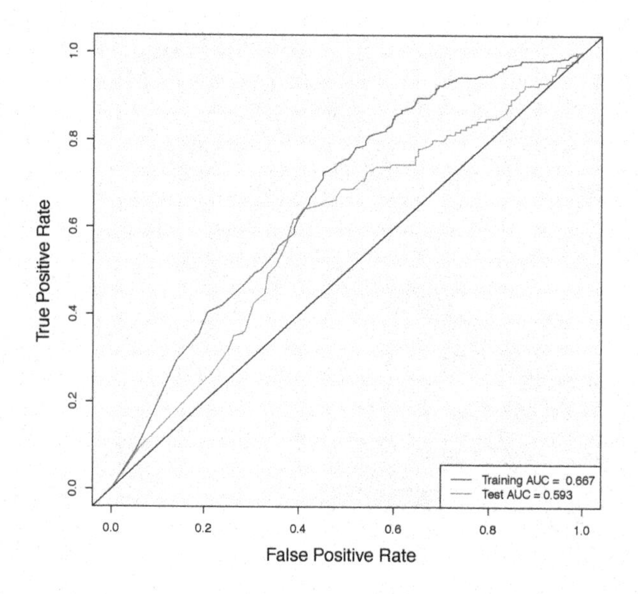


*Table 5.3. Evaluation of Classification Models for Customer Retention*

| | Area under ROC Curve | |
|---|---|---|
| | Training | Test |
| Logistic Regression | 0.651 | 0.602 |
| Naïve Bayes Classifier | 0.629 | 0.623 |
| Neural Network | 0.662 | 0.591 |
| Random Forests | 0.593 | 0.531 |
| Support Vector Machine | 0.667 | 0.593 |

Going back to our prior discussion of brand loyalty, we note that some customers (loyal customers) continue to buy the same brand year after year, while others are at risk of switching to competitive brands.

In the context of risk assessment, our objective may be to predict a binary response (stay on the job or quit, stay with the brand or switch, pay off or default on a loan). Logistic regression and predictive modeling are relevant for binary predictions. Alternatively, we may want to predict how long a customer or client will stay with a particular product, service, or brand.

Increasingly, firms look at the entire range of customer experience from initial contact and sale to subsequent product support. They identify customer touch-points or the times and places at which they have contact with the customer. Satisfaction research can be tied to programs for total quality improvement. Call centers and customer support programs benefit from satisfaction research. Many firms use customer satisfaction tracking studies as a basis for evaluating employees and for determining the size of financial incentives for management.

Brandt, Gupta, and Roberts (2004) cite results of a nationwide study of managers responsible for making decisions about computer hardware and software. Their results showed a strong relationship between expressed degrees of loyalty/satisfaction and customer retention.

Hayes (1998) provides an overview of survey methods and rating scales for measuring satisfaction. Fournier and Mick (1999) review qualitative approaches. Satisfaction and customer experience are especially relevant to services marketing (Zeithaml, Bitner, and Gremler 2005; Lovelock and Wirtz 2006).

Methods of survival analysis are appropriate for problems in customer retention. Le (1997) and Therneau and Grambsch (2000) discuss survival methods. Recurrent events methods represent a useful alternative to survival methods (Nelson 2003), but these have seen limited application in consumer research to date.

Reviews of classification methods have been provided by Tan, Steinbach, and Kumar (2006), Izenman (2008), Hastie, Tibshirani, and Friedman (2009), and Murphy (2012). See Hand (1997) for discussion of classifier performance. Duda, Hart, and Stork (2001) review naïve Bayes classifiers. Neural networks are discussed in Maren, Harston, and Pap (1990), Bishop (1995), and Ripley (1996).

The method of random forests represents a committee or ensemble technique that uses thousands of tree-structured classifiers to arrive at a single prediction. The method employs bootstrap resampling techniques. The components of a random forest are tree-structured classifiers, which build on classification and regression trees (CART) from Breiman et al. (1984). Useful introductions to trees and random forests may be found in Izenman (2008) and Hastie, Tibshirani, and Friedman (2009).

Support vector machines utilize a machine learning algorithm that has been shown to be an effective technique in many classification problems. Most closely identified with Vladimir Vapnik (Boser, Guyon, and Vapnik 1992; Vapnik 1998; Vapnik 2000), discussion of support vector machines may be found in Cristianini and Shawe-Taylor (2000), Izenman (2008), and Hastie, Tibshirani, and Friedman (2009).

Exhibit 5.1 shows an R program for analyzing data from the AT&T Choice Study. The program draws on R packages provided by Sarkar (2008, 2014), Sarkar and Andrews (2014), Hastie (2013), Feinerer (2014), Liaw and Wiener (2014), Therneau, Atkinson, and Ripley (2014), Milborrow (2014), Meyer et al. (2014), Ripley and Venables (2015), and and Sing et al. (2015). The corresponding Python program is on the website for the book.

***Exhibit 5.1.*** *Predicting Customer Retention (R)*

```
# Predicting Customer Retention (R)

library(lattice)  # lattice plot
library(vcd)  # mosaic plots
library(gam)  # generalized additive models for probability smooth
library(rpart)  # tree-structured modeling
library(e1071)  # support vector machines
library(randomForest)  # random forests
library(nnet)  # neural networks
library(rpart.plot)  # plot tree-structured model information
library(ROCR)  # ROC curve objects for binary classification

# user-defined function for plotting ROC curve using ROC objects from ROCR
plot_roc <- function(train_roc, train_auc, test_roc, test_auc) {
    plot(train_roc, col = "blue", lty = "solid", main = "", lwd = 2,
        xlab = "False Positive Rate",
        ylab = "True Positive Rate")
        plot(test_roc, col = "red", lty = "dashed", lwd = 2, add = TRUE)
        abline(c(0,1))
       # Draw a legend.
       train.legend <- paste("Training AUC = ", round(train_auc, digits=3))
       test.legend <- paste("Test AUC =", round(test_auc, digits=3))
       legend("bottomright", legend = c(train.legend, test.legend),
           lty = c("solid", "dashed"), lwd = 2, col = c("blue", "red"))
    }

# read in comma-delimited text file and create data frame
# there are blank character fields for missing data
# read them as character fields initially
att <- read.csv("att.csv", stringsAsFactors = FALSE)
print(str(att))

# convert blank character fields to missing data codes
att[att == ""] <- NA

# convert character fields to factor fields
att$pick <- factor(att$pick)
att$income <- factor(att$income)
att$moves <- factor(att$moves)
att$age <- factor(att$age)
att$education <- factor(att$education)
att$employment <- factor(att$employment)
att$nonpub <- factor(att$nonpub)
att$reachout <- factor(att$reachout)
att$card <- factor(att$card)

# check revised structure of att data frame
print(str(att))
# select usage and AT&T marketing plan factors
attwork <- subset(att, select = c("pick", "usage", "reachout", "card"))
attwork <- na.omit(attwork)
```

```
# listwise case deletion for usage and marketing factors
attwork <- na.omit(attwork)
print(summary(attwork))

# provide overview of data
print(summary(att))

# -----------------
# usage and pick
# -----------------
# examine relationship between age and response to promotion
pdf(file = "fig_retaining_customers_usage_lattice.pdf",
    width = 8.5, height = 8.5)
lattice_plot_object <- histogram(~usage | pick, data = att,
    type = "density", xlab = "Telephone Usage (Minutes per Month)",
    layout = c(1,2))
print(lattice_plot_object)  # switchers tend to have lower usage
dev.off()

att_gam_model <- gam(pick == "OCC"  ~ s(usage), family=binomial,data=att)
# probability smooth for usage and switching
pdf(file = "fig_retaining_customers_usage_probability_smooth.pdf",
    width = 8.5, height = 8.5)
plot(att$usage, att$pick == "OCC", type="n",
     ylim=c(-0.1,1.1), yaxt="n",
     ylab="Estimated Probability of Switching",
     xlab="Telephone Usage (Minutes per Month)")
     axis(2, at=c(0,.5,1))
     points(jitter(att$usage),
     att$pick=="OCC",pch="|")
     o <- order(att$usage)
     lines(att$usage[o],fitted(att_gam_model)[o])
dev.off()

# -----------------
# reachout and pick
# -----------------
# create a mosaic plot in using vcd package
pdf(file = "fig_retaining_customers_reachout_mosaic.pdf",
    width = 8.5, height = 8.5)
mosaic( ~ pick + reachout, data = attwork,
  labeling_args = list(set_varnames = c(pick = "Service Provider Choice",
  reachout = "AT&T Reach Out America Plan")),
  highlighting = "reachout",
  highlighting_fill = c("cornsilk","violet"),
  rot_labels = c(left = 0, top = 0),
  pos_labels = c("center","center"),
  offset_labels = c(0.0,0.6))
dev.off()

# create a mosaic plot in using vcd package
pdf(file = "fig_retaining_customers_card_mosaic.pdf",
    width = 8.5, height = 8.5)
```

```
mosaic( ~ pick + card, data = attwork,
  labeling_args = list(set_varnames = c(pick = "Service Provider Choice",
  card = "AT&T Credit Card")),
  highlighting = "card",
  highlighting_fill = c("cornsilk","violet"),
  rot_labels = c(left = 0, top = 0),
  pos_labels = c("center","center"),
  offset_labels = c(0.0,0.6))
dev.off()

# ----------------------------------
# fit logistic regression model
# ----------------------------------
att_spec <- {pick ~ usage + reachout + card}
att_fit <- glm(att_spec, family=binomial, data=attwork)
print(summary(att_fit))
print(anova(att_fit, test="Chisq"))

# compute predicted probability of switching service providers
attwork$Predict_Prob_Switching <- predict.glm(att_fit, type = "response")

pdf(file = "fig_retaining_customers_log_reg_density_evaluation.pdf",
    width = 8.5, height = 8.5)
plotting_object <- densityplot( ~ Predict_Prob_Switching | pick,
               data = attwork,
               layout = c(1,2), aspect=1, col = "darkblue",
               plot.points = "rug",
               strip=function(...) strip.default(..., style=1),
               xlab="Predicted Probability of Switching")
print(plotting_object)
dev.off()

# use a 0.5 cut-off in this problem
attwork$Predict_Pick <-
    ifelse((attwork$Predict_Prob_Switching > 0.5), 2, 1)
attwork$Predict_Pick <- factor(attwork$Predict_Pick,
    levels = c(1, 2), labels = c("AT&T", "OCC"))
confusion_matrix <- table(attwork$Predict_Pick, attwork$pick)
cat("\nConfusion Matrix (rows=Predicted Service Provider,",
   "columns=Actual Service Provider\n")
print(confusion_matrix)
predictive_accuracy <- (confusion_matrix[1,1] + confusion_matrix[2,2])/
                        sum(confusion_matrix)
cat("\nPercent Accuracy: ", round(predictive_accuracy * 100, digits = 1))
# mosaic rendering of the classifier with 0.10 cutoff
with(attwork, print(table(Predict_Pick, pick, useNA = c("always"))))
pdf(file = "fig_retaining_customers_confusion_mosaic_50_percent.pdf",
    width = 8.5, height = 8.5)
mosaic( ~ Predict_Pick + pick, data = attwork,
  labeling_args = list(set_varnames =
  c(Predict_Pick =
      "Predicted Service Provider (50 percent cut-off)",
       pick = "Actual Service Provider")),
```

```
  highlighting = c("Predict_Pick", "pick"),
  highlighting_fill = c("green","cornsilk","cornsilk","green"),
  rot_labels = c(left = 0, top = 0),
  pos_labels = c("center","center"),
  offset_labels = c(0.0,0.6))
dev.off()

# ------------------------------------------
# example of tree-structured classification
# ------------------------------------------
att_tree_fit <- rpart(att_spec, data = attwork,
    control = rpart.control(cp = 0.0025))
# plot classification tree result from rpart
pdf(file = "fig_retaining_customers_tree_classifier.pdf",
    width = 8.5, height = 8.5)
prp(att_tree_fit, main="",
    digits = 3,  # digits to display in terminal nodes
    nn = TRUE,  # display the node numbers
    branch = 0.5,  # change angle of branch lines
    branch.lwd = 2,  # width of branch lines
    faclen = 0,  # do not abbreviate factor levels
    trace = 1,  # print the automatically calculated cex
    shadow.col = 0,  # no shadows under the leaves
    branch.lty = 1,  # draw branches using dotted lines
    split.cex = 1.2,  # make the split text larger than the node text
    split.prefix = "is ",  # put "is" before split text
    split.suffix = "?",  # put "?" after split text
    split.box.col = "blue",  # lightgray split boxes (default is white)
    split.col = "white",  # color of text in split box
    split.border.col = "blue",  # darkgray border on split boxes
    split.round = .25)  # round the split box corners a tad
dev.off()
# ----------------------------------------------
# example of random forest model for importance
# ----------------------------------------------
# fit random forest model to the training data
set.seed (9999)  # for reproducibility
attwork_rf_fit <- randomForest(att_spec, data = attwork,
   mtry=3, importance=TRUE, na.action=na.omit)
# check importance of the individual explanatory variables
pdf(file = "fig_retaining_customers_random_forest_importance.pdf",
width = 11, height = 8.5)
varImpPlot(attwork_rf_fit, main = "", pch = 20, cex = 1.25)
dev.off()

# ------------------------------------------------------------
# training-and-test for evaluating alternative modeling methods
# ------------------------------------------------------------
set.seed(2020)
partition <- sample(nrow(attwork)) # permuted list of row index numbers
attwork$group <- ifelse((partition < nrow(attwork)/(3/2)),1,2)
attwork$group <- factor(attwork$group, levels=c(1,2),
    labels=c("TRAIN","TEST"))
```

```
train <- subset(attwork, subset = (group == "TRAIN"),
    select = c("pick", "usage", "reachout", "card"))
test <- subset(attwork, subset = (group == "TEST"),
    select = c("pick", "usage", "reachout", "card"))
# ensure complete data in both partitions
train <- na.omit(train)
test <- na.omit(test)
# check partitions for no-overlap and correct pick frequencies
if(length(intersect(rownames(train), rownames(test))) != 0)
    print("\nProblem with partition")
print(table(attwork$pick))
print(table(test$pick))
print(table(train$pick))
# --------------------------------------
# Logistic regression training-and-test
# auc = area under ROC curve
# --------------------------------------
# fit logistic regression model to the training set
train_lr_fit <- glm(att_spec, family=binomial, data=train)
train$lr_predict_prob <- predict(train_lr_fit, type = "response")
train_lr_prediction <- prediction(train$lr_predict_prob, train$pick)
train_lr_auc <- as.numeric(performance(train_lr_prediction, "auc")@y.values)
# use model fit to training set to evaluate on test data
test$lr_predict_prob <- as.numeric(predict(train_lr_fit, newdata = test,
    type = "response"))
test_lr_prediction <- prediction(test$lr_predict_prob, test$pick)
test_lr_auc <- as.numeric(performance(test_lr_prediction, "auc")@y.values)
# ----------------------------------------
# ROC for logistic regression
# ----------------------------------------
train_lr_roc <- performance(train_lr_prediction, "tpr", "fpr")
test_lr_roc <- performance(test_lr_prediction, "tpr", "fpr")
pdf(file = "fig_retaining_customers_logistic_regression_roc.pdf",
    width = 8.5, height = 8.5)
plot_roc(train_roc = train_lr_roc,
    train_auc = train_lr_auc,
    test_roc = test_lr_roc,
    test_auc = test_lr_auc)
dev.off()

# ----------------------------------------
# Support vector machine training-and-test
# ----------------------------------------
set.seed (9999)  # for reproducibility
# determine tuning parameters prior to fitting model to training set
train_svm_tune <- tune(svm, att_spec, data = train,
                  ranges = list(gamma = 2^(-8:1), cost = 2^(0:4)),
                  tunecontrol = tune.control(sampling = "fix"))
# fit the support vector machine to the training set using tuning parameters
train_svm_fit <- svm(att_spec, data = train,
    cost = train_svm_tune$best.parameters$cost,
    gamma=train_svm_tune$best.parameters$gamma,
    probability = TRUE)
```

```
train_svm_predict <- predict(train_svm_fit, train, probability = TRUE)
train$svm_predict_prob <- attr(train_svm_predict, "probabilities")[,1]
train_svm_prediction <- prediction(train$svm_predict_prob, train$pick)
train_svm_auc <- as.numeric(performance(train_svm_prediction, "auc")@y.values)
# use model fit to training data to evaluate on test set
test_svm_predict <- predict(train_svm_fit, test, probability = TRUE)
test$svm_predict_prob <- attr(test_svm_predict, "probabilities")[,1]
test_svm_prediction <- prediction(test$svm_predict_prob, test$pick)
test_svm_auc <- as.numeric(performance(test_svm_prediction, "auc")@y.values)
# ------------------------------------------
# ROC for support vector machines
# ------------------------------------------
train_svm_roc <- performance(train_svm_prediction, "tpr", "fpr")
test_svm_roc <- performance(test_svm_prediction, "tpr", "fpr")
pdf(file = "fig_retaining_customers_support_vector_machine_roc.pdf",
    width = 8.5, height = 8.5)
plot_roc(train_roc = train_svm_roc,
    train_auc = train_svm_auc,
    test_roc = test_svm_roc,
    test_auc = test_svm_auc)
dev.off()

# ----------------------------------------
# Random forest training-and-test
# ----------------------------------------
set.seed (9999)  # for reproducibility
train_rf_fit <- randomForest(att_spec, data = train,
   mtry=3, importance=FALSE, na.action=na.omit)
train$rf_predict_prob <- as.numeric(predict(train_rf_fit, type = "prob")[,2])
train_rf_prediction <- prediction(train$rf_predict_prob, train$pick)
train_rf_auc <- as.numeric(performance(train_rf_prediction, "auc")@y.values)
# use model fit to training set to evaluate on test data
test$rf_predict_prob <- as.numeric(predict(train_rf_fit, newdata = test,
    type = "prob")[,2])
test_rf_prediction <- prediction(test$rf_predict_prob, test$pick)
test_rf_auc <- as.numeric(performance(test_rf_prediction, "auc")@y.values)
# ------------------------------------------
# ROC for random forest
# ------------------------------------------
train_rf_roc <- performance(train_rf_prediction, "tpr", "fpr")
test_rf_roc <- performance(test_rf_prediction, "tpr", "fpr")
pdf(file = "fig_retaining_customers_random_forest_roc.pdf",
    width = 8.5, height = 8.5)
plot_roc(train_roc = train_rf_roc,
    train_auc = train_rf_auc,
    test_roc = test_rf_roc,
    test_auc = test_rf_auc)
dev.off()

# ----------------------------------------
# Naive Bayes training-and-test
# ----------------------------------------
set.seed (9999)  # for reproducibility
```

```
train_nb_fit <- naiveBayes(att_spec, data = train)
train$nb_predict_prob <- as.numeric(predict(train_nb_fit, newdata = train,
    type = "raw")[,2])
train_nb_prediction <- prediction(train$nb_predict_prob, train$pick)
train_nb_auc <- as.numeric(performance(train_nb_prediction, "auc")@y.values)
# use model fit to training set to evaluate on test data
test$nb_predict_prob <- as.numeric(predict(train_nb_fit, newdata = test,
    type = "raw")[,2])
test_nb_prediction <- prediction(test$nb_predict_prob, test$pick)
test_nb_auc <- as.numeric(performance(test_nb_prediction, "auc")@y.values)
# ---------------------------------------------
# ROC for Naive Bayes
# ---------------------------------------------
train_nb_roc <- performance(train_nb_prediction, "tpr", "fpr")
test_nb_roc <- performance(test_nb_prediction, "tpr", "fpr")
pdf(file = "fig_retaining_customers_naive_bayes_roc.pdf",
    width = 8.5, height = 8.5)
plot_roc(train_roc = train_nb_roc,
    train_auc = train_nb_auc,
    test_roc = test_nb_roc,
    test_auc = test_nb_auc)
dev.off()
# ---------------------------------------
# Neural Network training-and-test
# ---------------------------------------
set.seed (9999)  # for reproducibility
train_nnet_fit <- nnet(att_spec, data = train, size = 3, decay = 0.0,
    probability = TRUE, trace = FALSE)
train$nnet_predict_prob <- as.numeric(predict(train_nnet_fit, newdata = train))
train_nnet_prediction <- prediction(train$nnet_predict_prob, train$pick)
train_nnet_auc <- as.numeric(performance(train_nnet_prediction, "auc")@y.values)
# use model fit to training set to evaluate on test data
test$nnet_predict_prob <- as.numeric(predict(train_nnet_fit, newdata = test))
test_nnet_prediction <- prediction(test$nnet_predict_prob, test$pick)
test_nnet_auc <- as.numeric(performance(test_nnet_prediction, "auc")@y.values)
# ---------------------------------------------
# ROC for Neural Network
# ---------------------------------------------
train_nnet_roc <- performance(train_nnet_prediction, "tpr", "fpr")
test_nnet_roc <- performance(test_nnet_prediction, "tpr", "fpr")
pdf(file = "fig_retaining_customers_neural_network_roc.pdf",
    width = 8.5, height = 8.5)
plot_roc(train_roc = train_nnet_roc, train_auc = train_nnet_auc,
    test_roc = test_nnet_roc, test_auc = test_nnet_auc)
dev.off()

# Suggestions for the student: Build a more complete model for customer
# retention using additional information in the AT&T Choice Study.
# Take care to deal with missing data issues in customer demographics,
# perhaps using technologies for missing data imputation.
# Employ additional classification methods and evaluate them using
# multi-fold cross-validation, another training-and-test regimen.
```

# 6 Positioning Products

"...I don't want to get any reports saying we are holding our position. We'll let the enemy do that. We are not holding on to anything except the enemy. We're going to hold on to him all the time and we're gonna kick him in the ass, we're gonna kick him all the time, and we're gonna go through him like crap through a goose. Now, I will be proud to lead you wonderful men into battle anywhere, anytime. That is all."

—George C. Scott as General George S. Patton, Jr.
in *Patton* (1970)

Being a fan of movies, I thought it only appropriate to introduce product positioning with a similarity ranking task. I selected five Gene Hackman movies and ranked all possible pairs of these movies by similarity, as shown in figure 6.1.

Similarity rankings may be used to construct a distance matrix showing how far each movie is from every other movie. When studying small sets of products or brands, we can seek direct ratings or rankings across all pairs of products or brands. With $k$ products or brands there will be $k(k-1)/2$ pairs or combinations of two movies at a time. With five movies, there are $\frac{5(5-1)}{2} = 10$ pairs of movies.

*Figure 6.1. A Product Similarity Ranking Task*

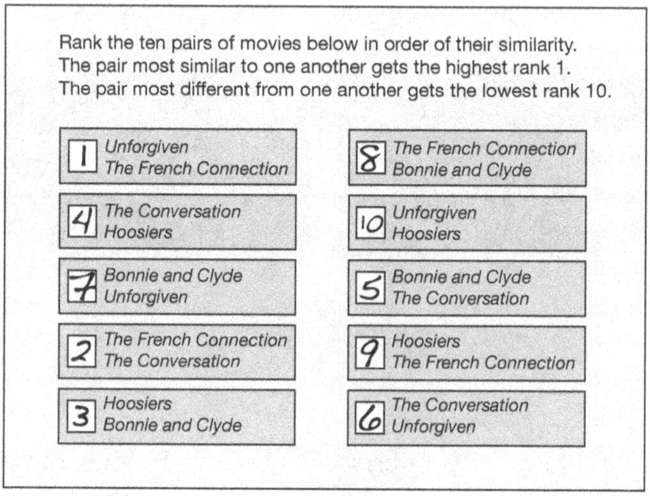

Rank the ten pairs of movies below in order of their similarity.
The pair most similar to one another gets the highest rank 1.
The pair most different from one another gets the lowest rank 10.

| Rank | Pair | Rank | Pair |
|---|---|---|---|
| 1 | *Unforgiven* / *The French Connection* | 8 | *The French Connection* / *Bonnie and Clyde* |
| 4 | *The Conversation* / *Hoosiers* | 10 | *Unforgiven* / *Hoosiers* |
| 7 | *Bonnie and Clyde* / *Unforgiven* | 5 | *Bonnie and Clyde* / *The Conversation* |
| 2 | *The French Connection* / *The Conversation* | 9 | *Hoosiers* / *The French Connection* |
| 3 | *Hoosiers* / *Bonnie and Clyde* | 6 | *The Conversation* / *Unforgiven* |

Table 6.2 shows how the movie similarity ranks translate into a distance matrix for the five movies. Because larger numbers correspond to larger differences between movies, we call this a dissimilarity matrix. And figure 6.3 shows the corresponding map of these movies in space.

We call this a perceptual map because it is a rendering of consumer perceptions of products or brands—in this case one consumer's perceptions. And it is not too hard to imagine how making a map of products in space for one consumer would translate into making a map for many consumers. We would ask many consumers to rank the products and place average ranks in the dissimilarity matrix.

Similarity judgments are especially useful in product categories for which attributes are difficult to identify or describe, such as categories defined by style, look, odor, or flavor. The resulting distance or dissimilarity matrix can serve as input to cluster analysis as well as product positioning.

The algorithm used to create the product positioning map is called multidimensional scaling, and another way to see how it works is to construct a map of physical distances between geographical objects—a mileage chart from a roadmap will serve the purpose. Then we use metric multidimensional scaling to draw the map. The process is sometimes called *unfolding*.

***Figure 6.2.*** *Rendering Similarity Judgments as a Matrix*

| | Bonnie and Clyde | The Conversation | The French Connection | Hoosiers | Unforgiven |
|---|---|---|---|---|---|
| Bonnie and Clyde | 0 | 5 | 8 | 3 | 7 |
| The Conversation | 5 | 0 | 2 | 4 | 6 |
| The French Connection | 8 | 2 | 0 | 9 | 1 |
| Hoosiers | 3 | 4 | 9 | 0 | 10 |
| Unforgiven | 7 | 6 | 1 | 10 | 0 |

***Figure 6.3.*** *Turning a Matrix of Dissimilarities into a Perceptual Map*

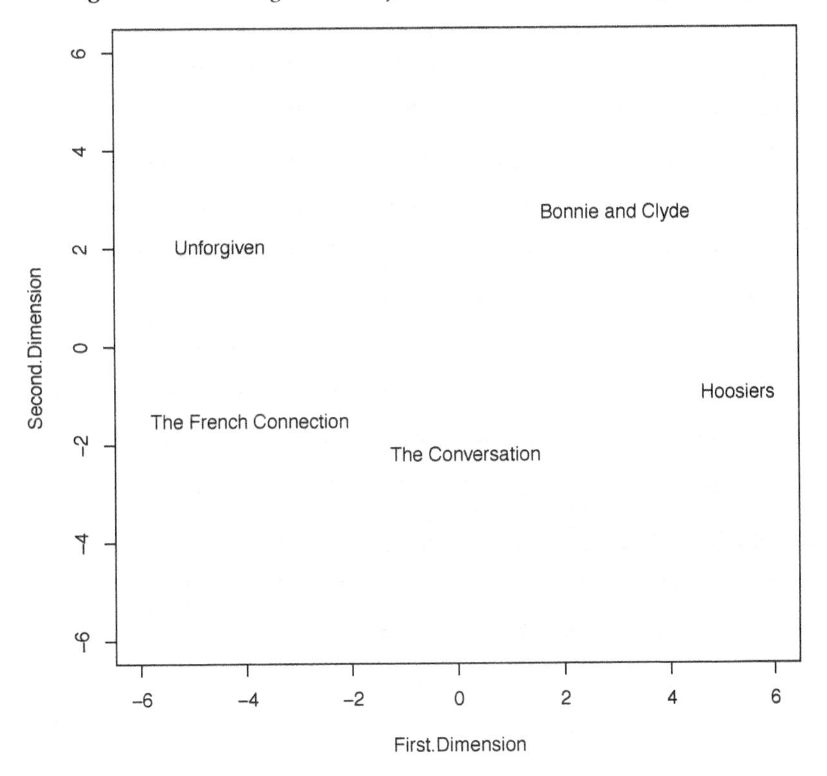

To study product positioning, data are collected for numerous products within a category, making sure that the primary competitive products are included. Data can relate to actual product specifications. Data can represent consumer ratings of product attributes. We can also use market basket data as input for multidimensional scaling. That is, products purchased within the same context or buying experience may be viewed as situationally similar or similar in usage.

To show an example of market basket product positioning, we draw on data from the Wisconsin Dells case (page 384). Activities of selected visitors to the Dells provide market baskets for input to a distance metric, which in turn provides input to multidimensional scaling. To carry out the analysis, we must first define the distance metric.

The Wisconsin Dells case draws data from more than a thousand face-to-face interviews conducted with visitors to the Dells. As part of the interviews, visitors were asked whether they had participated in or were likely to participate in any of thirty-three entertainment activities. The resulting data for these activities constitute a binary response matrix, much like we would obtain in a market basket study. Entries in the matrix are one if the visitor group participated in or planned to participate in the activity and zero if not.[1]

The multidimensional scaling solution in figure 6.5 shows a group of activities in close proximity: *Boating/Swimming*, *Shopping*, and *Casual Dining*, among others. These are activities in which most Dells visitors participate. They have low Jaccard dissimilarities, so they appear close to one another on the map.

In many research contexts, activities close together on maps are similar to one another and, hence, potential substitutes for one another. And then there is *Bungee Jumping*, which seems to have no close substitute among the set of Dells activities.

[1] To prepare the Dells training data for input to multidimensional scaling, we begin by converting the $1000 \times 33$ data matrix $\mathbf{X}$ to a distance matrix $\mathbf{D}$ that shows the distance or dissimilarity between each pair of activities. There are many ways to compute distances or dissimilarities between pairs of binary variables. We use the Jaccard dissimilarity index, which seems to work well in market basket data. (See figure 6.4.) The resulting matrix $\mathbf{D}$ is a symmetric $33 \times 33$ matrix, with each item in the matrix being a number between zero and one representing the degree of dissimilarity between a pair of activities.

***Figure 6.4.*** *Indices of Similarity and Dissimilarity between Pairs of Binary Variables*

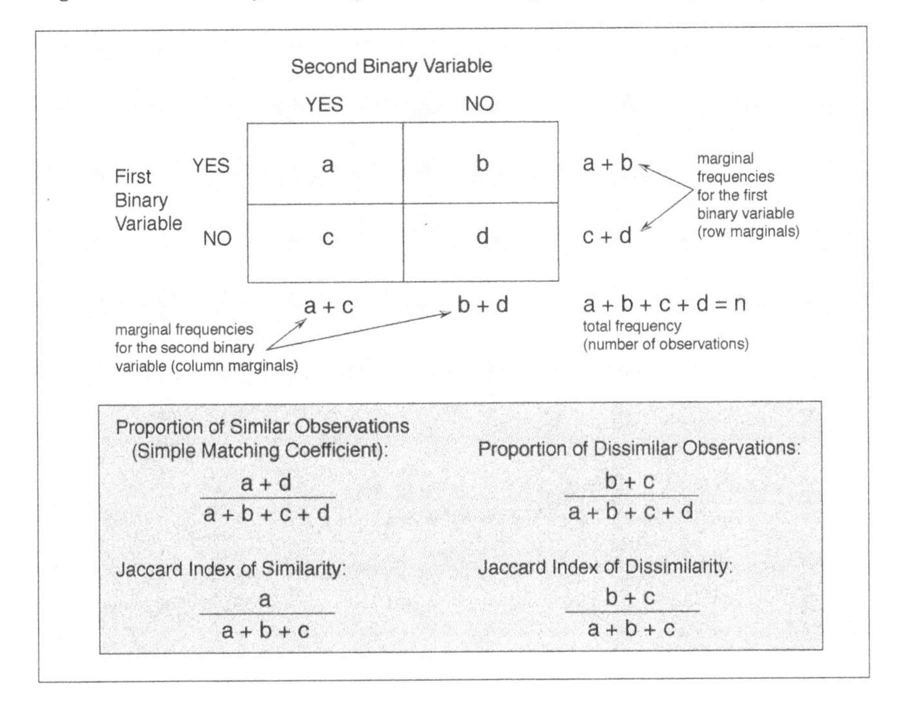


The Jaccard index of dissimilarity is obtained from a two-by-two table for binary variables. The table above shows how to compute the Jaccard indices of similarity and dissimilarity from the frequencies $a$, $b$, $c$, and $d$ in the two-by-two table. In many research problems the Jaccard index of dissimilarity makes more sense as an index of dissimilarity than the proportion of dissimilar observations. For pairs of low-incidence binary variables, the proportion of dissimilar observations is low because most observations are in the *NO/NO* cell of the two-by-two table. These *NO/NO* observations are more a reflection of low-incidence than similarity between the binary variables. In computing the Jaccard dissimilarity index, we drop the *NO/NO* cell frequency, thus avoiding problems with low-incidence binary variables. Similarity and dissimilarity indices play an important role in methods like multidimensional scaling, cluster analysis, and nearest neighbor analysis. Market basket analysis often begins with consumers or individual market baskets defining the rows of a binary data matrix and products or services defining the columns. Indices of similarity show how likely it is that two products will be included in a shopper's market basket. Further discussion of indices of similarity and dissimilarity may be found in Kaufman and Rousseeuw (1990).

*Figure 6.5. A Map of Wisconsin Dells Activities Produced by Multidimensional Scaling*

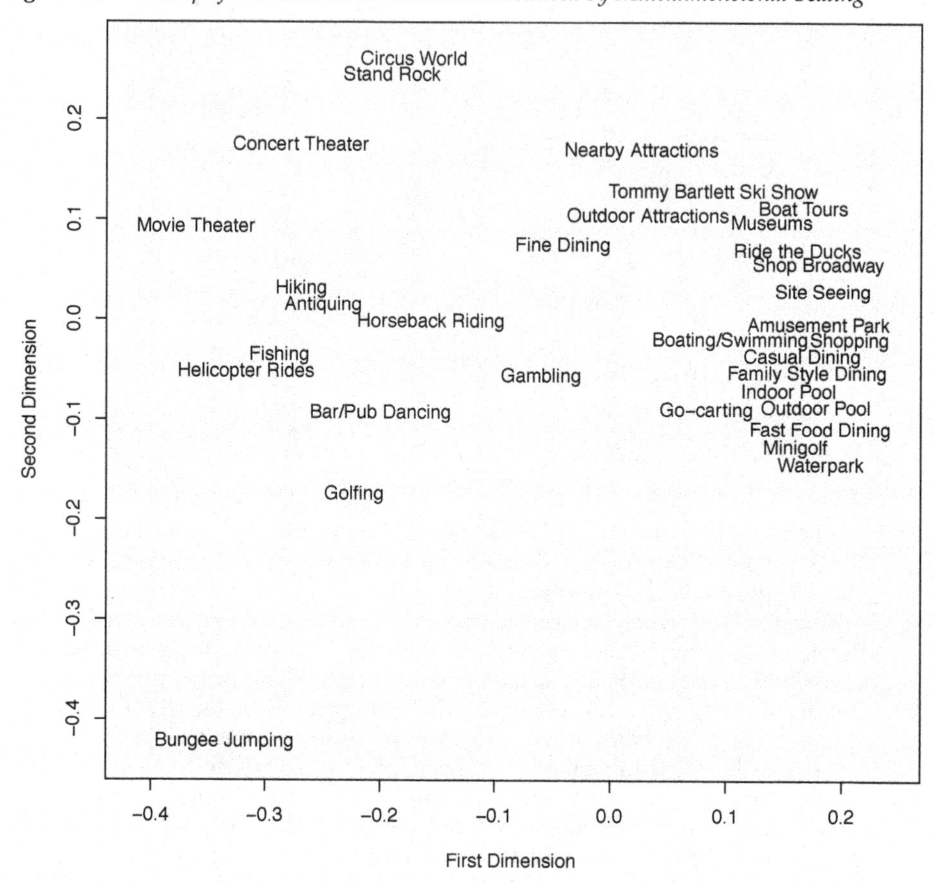

Multidimensional scaling uses dissimilarities as inputs, providing a map as output. The orientation of the map is arbitrary. We can rotate the map without changing the solution. We can develop a reflection of the map without changing the solution. We can exchange the top for the bottom or the left for the right without changing the solution.

We use multivariate methods to study relationships among many variables. The language of principal components, factor analysis, multidimensional scaling, and other multivariate methods suggests that we are trying to identify underlying or latent dimensions in the data. In practical terms, we want to reduce the amount of data and the number of dimensions being considered. A plot of points in two dimensions is easier to understand than a large table of numbers. It is hard to see relationships in large matrices of correlations or dissimilarities, but it is not so hard to see relationships in maps.

When different multivariate methods yield decidedly different results, as they often do, we use our best judgment to choose the method that will make the most sense to management. We choose methods of analysis appropriate for the data and the business problem at hand. And we choose methods for presenting results that are both appropriate for the data and meaningful to management. Some analyses and maps make more sense to management than others.

Regarding the Wisconsin Dells case, there are many alternative methods of analysis and presentation. One obvious alternative would be to employ market basket analysis. Another would be hierarchical cluster analysis.

As we have seen earlier with the analysis of the Bank Marketing Study (pages 52 through 54), cluster analysis may also be used to examine similarities and differences among customers.

Market segmentation involves identifying groups or segments of customers. We could employ segmentation of visitor groups in the Wisconsin Dells case, of course, but that is not our mission here. We want to see how entertainment activities or tourist products relate to one another, so we cluster the products rather than the visitors to the Dells.

There are two general types of cluster analysis: hierarchical clustering and non-hierarchical clustering (partitioning). We have seen partitioning previously. Here we will use hierarchical clustering, with the input being the Jaccard dissimilarity matrix we had defined for multidimensional scaling.

Cluster analysis is appropriate for classification problems in which classes are not known in advance. We use interrelationships among objects to define the classes. Objects that are similar to one another in terms of their values on measured attributes in the data set get placed in the same cluster. Objects within a cluster should be more similar to one another than they are to objects in other clusters.

Our analysis for the Dells data involves clustering of products or activities. Along with the product positioning maps we drew earlier, the hierarchical cluster analysis output, shown in Figure 6.6, helps us understand the product space of Dells activities as well as the competitive environment among Dells businesses. Quite often, products or activities within the same cluster are potential substitutes for one another.

Whether concerned with new or current products and services, product positioning is an important part of business and marketing strategy. There are basic questions to address: What constitutes the category of products? How do products within the category compare with one another? What is the nature of the product space? Which products serve as close substitutes for one another? What can a firm do to distinguish its products and services from those offered by other firms?

Product positioning maps are especially useful in product planning and competitive analysis. Products close to one another in space may be thought of as substitute products or close competitors. Open areas in the product space may represent opportunities for new, differentiated products. These same technologies may be applied to brands to obtain information about brand positioning and to guide branding strategy.

Product or brand positioning may be studied in concert with product and brand preferences, yielding a joint perception/preference mapping of products or brands in space. Critical to strategic product positioning are areas of the product space most desirable to consumers. Product managers like to find areas of the product space where there are many potential customers and few competitive brands or products.

***Figure 6.6.*** *Hierarchical Clustering of Wisconsin Dells Activities*

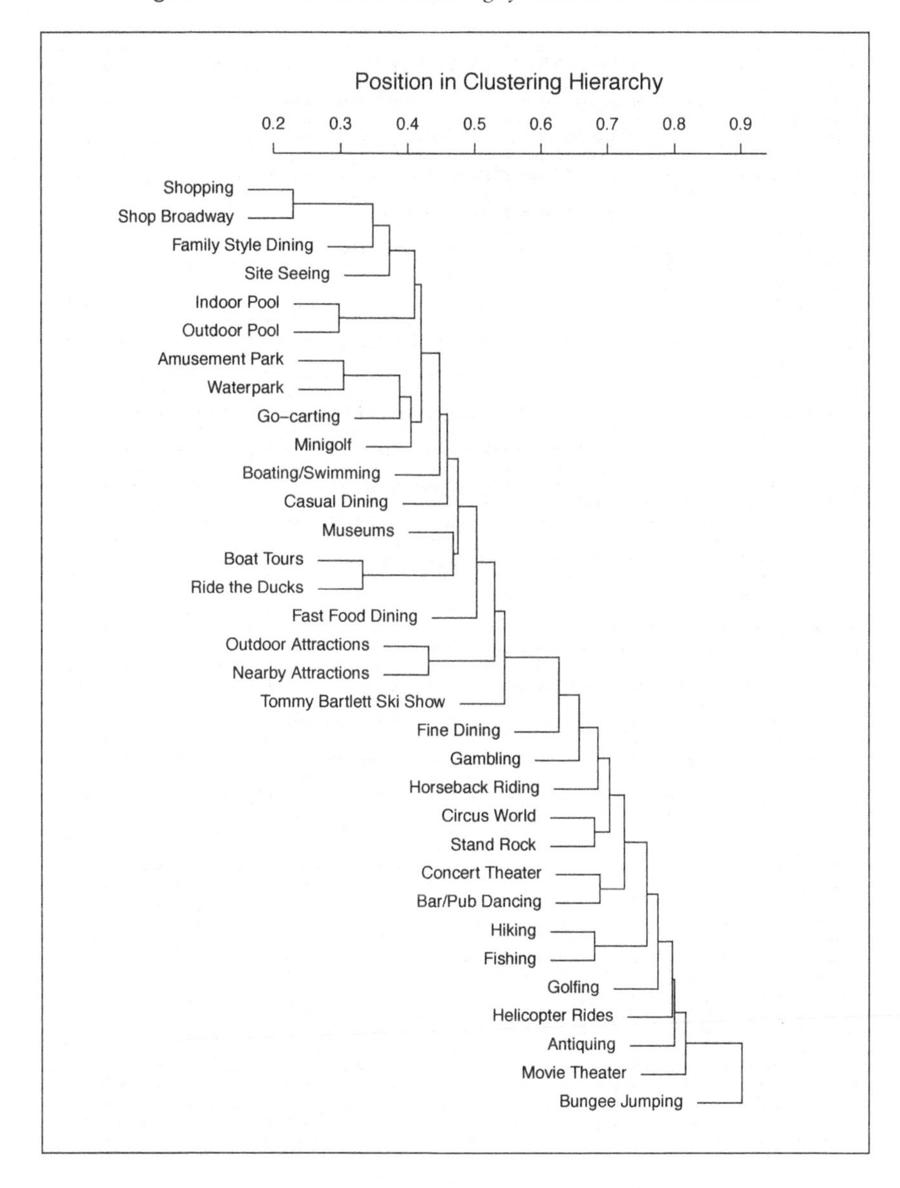

Whatever the product category, a firm wants consumers to prefer its products over the products of others. Given the variety of products and the complexity of many product offerings, it can be difficult to assess consumer preferences. Typical objects in consumer research represent alternative products or services, features of products, or names of products and brands. Assuming that consumer preferences for objects follow a normal distribution, mean proportions from the columns of a paired-comparisons preference matrix may be converted to scale scores using standard normal quantiles. This is traditional unidimensional scaling. It results in scale scores that can be easily understood—products or product features positioned along a real number line.

Preference scaling has a long history, dating back to early work in psychometrics by Thurstone (1927), Guilford (1954, first published in 1936), and Torgerson (1958). Traditional univariate methods, building as they do upon a paired comparison preference matrix, may be used for data arising from actual paired comparisons, rank orders, and multiple rank orders, as well as from best-worst scaling items, choice studies, pick lists, and elimination pick lists. Alternative survey methods are reviewed in appendix B (page 312).

Early thinking about multidimensional scaling was published in psychology journals (Young and Householder 1938; Richardson 1938). Torgerson (1958) provided an in-depth treatment. There are now many useful sources for learning about multidimensional scaling (Davison 1992; Cox and Cox 1994; Carroll and Green 1997; Borg and Groenen 2010). Lilien and Rangaswamy (2003) review joint perception/preference mappings. R software for multidimensional scale is described in Venables and Ripley (2002).

In addition to multidimensional scaling, multivariate procedures such as principal component analysis or factor analysis may be used in a search for dimensions that summarize the product attribute space. Multivariate methods are reviewed by Seber (2000), Manly (1994), Sharma (1996), Gnanadesikan (1997), Johnson and Wichern (1998), and Izenman (2008). Principal component biplots represent an alternative to multidimensional scaling plots (Gabriel 1971; Gower and Hand 1996) for product positioning. Biplots allow us to plot consumers and products/brands in the same space.

Exhibit 6.1 shows an R program for product positioning using the movie example from the beginning of the chapter. The R program draws on the work of Ripley et al. (2015). The corresponding Python program is provided in exhibit 6.2.

Exhibit 6.3 provides a demonstration of metric multidimensional scaling using distances between U.S. cities as input. The R program draws on the work of Ripley et al. (2015). The corresponding Python program is provided in exhibit 6.4.

Exhibit 6.5 is an R program for analyzing the Wisconsin Dells case data using multidimensional scaling. It uses activities market baskets as input and draws on R programs from Ripley et al. (2015) and Fellows (2014). Dissimilarity measures build on the work of Kaufman and Rousseeuw (1990). The corresponding Python program is provided in exhibit 6.6.

Exhibit 6.7 is an R program for analyzing the Wisconsin Dells case data using hierarchical cluster analysis and programs from Maechler (2014a) and Kaufman and Rousseeuw (1990). The corresponding Python program is on the website for the book.

**Exhibit 6.1.** *Product Positioning of Movies (R)*

```
# Product Positioning of Movies (R)

library(MASS) # need MASS package for isoMDS() function for non-metric MDS

# We define a utility function for converting a distance structure
# to a distance matrix as required for some routines and
# for printing of the complete matrix for visual inspection.
make.distance_matrix <- function(distance_structure)
{ n <- attr(distance_structure, "Size")
  full <- matrix(0,n,n)
  full[lower.tri(full)] <- distance_structure
  full+t(full)
}

# These are the original data from one respondent
# Pairs of movies are judged on their similarity
# Smaller numbers are more similar to one another
# Zero on the diagonal means no difference
#
#  0  5  8  3  7    Bonnie and Clyde
#  5  0  2  4  6    The Conversation
#  8  2  0  9  1    The French Connection
#  3  4  9  0 10    Hoosiers
#  7  6  1 10  0    Unforgiven
# We enter these into a distance structure as required for various
# distance-based routines. That is, we enter the upper triangle
# of the distance matrix as a single vector of distances.
distance_structure <- as.single(c(5, 8, 3, 7,
                                  2, 4, 6,
                                  9, 1,
                                  10))

# We also provide a character vector of movie names.
movie.names <- c("Bonnie and Clyde",
    "The Conversation",
    "The French Connection",
     "Hoosiers",
     "Unforgiven")
attr(distance_structure, "Size") <- length(movie.names)  # set size attribute

# We can check to see that the distance structure has been entered correctly
# by converting the distance structure to a distance matrix using the utility
# function make.distance_matrix, which we had defined above.
distance_matrix <- unlist(make.distance_matrix(distance_structure))
cat("\n", "Distance Matrix for Five Movies", "\n")
print(distance_matrix)

# apply the multidimensional scaling algorithm and plot the map
nonmetric.mds.solution <- isoMDS(distance_matrix, k=2, trace = FALSE)
First.Dimension <- nonmetric.mds.solution$points[,1]
Second.Dimension <- nonmetric.mds.solution$points[,2]
```

```
pdf(file = "fig_positioning_products_movies_mds.pdf",
    width = 8.5, height=8.5)  # opens pdf plotting device
# use par(mar = c(bottom, left, top, right)) to set up margins on the plot
par(mar=c(7.5, 7.5, 7.5, 5))

# We set up the plot but do not plot points... we use names for points.
plot(First.Dimension,Second.Dimension,xlim=c(-6,6),ylim=c(-6,6),type="n")

# We plot the movie names in the locations where points normally go.
text(First.Dimension, Second.Dimension, labels=movie.names, offset = 0.0)
dev.off()  # closes the pdf plotting device

# Suggestions for the student: Here is another set of movies.
# These movies are similar in having simple two-word titles: "The ______."
# Again, pairs of movies are judged on their similarity.
# Candidate Godfather Hustler Passenger Stranger
#  0  6  2  4  9     Candidate
#  6  0  5  7 10     Godfather
#  2  5  0  3  8     Hustler
#  4  7  3  0  1     Passenger
#  9 10  8  1  0     Stranger
#
# Try your own set of five movies.
#
# Modify the program to accommodate larger sets of movies.
```

***Exhibit 6.2.*** *Product Positioning of Movies (Python)*

```
# Product Positioning of Movies (Python)

# prepare for Python version 3x features and functions
from __future__ import division, print_function

# import packages for multivariate analysis
import numpy as np  # arrays and numerical processing
import matplotlib.pyplot as plt  # 2D plotting

# alternative distance metrics for multidimensional scaling
from sklearn.metrics import euclidean_distances
from sklearn.metrics.pairwise import linear_kernel as cosine_distances
from sklearn.metrics.pairwise import manhattan_distances as manhattan_distances

from sklearn import manifold  # multidimensional scaling

# define a numpy array for the movie data from one respondent
distance_matrix = np.array([[0,  5,  8,  3,  7],
    [5,  0,  2,  4,  6],
    [8,  2,  0,  9,  1],
    [3,  4,  9,  0, 10],
    [7,  6,  1, 10,  0]])
# check to see that the distance structure has been entered correctly
print(distance_matrix)
print(type(distance_matrix))

# apply the multidimensional scaling algorithm and plot the map
mds_method = manifold.MDS(n_components = 2, random_state = 9999,\
    dissimilarity = 'precomputed')
mds_fit = mds_method.fit(distance_matrix)
mds_coordinates = mds_method.fit_transform(distance_matrix)

movie_label = ["Bonnie and Clyde", "The Conversation",
    "The French Connection", "Hoosiers", "Unforgiven"]
# plot mds solution in two dimensions using movie labels
# defined by multidimensional scaling
plt.figure()
plt.scatter(mds_coordinates[:,0],mds_coordinates[:,1],\
    facecolors = 'none', edgecolors = 'none')  # points in white (invisible)
labels = movie_label
for label, x, y in zip(labels, mds_coordinates[:,0], mds_coordinates[:,1]):
    plt.annotate(label, (x,y), xycoords = 'data')
plt.xlabel('First Dimension')
plt.ylabel('Second Dimension')
plt.show()
plt.savefig('fig_positioning_products_mds_movies_python.pdf',
    bbox_inches = 'tight', dpi=None, facecolor='w', edgecolor='b',
    orientation='landscape', papertype=None, format=None,
    transparent=True, pad_inches=0.25, frameon=None)
```

***Exhibit 6.3.*** *Multidimensional Scaling Demonstration: US Cities (R)*

```
# Multidimensional Scaling Demonstration: US Cities (R)

library(MASS) # need MDS functions

# We define a utility function for converting a distance structure
# to a distance matrix as required for some routines and
# for printing of the complete matrix for visual inspection.
make.distance.matrix <- function(distance.structure)
{ n <- attr(distance.structure, "Size")
  full <- matrix(0,n,n)
  full[lower.tri(full)] <- distance.structure
  full+t(full)
}

# These are the original data from a chart of U.S. airline miles.
# Atla Chic  Den Hous   LA Miam   NY SanF Seat WaDC
#    0  587 1212  701 1936  604  748 2139 2182  543     Atlanta
#  587    0  920  940 1745 1188  713 1858 1737  597     Chicago
# 1212  920    0  878  831 1726 1631  949 1021 1494     Denver
#  701  940  878    0 1374  968 1420 1645 1891 1220     Houston
# 1936 1745  831 1374    0 2339 2451  347  959 2300     Los.Angeles
#  604 1188 1726  968 2339    0 1092 2594 2734  923     Miami
#  748  713 1631 1420 2451 1092    0 2571 2408  205     New.York
# 2139 1858  949 1645  347 2594 2571    0  678 2442     San.Francisco
# 2182 1737 1021 1891  959 2734 2408  678    0 2329     Seattle
#  543  597 1494 1220 2300  923  205 2442 2329    0     Washington.D.C.

# We enter these into a distance structure as required for various
# distance-based routines. That is, we enter the upper triangle
# of the distance matrix as a single vector of distances.
# We enter these into a distance structure as required for various
# distance-based routines. That is, we enter the upper triangle
# of the distance matrix as a single vector of distances.
distance.structure <-
    as.single(c(587, 1212,  701, 1936,  604,  748, 2139, 2182,  543,
                      920,  940, 1745, 1188,  713, 1858, 1737,  597,
                            878,  831, 1726, 1631,  949, 1021, 1494,
                                 1374,  968, 1420, 1645, 1891, 1220,
                                       2339, 2451,  347,  959, 2300,
                                             1092, 2594, 2734,  923,
                                                   2571, 2408,  205,
                                                          678, 2442,
                                                               2329))

# We also provide a character vector of city names.
city.names <- c("Atlanta", "Chicago", "Denver", "Houston", "LA",
    "Miami","NY","SanFran","Seattle","WashDC")

attr(distance.structure, "Size") <- length(city.names)  # set size attribute

# We can check to see that the distance structure has been entered correctly
```

```
# by converting the distance structure to a distance matrix
# using the utility function make.distance.matrix, which we had defined
distance.matrix <- unlist(make.distance.matrix(distance.structure))
cat("\n","Distance Matrix for U.S. Airline Miles","\n")
print(distance.matrix)

# apply the multidimensional scaling algorithm and plot the map
mds.solution <- cmdscale(distance.structure, k=2, eig=T)
First.Dimension <- mds.solution$points[,1]
Second.Dimension <- mds.solution$points[,2]

pdf(file = "plot_metric_mds_airline_miles.pdf",
    width=11, height=8.5) # opens pdf plotting device

# use par(mar = c(bottom, left, top, right)) to set up margins on the plot
par(mar=c(7.5, 7.5, 7.5, 5))

# We set up the plot but do not plot points... use names for points.
plot(First.Dimension,Second.Dimension,type="n") # first page of pdf plots
# We plot the city names in the locations where points normally go.
text(First.Dimension,Second.Dimension,labels=city.names,offset = 0.0)
title("Multidimensional Scaling of U.S. Airline Miles (First Draft)")

# a review of the plot shows that the horizontal dimension should be reflected
# multiply the first dimension by -1 to get closer to desired map of the US
First.Dimension <- mds.solution$points[,1] * -1
Second.Dimension <- mds.solution$points[,2]
plot(First.Dimension,Second.Dimension,type="n")  # second page of pdf plots
text(First.Dimension,Second.Dimension,labels=city.names,offset = 0.0)
title("Multidimensional Scaling of U.S. Airline Miles (Second Draft)")

# a review of the plot shows vertical dimension should also be reflected
# so we multiply the first and second dimensions by -1
# this gives us the desired map of the US
First.Dimension <- mds.solution$points[,1] * -1
Second.Dimension <- mds.solution$points[,2] * -1
plot(First.Dimension,Second.Dimension,type="n")  # third page of pdf plots
text(First.Dimension,Second.Dimension,labels=city.names,offset = 0.0)
title("Multidimensional Scaling of U.S. Airline Miles")
dev.off()  # closes the pdf plotting device

# Suggestions for the student: Try another geographic map.
# Try using distance in highway miles in place of distance in miles by air.
# Try non-metric MDS in place of metric MDS. See if there is a difference
# in the plotted solutions.
```

***Exhibit 6.4.*** *Multidimensional Scaling Demonstration: US Cities (Python)*

```
# Multidimensional Scaling Demonstration: US Cities (Python)

# prepare for Python version 3x features and functions
from __future__ import division, print_function

# import packages for multivariate analysis
import numpy as np  # arrays and numerical processing
import scipy
import matplotlib.pyplot as plt  # 2D plotting

# alternative distance metrics for multidimensional scaling
from sklearn.metrics import euclidean_distances
from sklearn.metrics.pairwise import linear_kernel as cosine_distances
from sklearn.metrics.pairwise import manhattan_distances as manhattan_distances
from sklearn import manifold  # multidimensional scaling

# These are the original data from a chart of U.S. airline miles.
# Atla Chic  Den Hous   LA Miam   NY SanF Seat WaDC
#    0  587 1212  701 1936  604  748 2139 2182  543     Atlanta
#  587    0  920  940 1745 1188  713 1858 1737  597     Chicago
# 1212  920    0  878  831 1726 1631  949 1021 1494     Denver
#  701  940  878    0 1374  968 1420 1645 1891 1220     Houston
# 1936 1745  831 1374    0 2339 2451  347  959 2300     Los.Angeles
#  604 1188 1726  968 2339    0 1092 2594 2734  923     Miami
#  748  713 1631 1420 2451 1092    0 2571 2408  205     New.York
# 2139 1858  949 1645  347 2594 2571    0  678 2442     San.Francisco
# 2182 1737 1021 1891  959 2734 2408  678    0 2329     Seattle
#  543  597 1494 1220 2300  923  205 2442 2329    0     Washington.D.C.

# we enter these into a distance matrix for multidimensional scaling
# defining a numpy array for these data
distance_matrix = \
np.array([[ 0,  587, 1212,  701, 1936,  604,  748, 2139, 2182,  543],
    [587,    0,  920,  940, 1745, 1188,  713, 1858, 1737,  597],
    [1212,  920,    0,  878,  831, 1726, 1631,  949, 1021, 1494],
    [701,  940,  878,    0, 1374,  968, 1420, 1645, 1891, 1220],
    [1936, 1745,  831, 1374,    0, 2339, 2451,  347,  959, 2300],
    [604, 1188, 1726,  968, 2339,    0, 1092, 2594, 2734,  923],
    [748,  713, 1631, 1420, 2451, 1092,    0, 2571, 2408,  205],
    [2139, 1858,  949, 1645,  347, 2594, 2571,    0,  678, 2442],
    [2182, 1737, 1021, 1891,  959, 2734, 2408,  678,    0, 2329],
    [543,  597, 1494, 1220, 2300,  923,  205, 2442, 2329,    0]])
# check to see that the distance structure has been entered correctly
print(distance_matrix)
print(type(distance_matrix))

# apply the multidimensional scaling algorithm and plot the map
mds_method = manifold.MDS(n_components = 2, random_state = 9999,\
    dissimilarity = 'precomputed')
mds_fit = mds_method.fit(distance_matrix)
mds_coordinates = mds_method.fit_transform(distance_matrix)
```

```
city_label = ['Atlanta', 'Chicago', 'Denver', 'Houston', 'Los Angeles',
    'Miami', 'New York', 'San Francisco', 'Seattle', 'Washington D.C.']

# plot mds solution in two dimensions using city labels
# defined by multidimensional scaling
plt.figure()
plt.scatter(mds_coordinates[:,0],mds_coordinates[:,1],\
    facecolors = 'none', edgecolors = 'none')  # points in white (invisible)
labels = city_label
for label, x, y in zip(labels, mds_coordinates[:,0], mds_coordinates[:,1]):
    plt.annotate(label, (x,y), xycoords = 'data')
plt.xlabel('First Dimension')
plt.ylabel('Second Dimension')
plt.show()
plt.savefig('fig_positioning_products_mds_cities_python.pdf',
    bbox_inches = 'tight', dpi=None, facecolor='w', edgecolor='b',
    orientation='landscape', papertype=None, format=None,
    transparent=True, pad_inches=0.25, frameon=None)
```

**Exhibit 6.5.** *Using Activities Market Baskets for Product Positioning (R)*

```
# Using Activities Market Baskets for Product Positioning (R)

library(MASS)  # MASS package for isoMDS() function for non-metric MDS
library(cluster)  # cluster analysis algorithms
library(wordcloud)  # provides function textplot() for non-overlapping text

# We define a utility function for converting a distance structure
# to a distance matrix as required for some routines and
# for printing of the complete matrix for visual inspection.
make.distance_matrix <- function(distance_structure)
{ n <- attr(distance_structure, "Size")
  full <- matrix(0,n,n)
  full[lower.tri(full)] <- distance_structure
  full+t(full)
}

wisconsin_dells_data_frame <- read.csv("wisconsin_dells.csv") # read in data
print(str(wisconsin_dells_data_frame))  # show structure of data frame

binary_variable_names <- c("shopping","antiquing",
"scenery","eatfine","eatcasual","eatfamstyle","eatfastfood","museums",
"indoorpool","outdoorpool","hiking","gambling","boatswim","fishing",
"golfing","boattours","rideducks","amusepark","minigolf","gocarting",
"waterpark","circusworld","tbskishow","helicopter","horseride","standrock",
"outattract","nearbyattract","movietheater","concerttheater","barpubdance",
"shopbroadway","bungeejumping")

binary_activity_data_frame <-
    wisconsin_dells_data_frame[,binary_variable_names]

# ifelse() converts YES/NO to 1/0 as needed for distance calculations
# then we form a matrix and take the transpose of the matrix
# which will be needed to compute distances between activities
binary_activity_matrix <-
    t(as.matrix(ifelse((binary_activity_data_frame == "YES"), 1, 0)))

# We compute the distance structure using the dist() function
# We specify that we want to use a distance metric for binary input data
# this takes the number of occurrences of zero and one or one and zero
# divided by the number of times at least one variable has a one
# The distance structure is for pairs of activities in this case study.
# A distance structure is needed for many distance-guided routines.
distance_structure <- dist(binary_activity_matrix, method="binary")

# To see this structure in the form of a distance matrix
# we can run the utility function for converting a distance structure
# to a distance matrix, which is make.distance_matrix
distance_matrix <- unlist(make.distance_matrix(distance_structure))
cat("\n", "Distance Matrix for Activities","\n")
print(str(distance_matrix))
```

```
# use par(mar = c(bottom, left, top, right)) to set up margins on the plot
par(mar=c(7.5, 7.5, 7.5, 5))

# The non-metric solution for this case was not especially useful,
# so we use metric multidimensional scaling for this problem.
mds_solution <- cmdscale(distance_structure, k=2, eig=T)
First_Dimension <- mds_solution$points[,1]
Second_Dimension <- mds_solution$points[,2]
plot(First_Dimension,Second_Dimension, type="n")
# setting xpd = TRUE allows text labels to extend beyond plotting region
text(First_Dimension, Second_Dimension, labels=binary_variable_names,
    offset=0, xpd=TRUE)

# let's plot all of the activities on a single plot using the
# textplot() utility function from the wordcloud package
# to avoid overlapping text strings... and for presentation to
# management, let's spell out the activities by name
activity_names <- c("Shopping", "Antiquing",
"Site Seeing", "Fine Dining", "Casual Dining",
"Family Style Dining", "Fast Food Dining", "Museums",
"Indoor Pool", "Outdoor Pool", "Hiking", "Gambling",
"Boating/Swimming", "Fishing", "Golfing", "Boat Tours",
"Ride the Ducks", "Amusement Park", "Minigolf", "Go-carting",
"Waterpark", "Circus World", "Tommy Bartlett Ski Show",
"Helicopter Rides", "Horseback Riding", "Stand Rock",
"Outdoor Attractions", "Nearby Attractions",
"Movie Theater", "Concert Theater", "Bar/Pub Dancing",
"Shop Broadway", "Bungee Jumping")
pdf(file = "fig_positioning_products_dells_mds.pdf",
    width=8.5,height=8.5) # opens pdf plotting device
textplot(x = mds_solution$points[,1],
    y = mds_solution$points[,2],
    words = activity_names,
    show.lines = FALSE,
    xlim = range(mds_solution$points[,1]),
    ylim = range(mds_solution$points[,2]),
    xlab = "First Dimension",
    ylab = "Second Dimension")
dev.off()  # closes the pdf plotting device

# Suggestions for the student:
# Try alternative distance metrics to see how these change
# the multidimensional solution. Try reflection of the map
# and roation of the map to see if the resulting visualization
# may be more useful for management. Try nom-metric as well
# as metric MDS to see how the solutions compare.
# Having identified bungee jumping as an entertainment
# activity distinct from the set of other Dells activities,
# develop a classification model from visitor group demographics
# to find/target the bungee jumpers. Try alternative methods
# for this, including traditional and machine learning methods.
```

**Exhibit 6.6.** *Using Activities Market Baskets for Product Positioning (Python)*

```
# Using Activities Market Baskets for Product Positioning (Python)

# prepare for Python version 3x features and functions
from __future__ import division, print_function

# import packages for multivariate analysis
import numpy as np  # arrays and numerical processing
import matplotlib.pyplot as plt  # 2D plotting
import pandas as pd

# alternative distance metrics for multidimensional scaling
from sklearn.metrics import euclidean_distances
from sklearn.metrics.pairwise import linear_kernel as cosine_distances
from sklearn.metrics.pairwise import manhattan_distances as manhattan_distances

from sklearn import manifold  # multidimensional scaling

# read data from comma-delimited text file... create DataFrame object
wisconsin_dells_data_frame = pd.read_csv('wisconsin_dells.csv')
print(wisconsin_dells_data_frame.head)  # check the structure of the data frame
print(wisconsin_dells_data_frame.shape)

binary_variable_names = ['shopping','antiquing',
'scenery','eatfine','eatcasual','eatfamstyle','eatfastfood','museums',
'indoorpool','outdoorpool','hiking','gambling','boatswim','fishing',
'golfing','boattours','rideducks','amusepark','minigolf','gocarting',
'waterpark','circusworld','tbskishow','helicopter','horseride','standrock',
'outattract','nearbyattract','movietheater','concerttheater','barpubdance',
'shopbroadway','bungeejumping']

# let's focus on activities data
dells_activities_data_frame_preliminary = \
    pd.DataFrame(wisconsin_dells_data_frame, columns = binary_variable_names)
# remove any records with missing data
dells_activities_data_frame = dells_activities_data_frame_preliminary.dropna()
print(dells_activities_data_frame.shape)

# use dictionary object for mapping the response/target variable
activity_to_binary = {'NO' : 0, 'YES' : 1, '': 0}
for iname in binary_variable_names:
    dells_activities_data_frame[iname] = \
        dells_activities_data_frame[iname].map(activity_to_binary)
print(dells_activities_data_frame[0:10])  # examine the first 10 rows of data

# convert DataFrame to numpy array representation of activities matrix
activities_binary_matrix = dells_activities_data_frame.as_matrix().transpose()
print(type(activities_binary_matrix))
print(activities_binary_matrix.shape)
# compute distance matrix
distance_matrix = manhattan_distances(activities_binary_matrix)
print(distance_matrix.shape)
```

```
# apply the multidimensional scaling algorithm and plot the map
mds_method = manifold.MDS(n_components = 2, random_state = 9999,\
    dissimilarity = 'precomputed')
mds_fit = mds_method.fit(distance_matrix)
mds_coordinates = mds_method.fit_transform(distance_matrix)

activity_names = ['Shopping', 'Antiquing',
'Site Seeing', 'Fine Dining', 'Casual Dining',
'Family Style Dining', 'Fast Food Dining', 'Museums',
'Indoor Pool', 'Outdoor Pool', 'Hiking', 'Gambling',
'Boating/Swimming', 'Fishing', 'Golfing', 'Boat Tours',
'Ride the Ducks', 'Amusement Park', 'Minigolf', 'Go-carting',
'Waterpark', 'Circus World', 'Tommy Bartlett Ski Show',
'Helicopter Rides', 'Horseback Riding', 'Stand Rock',
'Outdoor Attractions', 'Nearby Attractions',
'Movie Theater', 'Concert Theater', 'Bar/Pub Dancing',
'Shop Broadway', 'Bungee Jumping']

# plot mds solution in two dimensions using activity labels
# defined by multidimensional scaling
plt.figure()
plt.scatter(mds_coordinates[:,0],mds_coordinates[:,1],\
    facecolors = 'none', edgecolors = 'none')  # points in white (invisible)
labels = activity_names
for label, x, y in zip(labels, mds_coordinates[:,0], mds_coordinates[:,1]):
    plt.annotate(label, (x,y), xycoords = 'data')
plt.xlabel('First Dimension')
plt.ylabel('Second Dimension')
plt.show()
plt.savefig('fig_positioning_products_mds_dells_python.pdf',
    bbox_inches = 'tight', dpi=None, facecolor='w', edgecolor='b',
    orientation='landscape', papertype=None, format=None,
    transparent=True, pad_inches=0.25, frameon=None)
```

*Exhibit 6.7. Hierarchical Clustering of Activities (R)*

```
# Hierarchical Clustering of Activities (R)

library(cluster)  # cluster analysis algorithms

# We define a utility function for converting a distance structure.
make.distance_matrix <- function(distance_structure)
{ n <- attr(distance_structure, "Size")
  full <- matrix(0,n,n)
  full[lower.tri(full)] <- distance_structure
  full+t(full)
}

wisconsin_dells_data_frame <- read.csv("wisconsin_dells.csv") # read in data
print(str(wisconsin_dells_data_frame))  # show structure of data frame

binary_variable_names <- c("shopping","antiquing",
"scenery","eatfine","eatcasual","eatfamstyle","eatfastfood","museums",
"indoorpool","outdoorpool","hiking","gambling","boatswim","fishing",
"golfing","boattours","rideducks","amusepark","minigolf","gocarting",
"waterpark","circusworld","tbskishow","helicopter","horseride","standrock",
"outattract","nearbyattract","movietheater","concerttheater","barpubdance",
"shopbroadway","bungeejumping")
binary_activity_data_frame <-
    wisconsin_dells_data_frame[,binary_variable_names]
# ifelse() converts YES/NO to 1/0 as needed for distance calculations
# then we form a matrix and take the transpose of the matrix
# which will be needed to compute distances between activities
binary_activity_matrix <-
    t(as.matrix(ifelse((binary_activity_data_frame == "YES"), 1, 0)))
# We compute the distance structure using the dist() function
distance_structure <- dist(binary_activity_matrix, method="binary")

# To see this structure in the form of a distance matrix
# we can run the utility function for converting a distance structure
# to a distance matrix, which is make.distance_matrix
distance_matrix <- unlist(make.distance_matrix(distance_structure))

activity_names <- c("Shopping", "Antiquing",
"Site Seeing", "Fine Dining", "Casual Dining",
"Family Style Dining", "Fast Food Dining", "Museums",
"Indoor Pool", "Outdoor Pool", "Hiking", "Gambling",
"Boating/Swimming", "Fishing", "Golfing", "Boat Tours",
"Ride the Ducks", "Amusement Park", "Minigolf", "Go-carting",
"Waterpark", "Circus World", "Tommy Bartlett Ski Show",
"Helicopter Rides", "Horseback Riding", "Stand Rock",
"Outdoor Attractions", "Nearby Attractions",
"Movie Theater", "Concert Theater", "Bar/Pub Dancing",
"Shop Broadway", "Bungee Jumping")
rownames(distance_matrix) <- activity_names
colnames(distance_matrix) <- activity_names
print(str(distance_matrix))
```

```
clustering_solution <- agnes(distance_matrix, diss = TRUE,
    metric = "manhattan", method = "average")

pdf(file = "fig_positioning_products_dells_clustering.pdf",
  width = 11, height = 8.5)
plot(clustering_solution)
dev.off()

# Suggestions for the student:
# Try other clustering solutions by varying the distance metric
# and hierarchical clustering algorithms.
```

# 7 Developing New Products

Stock Boy: "Excuse me, sir, what are you doing?"

George: "I'll tell you what I'm doing. I want to buy eight hot dogs and eight hot dog buns to go with them. But no one sells eight hot dog buns. They only sell twelve hot dog buns. So I end up paying for four buns I don't need. So I am removing the superfluous buns."

Stock Boy: "I'm sorry, sir. But you're going to have to pay for all twelve buns. They're not marked individually."

George: "Yeah. And you want to know why? Because some big-shot over at the wiener company got together with some big-shot over at the bun company and decided to rip off the American public. Because they think the American public is a bunch of trusting nit-wits who will pay for everything they don't need rather than make a stink. Well they are not ripping off this nit-wit anymore because I'm not paying for one more thing I don't need. George Banks is saying 'no.' "

Stock Boy: "Who's George Banks?"

George: "Me!"

—Ira Heiden as Stock Boy and Steve Martin as George Banks
in *Father of the Bride* (1991)

*Figure 7.1. The Precarious Nature of New Product Development*

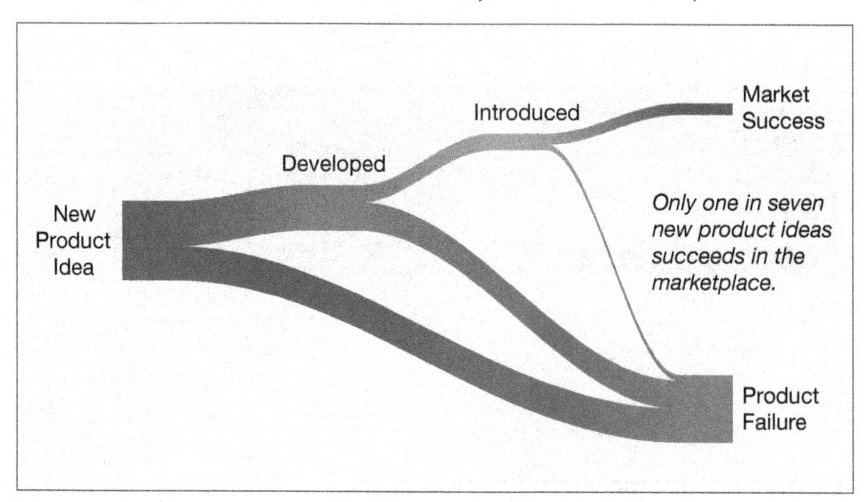


New products are essential to business success, and marketing data science has much to contribute to the product development process. It is common to identify stages in product development as opportunity identification, design, testing, introduction, and product management. At each stage in the evaluation process, new product ideas are discarded.

Developing new products is a precarious process. Cooper (2011) cites studies from the Product Development and Management Association suggesting that for every seven new product ideas, four enter development, one or two are introduced as new products, and only one ultimately succeeds. Figure 7.1 shows the degree of drop-off at these checkpoints in the new product development process.

Marketplace evaluation involves a review of the product category and a company's position in that category. This is a matter of screening and evaluation. We must understand our position relative to the competitors. Business strategy questions referred to as SWOT are key here. What are our strengths and weaknesses? What are the opportunities and threats?

Business analysis concerns the capabilities of the company for new product development. There are financial and human capital considerations, limitations regarding budgets and time frames, as well as requirements for the physical plant and information infrastructure.

Product design is a major component of the new development process and a place where data scientists can provide great value. Experiments may be designed to provide answers about the particular products or product bundles that will be best received by consumers.

Imagine that we are helping an automobile product manager make decisions about next year's sports car line. There are many decisions to make, decisions about interior and exterior color options, engine horsepower or torque, fuel efficiency, electronics (including GPS navigation systems), manual or automatic transmission, and warrantee options. All of these decisions must be made with an eye toward the competition in the sports car market. After product design and pricing decisions have been made, there is the need to draft advertising copy for each target market. There may be questions about market segments. Who are the potential buyers, and what are their numbers? And, of course, there are questions about the pricing of new product offerings.

The product manager needs to decide which product characteristics to include in all cars of the brand, which product characteristics to include in product bundles or car models, and which product characteristics to include as consumer options.

The selection of attributes to include in a product design study will depend on the purpose or objective of the study. If the purpose of a study is product design for a particular brand or product line, then it makes little sense to include brand in the design.

When it comes to product design studies, questions abound, and the answers are not always easy to come by. How many attributes can be tested in a single study? How many levels of each attribute make sense? Should levels of quantitative attributes be defined at equally spaced intervals, or should we employ a logarithmic scale?

What about price? Due to its importance in consumer choice, price can dominate a product design study, making consumer preferences across product characteristics difficult to understand. For this reason, research studies with a primary focus on product design often exclude price. If price is included, then we try to ensure that price does not dominate a study. We set an appropriate range for price levels vis à vis ranges for levels of product attributes. The importance of one attribute is measured relative to the importance of other attributes. In the design of conjoint studies, we want to ensure that price does not dominate in terms of importance, so we are careful to set a range of price levels that is not too wide.

Does it make sense to include brand? If the purpose of a study is pricing, then brand is a key factor to include in the study. This is because there is competition among brands in most product categories. In fact, some studies are designed to measure brand equity alone. These studies answer the question: "How much is our brand worth over and above other brands in the same product category?"

If we include both brand and price, which brands should we include and what range of prices should we use? Must all combinations of attribute levels be permitted, or can we have certain prohibited pairings of attributes?

The fundamental character of conjoint studies, what distinguishes these studies from other product design research methods, is their focus on tradeoffs. Curiously, there are many tradeoffs in the design of conjoint studies themselves. In addition to questions about product attributes, we must answer questions about the format of the design.

How long should the survey be? What is the most appropriate response modality: a rating, ranking, choice, or sequence of choices? If we decide to use choice-based conjoint, how many choice alternatives should be used in each item or set of choices and how many choice sets should be used?

Which method of analysis makes the most sense for the chosen response modality and research objective? If possible, shall we obtain measurements at the individual level, or is aggregate (group-level) analysis sufficient? How should we approach the question of market segments or consumer groups within the context of product design research?

After products have been designed, there is testing in the field. To demonstrate this process, we draw on a classic study, Procter & Gamble Laundry Soaps as described in appendix C (page 370).

Consumers in 1,008 households, some of whom were previous users of the original formula $M$, were given the opportunity to try formulas $X$ and $M$ in blind preference tests. At the end of the tests, consumers were asked to indicate their soap preferences by choosing either $X$ or $M$. Water temperature (cold or hot) and type (hard, medium, soft) were noted for each household.

The field test data, first reported by Reis and Smith (1963), were presented as cross-classified count data (that is, a multi-way contingency or grouped frequency table). Fienberg (2007) presented an analysis of these data using log-linear models. But there is doubt as to whether log-linear modeling is the most effective way to learn about these data. One could argue that logistic regression provides a more straightforward approach. The soap choice in the Procter & Gamble study is a binary response, much like we have seen in choice studies reviewed earlier.

In order to fit a logistic regression model to data from the laundry soap experiment, we begin by converting the grouped frequency table to a set of records for individual respondents. Then we fit a complete logistic regression model with main effects and interactions among the factors in the experiment. The analysis of deviance for the fitted logistic regression is shown in table 7.1.

What are the implications of the study for Procter & Gamble. Of special note is the fact that $M$ and non-$M$ users, although engaging in a blind consumer study, had differential preferences regarding the laundry soaps. There were also slight differences across users by the water temperature being used for laundry. There is no evidence of an effect for water type.

We return to the original field test data to see what these experimental effects mean for product managers at Procter & Gamble. Figure 7.2 shows an interaction plot for the Procter & Gamble Laundry Soaps. The new soap formulation $X$ is preferred by consumers who are not currently soap $M$ users. Among these consumers who are not currently soap $M$ users, there is also a slight preference for soap $X$ when washing at high temperatures. An interaction plot like this is much easier for managers to understand than the initial contingency tables or the analysis of deviance results.

Interaction plots are useful even when interactions are nonsignificant. Like lattice and mosaic plots, they allow us to show results for many variables simultaneously.

*Table 7.1. Analysis of Deviance for New Product Field Test: Procter & Gamble Laundry Soaps*

| | df | Deviance | Residual df | Residual Deviance | P-Value |
|---|---|---|---|---|---|
| NULL | | | 1007 | 1397.3 | |
| M User | 1 | 20.5815 | 1006 | 1376.7 | 5.715e-06 |
| Water Temperature | 1 | 3.8002 | 1005 | 1372.9 | 0.05125 |
| Water Type | 2 | 0.2160 | 1003 | 1372.7 | 0.89763 |
| M User × Water Temperature | 1 | 2.7328 | 1002 | 1370.0 | 0.09831 |
| M User × Water Type | 2 | 4.5961 | 1000 | 1365.4 | 0.10045 |
| Water Temperature × Water Type | 2 | 0.1618 | 998 | 1365.2 | 0.92228 |
| M User × Water Temperature × Water Type | 2 | 0.7373 | 996 | 1364.5 | 0.69166 |

*Figure 7.2. Implications of a New Product Field Test: Procter & Gamble Laundry Soaps*

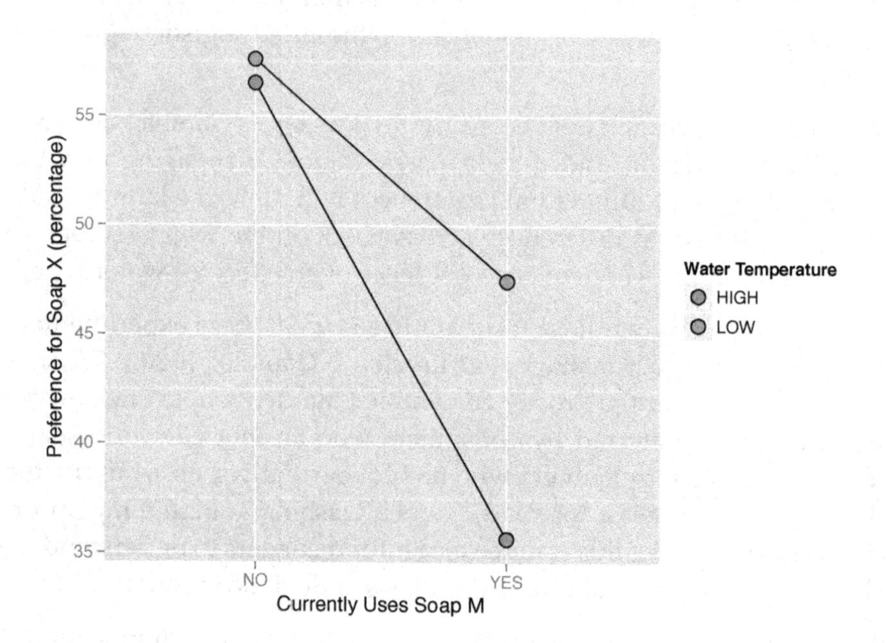

For an overview of the new product development process, see Cooper (2011), and for issues relating to the design and marketing of new products see Urban and Hauser (1993). Katz (2011) suggests that we need to look beyond commonly identified stages in product development. He argues that not enough time is spent in idea generation or discovery, which precedes all other stages. To do a good job of discovery, we must pursue exploratory research into markets and consumer needs.

A related field of research concerns marketplace response to new products or what is known as diffusion of innovations, a social network process. General models of buyer behavior help us understand trial purchase, repeat purchase, brand loyalty, and switching. Probability models are often used to represent regularities in aggregate buyer behavior over time, as is observed in the diffusion of new products and repeat buying models. Examples of applied probability models for understanding buyer behavior may be found in the work of Bass (1969), Ehrenberg (1988), and Fader and Hardie (1996, 2002).

Experimental design methods utilized in the analysis of the Procter & Gamble Laundry Soaps field test are of great value to those needing to engage in experimental research in marketing. What are sometimes referred to as $A/B$ tests in website design fall under the domain of experimental design. See appendix B (page 338) for additional discussion of experimental research in marketing and the relevant literature.

Conjoint methods are especially useful in designing new products. See Orme (2013) and Rao (2014) for a review of conjoint survey methods for new product design. We employ principles of experimental design in constructing conjoint tasks, and we use linear models in the analysis of the consumer responses. Methods for the design and analysis of conjoint studies have been discussed by Carroll and Green (1995), Marshall and Bradlow (2002), and authors in the edited volume by Gustafsson, Herrmann, and Huber (2000). Aizaki (2012, 2014) describes programs for constructing product profiles for conjoint and choice studies.

Exhibit 7.1 shows an R program for analyzing data for Procter & Gamble Laundry Soaps. The program utilizes a graphics package developed by Wickham and Chang (2014). The corresponding Python program is on the website for the book.

*Exhibit 7.1. Analysis for a Field Test of Laundry Soaps (R)*

```
# Analysis for a Field Test of Laundry Soaps (R)

library(ggplot2)  # graphics for interaction plotting

# set contrast for tests of significance with factors
options(contrasts=c(factor = "contr.treatment", ordered="contr.poly"))

# first import data from the comma-delimited file gsoaps.csv
# the response variable relates to brand choice
# choice is factor/binary response variable (M or X)
# the explanatory variables are factors (categorical variables):
#   wtemp = water temperature with levels LOW OR HIGH
#   wtype = water type with levels SOFT, MEDIUM, OR HARD
#   muser = is user of brand M with labels NO OR YES

# read grouped frequecy data file
gsoaps <- read.csv("gsoaps.csv")

# convert to individual observations as required for logistic regression
indices <- rep(1:nrow(gsoaps),gsoaps$freq)
soaps <- gsoaps[indices,-1]
soaps <- data.frame(soaps, row.names=NULL)

# check the data frame
print(str(soaps))
print(head(soaps))
print(tail(soaps))
with(soaps, print(table(wtemp, wtype, muser, choice)))

# specify complete experimental design with interactions
soaps_model <- choice ~ muser * wtemp * wtype

# fit the complete model
soaps_fit <- glm(soaps_model, family = binomial, data=soaps)

# summary of fit
print(soaps_fit)

# analysis of deviance for experimental factors
# providing likelihood ratio chi-square tests of each effect
print(anova(soaps_fit, test="Chisq"))

# -----------------------------
# Interaction Plotting
# -----------------------------
# code the choice as a 0/1 binary variable 1 = Brand X
soaps$response <- ifelse((soaps$choice == "X"), 1, 0)

# compute choice share cell means for use in interaction plot
response_mean <- aggregate(response ~ muser * wtemp,
  data = soaps, mean)
```

```
# generate an interaction plot for brand X choice as a percentage
pdf(file = "fig_developing_new_products_soaps_interaction_plot.pdf",
    width = 7, height = 5)
interaction_plot <- ggplot(response_mean,
  aes(x = muser, y = 100*response,
  group = wtemp, fill = wtemp)) +
  geom_line(linetype = "solid", size = 1, colour = "black") +
  geom_point(size = 4, shape = 21) +
  ylab("Preference for Soap X (percentage)") +
  xlab("Currently Uses Soap M") +
  labs(fill = "Water Temperature") +
  theme(plot.title = element_text(lineheight=.8, face="bold"))
print(interaction_plot)
dev.off()

# Suggestions for the student:
# Show that other effects in the experiment are of lesser
# importance by making additional interaction plots.
# Try alternative modeling methods for examining these
# field test data. Log-linear models are one possibility.
```

# 8 Promoting Products

Roy: "I'm not a criminal. I'm a con man."

Dr. Klein: "The difference being?"

Roy: "They give me their money."

—NICOLAS CAGE AS ROY AND BRUCE ALTMAN AS DR. KLEIN
IN *Matchstick Men* (2003)

It was a Thursday night in July. I was thinking about going to the ballpark. The Los Angeles Dodgers were playing the Colorado Rockies, and I was supposed to get an Adrian Gonzalez bobblehead with my ticket. Although I was not excited about the bobblehead, seeing a ball game at Dodger Stadium sounded like great fun. In April and May the Dodgers' record had not been the best, but things were looking better by July. I wondered if bobbleheads would bring additional fans to the park. Dodgers management may have been wondering the same thing, or perhaps making plans for Yasiel Puig bobbleheads.

Suppose we were working for the Dodgers and wanted to learn about promotions and their effect on attendance. We call this example *Bobbleheads and Dodger Dogs* or *Shaking Our Bobbleheads Yes and No*. The example draws on Major League Baseball data from the 2012 season.

Relevant data for Dodgers' home games are shown in table 8.1. Dodger Stadium, with a capacity of 56,000, is the largest ballpark in the world. We can see that Dodger Stadium was filled to capacity only twice in 2012. There were two cap promotions and three shirt promotions in 2012, not enough to draw meaningful inferences. Fireworks were used thirteen times on Friday nights, and once on the Fourth of July. The eleven bobblehead promotions occurred on night games, six of those being Tuesday nights.

Exploratory graphics help us find models that might work for predicting attendance and evaluating the effect of promotions on attendance. Figure 8.1 shows distributions of attendance across days of the week, and figure 8.2 shows attendance by month. Box plots like these reveal the overall values of the data, with the boxes covering the middle fifty percent or so of the distribution and with the center line representing the median. The dotted lines or whiskers extend to more extreme values in the distribution.[1] By looking at the box plots, we can make comparisons by day and by month across the distributions of attendance.

We can explore these data further using a lattice of scatter plots. In figure 8.3 we map the relationship between temperature and attendance, controlling for time of game (day or night) and clear or cloudy skies. On day games with clear skies, we see what appears to be a moderate inverse relationship between temperature and attendance. Day games are usually on Sunday, and in 2012 all but one of those games was played under clear skies—a benefit of being in Los Angeles.

More telling perhaps are strip plots of attendance by opponent or visiting team; these are the univariate scatter plots in figure 8.4. Opponents from the large metropolitan areas (the New York Mets, Chicago Cubs and White Sox, Los Angeles Angels, and Washington D.C. Nationals) are consistently associated with higher attendance. But there are seventeen visiting teams in this study, and only eighty-one games or observations. Accordingly, utilizing the visiting team as a categorical predictor presents problems.

[1] To determine the length of the whiskers, we first compute the interquartile range, which is the distance between the 25th percentile and the 75th percentile. The end-points of the whiskers are defined by what are called *adjacent values.* The upper whisker extends to the upper adjacent value, a point one-and-a-half times the interquartile range above the upper end of the box. Or, if the maximum value in the distribution is less than that, the upper whisker extends to that maximum value. We often think of outliers as being points outside the whiskers; these outlier points are plotted as open circles. Box plots were the invention of John Tukey (1977).

**Table 8.1.** *Bobbleheads and Dodger Dogs*

| month | day | attend | day_of_week | opponent | temp | skies | day_night | cap | shirt | fireworks | bobblehead |
|---|---|---|---|---|---|---|---|---|---|---|---|
| APR | 10 | 56000 | Tuesday | Pirates | 67 | Clear | Day | NO | NO | NO | NO |
| APR | 11 | 29729 | Wednesday | Pirates | 58 | Cloudy | Night | NO | NO | NO | NO |
| APR | 12 | 28328 | Thursday | Pirates | 57 | Cloudy | Night | NO | NO | NO | NO |
| APR | 13 | 31601 | Friday | Padres | 54 | Cloudy | Night | NO | NO | YES | NO |
| APR | 14 | 46549 | Saturday | Padres | 57 | Cloudy | Night | NO | NO | NO | NO |
| APR | 15 | 38359 | Sunday | Padres | 65 | Clear | Day | NO | NO | NO | NO |
| APR | 23 | 26376 | Monday | Braves | 60 | Cloudy | Night | NO | NO | NO | NO |
| APR | 24 | 44014 | Tuesday | Braves | 63 | Cloudy | Night | NO | NO | NO | NO |
| APR | 25 | 26345 | Wednesday | Braves | 64 | Cloudy | Night | NO | NO | NO | NO |
| APR | 27 | 44807 | Friday | Nationals | 66 | Clear | Night | NO | NO | YES | NO |
| APR | 28 | 54242 | Saturday | Nationals | 71 | Clear | Night | NO | NO | NO | YES |
| APR | 29 | 48753 | Sunday | Nationals | 74 | Clear | Day | NO | YES | NO | NO |
| MAY | 7 | 43713 | Monday | Giants | 67 | Clear | Night | NO | NO | NO | NO |
| MAY | 8 | 32799 | Tuesday | Giants | 75 | Clear | Night | NO | NO | NO | NO |
| MAY | 9 | 33993 | Wednesday | Giants | 71 | Clear | Night | NO | NO | NO | NO |
| MAY | 11 | 35591 | Friday | Rockies | 65 | Clear | Night | NO | NO | YES | NO |
| MAY | 12 | 33735 | Saturday | Rockies | 65 | Clear | Night | NO | NO | NO | NO |
| MAY | 13 | 49124 | Sunday | Rockies | 70 | Clear | Day | NO | NO | NO | NO |
| MAY | 14 | 24312 | Monday | Snakes | 67 | Clear | Night | NO | NO | NO | NO |
| MAY | 15 | 47077 | Tuesday | Snakes | 70 | Clear | Night | NO | NO | NO | YES |
| MAY | 18 | 40906 | Friday | Cardinals | 64 | Clear | Night | NO | NO | YES | NO |
| MAY | 19 | 39383 | Saturday | Cardinals | 67 | Clear | Night | NO | NO | NO | NO |
| MAY | 20 | 44005 | Sunday | Cardinals | 77 | Clear | Night | NO | NO | NO | NO |
| MAY | 25 | 36283 | Friday | Astros | 59 | Cloudy | Night | NO | NO | YES | NO |
| MAY | 26 | 36561 | Saturday | Astros | 61 | Cloudy | Night | NO | NO | NO | NO |
| MAY | 27 | 33306 | Sunday | Astros | 70 | Clear | Day | NO | NO | NO | NO |
| MAY | 28 | 38016 | Monday | Brewers | 73 | Clear | Night | NO | NO | NO | NO |
| MAY | 29 | 51137 | Tuesday | Brewers | 74 | Clear | Night | NO | NO | NO | YES |
| MAY | 30 | 25509 | Wednesday | Brewers | 69 | Clear | Night | NO | NO | NO | NO |
| MAY | 31 | 26773 | Thursday | Brewers | 70 | Clear | Night | NO | NO | NO | NO |
| JUN | 11 | 50559 | Monday | Angels | 68 | Clear | Night | NO | YES | NO | NO |
| JUN | 12 | 55279 | Tuesday | Angels | 66 | Cloudy | Night | NO | NO | NO | YES |
| JUN | 13 | 43494 | Wednesday | Angels | 67 | Clear | Night | NO | NO | NO | NO |
| JUN | 15 | 40432 | Friday | White Sox | 67 | Clear | Night | NO | NO | YES | NO |
| JUN | 16 | 45210 | Saturday | White Sox | 68 | Clear | Night | NO | NO | NO | NO |
| JUN | 17 | 53504 | Sunday | White Sox | 74 | Clear | Day | NO | NO | NO | NO |
| JUN | 28 | 49006 | Thursday | Mets | 75 | Clear | Night | NO | NO | NO | YES |
| JUN | 29 | 49763 | Friday | Mets | 72 | Clear | Night | NO | NO | YES | NO |
| JUN | 30 | 44217 | Saturday | Mets | 78 | Clear | Day | NO | NO | NO | NO |
| JUL | 1 | 55359 | Sunday | Mets | 75 | Clear | Night | NO | NO | NO | YES |
| JUL | 2 | 34493 | Monday | Reds | 70 | Clear | Night | NO | NO | NO | NO |
| JUL | 3 | 33884 | Tuesday | Reds | 70 | Cloudy | Night | YES | NO | NO | NO |
| JUL | 4 | 53570 | Wednesday | Reds | 70 | Clear | Night | NO | NO | YES | NO |
| JUL | 13 | 43873 | Friday | Padres | 76 | Cloudy | Night | NO | NO | YES | NO |
| JUL | 14 | 54014 | Saturday | Padres | 75 | Clear | Night | NO | NO | NO | YES |
| JUL | 15 | 39715 | Sunday | Padres | 77 | Clear | Day | NO | NO | NO | NO |
| JUL | 16 | 32238 | Monday | Phillies | 67 | Clear | Night | NO | NO | NO | NO |
| JUL | 17 | 53498 | Tuesday | Phillies | 70 | Clear | Night | NO | NO | NO | NO |
| JUL | 18 | 39955 | Wednesday | Phillies | 80 | Cloudy | Day | NO | NO | NO | NO |
| JUL | 30 | 33180 | Monday | Snakes | 73 | Clear | Night | NO | NO | NO | NO |
| JUL | 31 | 52832 | Tuesday | Snakes | 75 | Cloudy | Night | NO | NO | NO | YES |
| AUG | 1 | 36596 | Wednesday | Snakes | 79 | Clear | Day | NO | NO | NO | NO |
| AUG | 3 | 43537 | Friday | Cubs | 73 | Clear | Night | NO | NO | YES | NO |
| AUG | 4 | 46588 | Saturday | Cubs | 73 | Cloudy | Night | NO | NO | NO | NO |
| AUG | 5 | 42495 | Sunday | Cubs | 83 | Clear | Day | YES | NO | NO | NO |
| AUG | 6 | 32659 | Monday | Rockies | 79 | Clear | Night | NO | NO | NO | NO |
| AUG | 7 | 55024 | Tuesday | Rockies | 80 | Clear | Night | NO | NO | NO | YES |
| AUG | 8 | 37084 | Wednesday | Rockies | 84 | Clear | Night | NO | NO | NO | NO |
| AUG | 20 | 36878 | Monday | Giants | 80 | Clear | Night | NO | NO | NO | NO |
| AUG | 21 | 56000 | Tuesday | Giants | 75 | Clear | Night | NO | NO | NO | YES |
| AUG | 22 | 40173 | Wednesday | Giants | 75 | Clear | Night | NO | NO | NO | NO |
| AUG | 24 | 39805 | Friday | Marlins | 71 | Clear | Night | NO | NO | YES | NO |
| AUG | 25 | 40284 | Saturday | Marlins | 70 | Clear | Night | NO | NO | NO | NO |
| AUG | 26 | 41907 | Sunday | Marlins | 81 | Clear | Day | NO | NO | NO | NO |
| AUG | 30 | 54621 | Thursday | Snakes | 80 | Clear | Night | NO | NO | NO | YES |
| AUG | 31 | 37622 | Friday | Snakes | 77 | Clear | Night | NO | NO | YES | NO |
| SEP | 1 | 35992 | Saturday | Snakes | 81 | Clear | Night | NO | NO | NO | NO |
| SEP | 2 | 31607 | Sunday | Snakes | 89 | Clear | Day | NO | NO | NO | NO |
| SEP | 3 | 33540 | Monday | Padres | 84 | Cloudy | Night | NO | NO | NO | NO |
| SEP | 4 | 40619 | Tuesday | Padres | 78 | Clear | Night | NO | YES | NO | NO |
| SEP | 5 | 50560 | Wednesday | Padres | 77 | Cloudy | Night | NO | NO | NO | NO |
| SEP | 13 | 43309 | Thursday | Cardinals | 80 | Clear | Night | NO | NO | NO | NO |
| SEP | 14 | 40167 | Friday | Cardinals | 85 | Clear | Night | NO | NO | YES | NO |
| SEP | 15 | 42449 | Saturday | Cardinals | 95 | Clear | Night | NO | NO | NO | NO |
| SEP | 16 | 35754 | Sunday | Cardinals | 86 | Clear | Day | NO | NO | NO | NO |
| SEP | 28 | 37133 | Friday | Rockies | 77 | Clear | Night | NO | NO | YES | NO |
| SEP | 29 | 40724 | Saturday | Rockies | 84 | Cloudy | Night | NO | NO | NO | NO |
| SEP | 30 | 35607 | Sunday | Rockies | 95 | Clear | Day | NO | NO | NO | NO |
| OCT | 1 | 33624 | Monday | Giants | 86 | Clear | Night | NO | NO | NO | NO |
| OCT | 2 | 42473 | Tuesday | Giants | 83 | Clear | Night | NO | NO | NO | NO |
| OCT | 3 | 34014 | Wednesday | Giants | 82 | Cloudy | Night | NO | NO | NO | NO |

***Figure 8.1.*** *Dodgers Attendance by Day of Week*

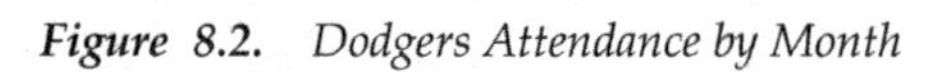

***Figure 8.2.*** *Dodgers Attendance by Month*

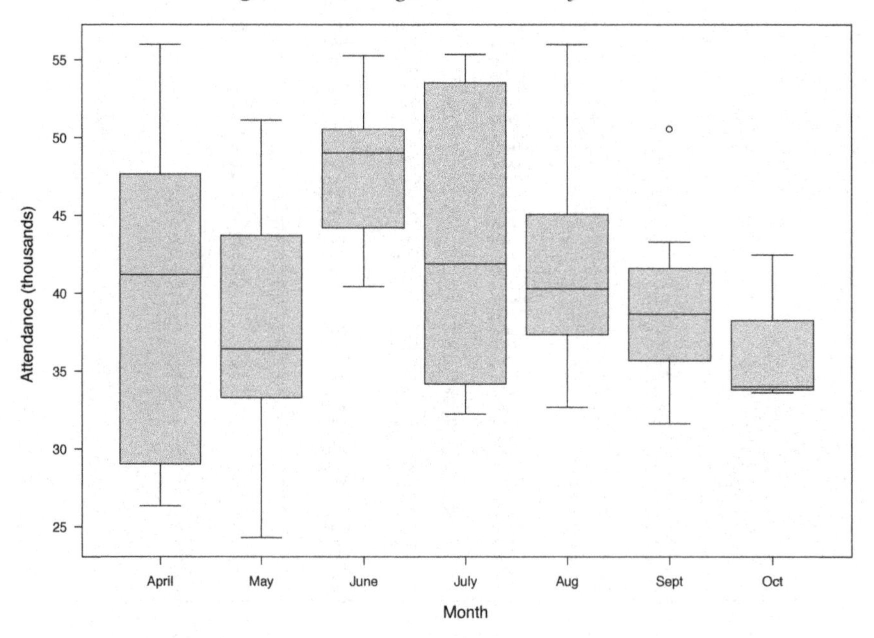

*Figure 8.3.* Dodgers Weather, Fireworks, and Attendance

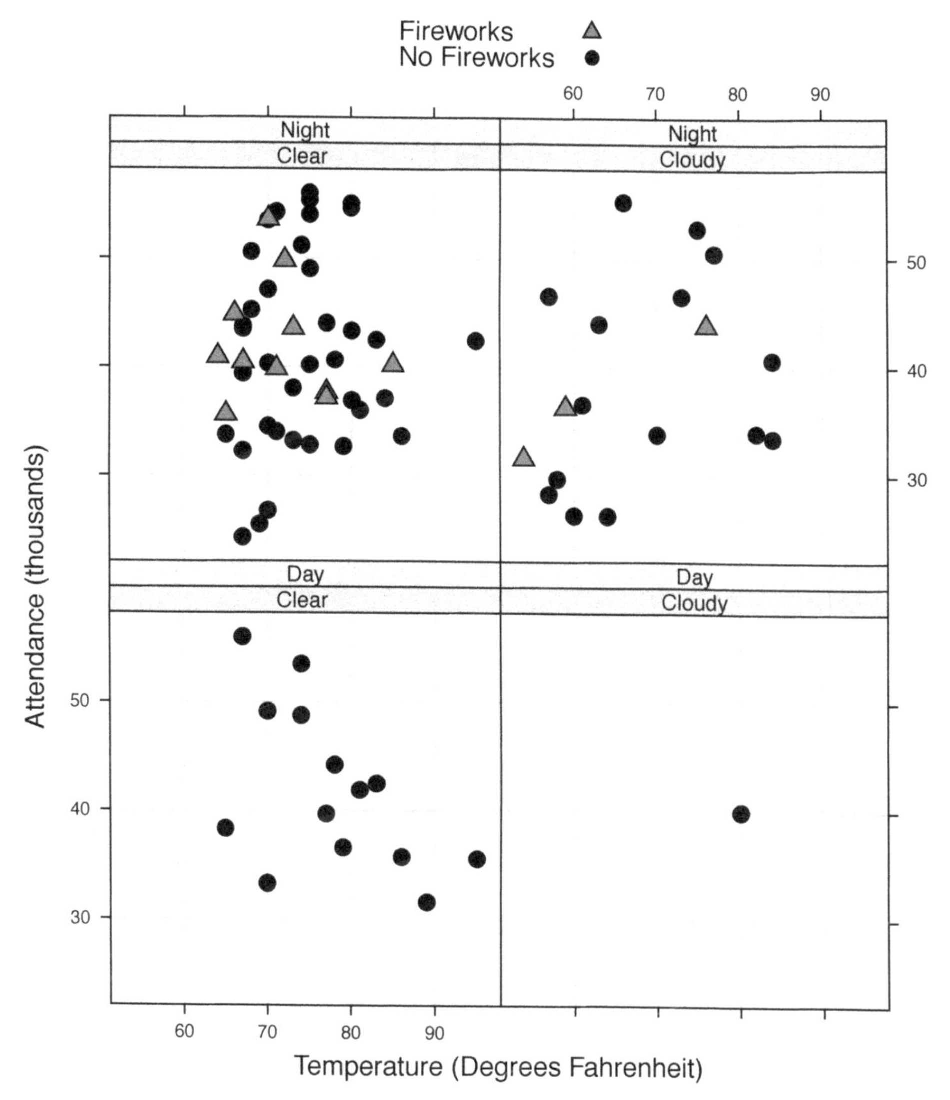

*Figure 8.4.* Dodgers Attendance by Visiting Team

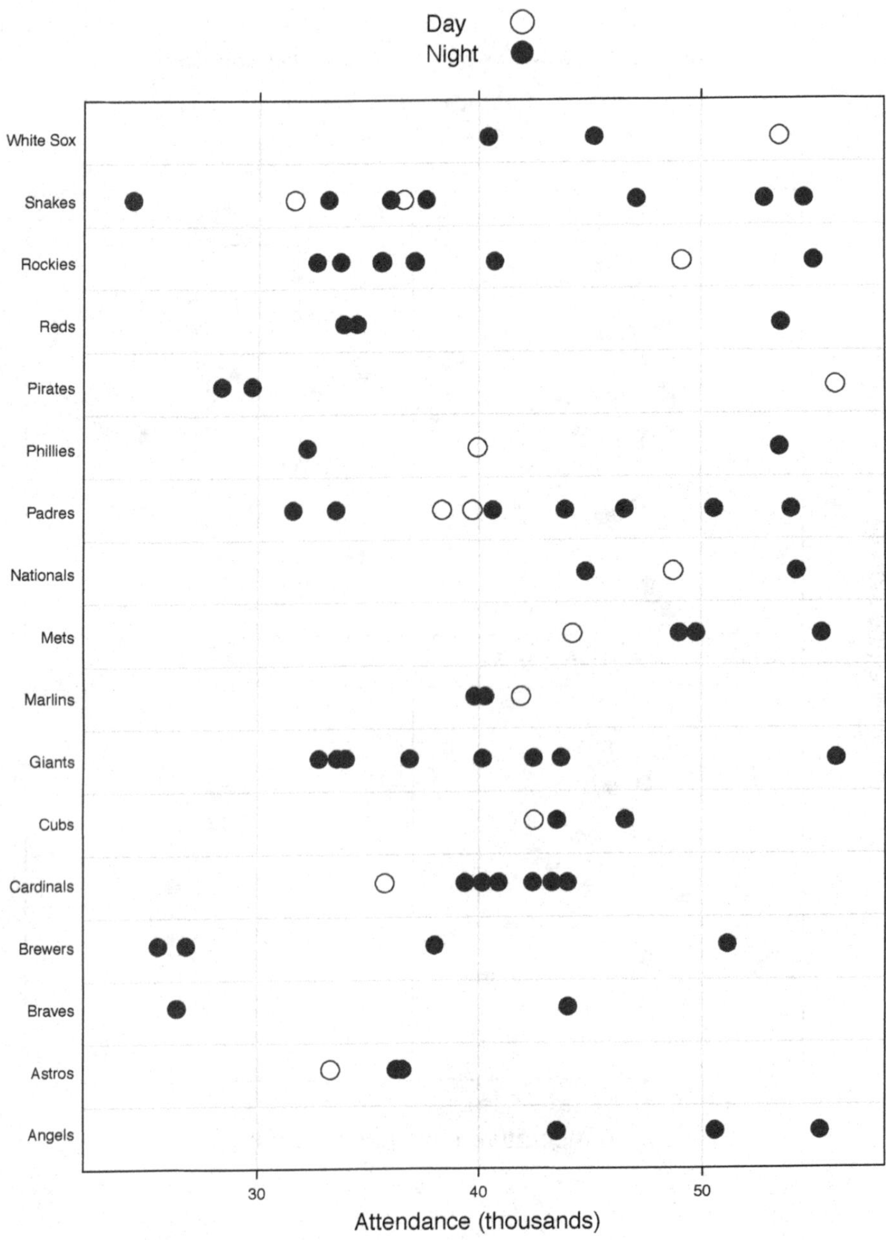

To advise management regarding promotions, we would like to know if promotions have a positive effect on attendance, and if they do have a positive effect, how much might that effect be. To provide this advice we build a linear model for predicting attendance using month, day of the week, and an indicator variable for the bobblehead promotion, and then we see how well it works. We enter these explanatory variables in a particular order so we can answer the basic question: *Do bobblehead promotions increase attendance, controlling for the date of the game (month and day of the week)?* Being data scientists, we employ a training-and-test regimen to provide an honest evaluation of the model's predictive performance.

For the Los Angeles Dodgers bobblehead promotion, the fitted model does a good job of predicting higher attendance at games when bobbleheads are distributed. Our computer programs can provide indices of goodness of fit, but, more importantly, they can provide predictions of attendance that we can display on scatter plots.

How does a training-and-test regimen play out for the model we have developed for the Dodgers? Figure 8.5 provides a picture of model performance that data scientists and business managers can understand. The model fit to the training set holds up when used with the test set.[2]

Running the code for the study on the complete set of home game data for the Los Angeles Dodgers in 2012 would yield a set of regression coefficients, estimates of the parameters in the linear model, as shown in table 8.2. A sequential analysis of variance shows a statistically significant effect for the bobblehead promotion, controlling for month and day of the week. A test of residuals from the model would identify any statistically significant outliers—there were none for this problem. Most importantly, the model can provide an assessment of the effect of the bobblehead promotion. In particular, we can see that bobblehead promotions have the potential of increasing attendance by 10,715 fans per game, all other things being equal.

---

[2] In figure 8.5, TRAIN refers to the training set, the data on which we fit the model and TEST refers to the hold-out-data on which we test our model. Running the code for this example, we would see that, in the test set, more than 45 percent of the variance in attendance is accounted for by the linear model—this is the square of the correlation of observed and predicted attendance. To explain a model to management, however, it is better to show a performance graph than to talk about squared correlation coefficients, mean squared errors of prediction, or other model summary statistics. This is one graph among many possible graphs that we could have produced for the Dodgers. It shows the results of one particular random splitting of the data into training and test.

*Figure 8.5. Regression Model Performance: Bobbleheads and Attendance*

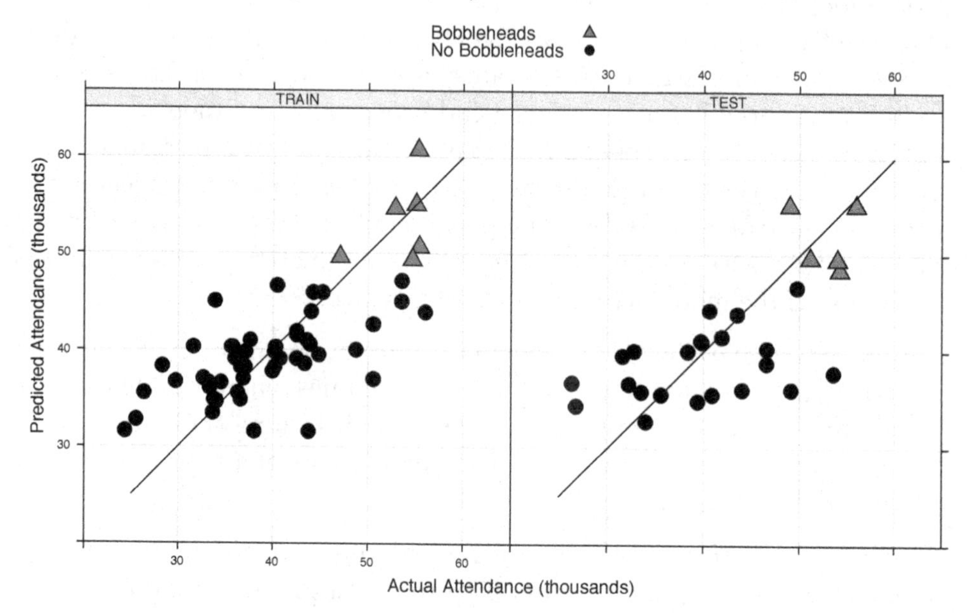


The baseball promotions example was chosen to be simple in structure, so that ordinary least squares regression could be employed. This is a cross-sectional study with the baseball game serving as the unit of analysis.

More complicated models are possible, and diagnostic plots for the model can provide additional information for the data scientist seeking to improve on the model we specified. Nonetheless, it is interesting to note how much information a linear regression model can provide. Predictive models like the one used in this small example, with results presented in graphical summaries, can help guide management decisions.

One of the things that distinguishes predictive analytics from statistics is its focus on business requirements. In evaluating the utility of a model, the analyst or data scientist considers financial criteria as well as statistical criteria. And, in presenting predictions to management, she provides financial analysis as well as a description of the statistical model itself.

*Table 8.2.* *Regression of Attendance on Month, Day of Week, and Bobblehead Promotion*

| | Response: Attendance |
|---|---|
| Month (May) | −2,385.625 |
| Month (June) | 7,163.234** |
| Month (July) | 2,849.828 |
| Month (August) | 2,377.924 |
| Month (September) | 29.030 |
| Month (October) | −662.668 |
| Day of Week (Tuesday) | 7,911.494*** |
| Day of Week (Wednesday) | 2,460.023 |
| Day of Week (Thursday) | 775.364 |
| Day of Week (Friday) | 4,883.818* |
| Day of Week (Saturday) | 6,372.056** |
| Day of Week (Sunday) | 6,724.003*** |
| Bobblehead Promotion (YES) | 10,714.900*** |
| Constant | 33,909.160*** |
| Observations | 81 |
| $R^2$ | 0.544 |
| Adjusted $R^2$ | 0.456 |
| Residual Std. Error | $6,120.158 (df = 67)$ |
| F statistic | $6.158^{***} (df = 13; 67)$ |
| *Notes:* | ***Significant at the 1 percent level.<br>**Significant at the 5 percent level.<br>*Significant at the 10 percent level. |

Using the fitted predictive model for the Dodgers bobblehead promotion, we can predict the attendance for each game in the forthcoming season, and we can predict this attendance with and without a bobblehead promotion. Knowing fixed and variable costs associated with a bobblehead promotion, as well as expected revenues from ticket sales and concessions, we can help the Los Angeles Dodgers assess the financial contribution of bobblehead promotions.

Considering costs for the forthcoming season in this example, the unit cost of a bobblehead doll is expected to be no more than $3 when ordered in quantities of at least 20,000. Bobbleheads are provided to the first 50,000 fans entering Dodger Stadium. To complete our work, then, we would use cost/volume/profit analysis[3] to assess the profit contribution of a bobblehead promotion for each game at Dodger Stadium. In this way, Dodger management could decide whether or not to use bobbleheads in the forthcoming season and which games most benefit from the use of bobbleheads.

Promotions like the bobblehead promotion do more than drive up attendance. They also reinforce the name of the brand in the minds of consumers. Advertising and promotion are the "promotion" part of the marketing mix or *the four Ps:* product, price, promotion, and place. Here "product" relates to product or service, and "price" is simply price. The word "place" refers to channels of distribution (face-to-face selling, wholesale, retail, brick-and-mortar, mail-order, online, or mobile). Advertising and promotion are thought of as distinct fields of study by marketing academics. Advertising refers to the message, the marketing communication, while promotion is what firms do in addition to the message.

As we have shown, traditional regression models are especially relevant to these areas of inquiry. Useful sources for regression modeling include Kutner, Nachtsheim, Neter, and Li (2004), Ryan (2008), and Chatterjee and Hadi (2012). For guidance in R programming, see Venables and Ripley (2002), Fox and Weisberg (2011), Matloff (2011), and Fox (2014).

[3] Cost/volume/profit analysis is a common technique in management accounting. It is sometimes called break even analysis or cost/benefit analysis. There are challenges in carrying out a financial analysis for Dodgers' promotions because ticket prices vary by the type of game and seating location. Ticket prices for (four-star) bobblehead games in 2013 varied from $20 for a top deck seat to $120 for a VIP field box seat. These prices were raised in 2014 to $25 for a top deck seat to $140 for a VIP field box seat. A portion of ticket revenues goes to support concessions and additional staff needed to distribute bobbleheads. We would obtain these cost estimates from Dodger management.

Moving beyond traditional linear models, we can consider modern, data-adaptive regression methods, as reviewed by Izenman (2008) and Hastie, Tibshirani, and Friedman (2009). We provide additional discussion of traditional and data-adaptive (machine learning) methods in appendix A.

For an overview of advertising and promotion, marketing management textbooks may be consulted (Dickson 1997; Kotler and Keller 2012). Market response models attempt to predict sales and market shares across products within categories. These build upon econometric and time series methods. Hanssens, Parsons, and Schultz (2001) discuss market response modeling. Lilien and Rangaswamy (2003) suggest applications of market response modeling in sales force and channel management. For an overview of marketing models, see Lilien, Kotler, and Moorthy (1992) and Leeflang et al. (2000).

The R program *Shaking Our Bobbleheads Yes and No* is shown in exhibit 8.1 and draws on packages for regression and graphics by Fox (2014) and Sarkar (2014), respectively. The corresponding Python program is shown in exhibit 8.2. For those wishing to explore models across all Major League Baseball teams, we describe additional data in the *Return of the Bobbleheads* case in appendix C on page 372.

*Exhibit 8.1. Shaking Our Bobbleheads Yes and No (R)*

```
# Predictive Model for Los Angeles Dodgers Promotion and Attendance (R)

library(car)  # special functions for linear regression
library(lattice)  # graphics package

# read in data and create a data frame called dodgers
dodgers <- read.csv("dodgers.csv")
print(str(dodgers))  # check the structure of the data frame

# define an ordered day-of-week variable
# for plots and data summaries
dodgers$ordered_day_of_week <- with(data=dodgers,
  ifelse ((day_of_week == "Monday"),1,
  ifelse ((day_of_week == "Tuesday"),2,
  ifelse ((day_of_week == "Wednesday"),3,
  ifelse ((day_of_week == "Thursday"),4,
  ifelse ((day_of_week == "Friday"),5,
  ifelse ((day_of_week == "Saturday"),6,7)))))))
dodgers$ordered_day_of_week <- factor(dodgers$ordered_day_of_week, levels=1:7,
labels=c("Mon", "Tue", "Wed", "Thur", "Fri", "Sat", "Sun"))

# exploratory data analysis with standard graphics: attendance by day of week
with(data=dodgers,plot(ordered_day_of_week, attend/1000,
xlab = "Day of Week", ylab = "Attendance (thousands)",
col = "violet", las = 1))

# when do the Dodgers use bobblehead promotions
with(dodgers, table(bobblehead,ordered_day_of_week)) # bobbleheads on Tuesday

# define an ordered month variable
# for plots and data summaries
dodgers$ordered_month <- with(data=dodgers,
  ifelse ((month == "APR"),4,
  ifelse ((month == "MAY"),5,
  ifelse ((month == "JUN"),6,
  ifelse ((month == "JUL"),7,
  ifelse ((month == "AUG"),8,
  ifelse ((month == "SEP"),9,10)))))))
dodgers$ordered_month <- factor(dodgers$ordered_month, levels=4:10,
labels = c("April", "May", "June", "July", "Aug", "Sept", "Oct"))

# exploratory data analysis with standard R graphics: attendance by month
with(data=dodgers,plot(ordered_month,attend/1000, xlab = "Month",
ylab = "Attendance (thousands)", col = "light blue", las = 1))

# exploratory data analysis displaying many variables
# looking at attendance and conditioning on day/night
# the skies and whether or not fireworks are displayed
library(lattice) # used for plotting
# let us prepare a graphical summary of the dodgers data
group.labels <- c("No Fireworks","Fireworks")
```

```
group.symbols <- c(21,24)
group.colors <- c("black","black")
group.fill <- c("black","red")
xyplot(attend/1000 ~ temp | skies + day_night,
    data = dodgers, groups = fireworks, pch = group.symbols,
    aspect = 1, cex = 1.5, col = group.colors, fill = group.fill,
    layout = c(2, 2), type = c("p","g"),
    strip=strip.custom(strip.levels=TRUE,strip.names=FALSE, style=1),
    xlab = "Temperature (Degrees Fahrenheit)",
    ylab = "Attendance (thousands)",
    key = list(space = "top",
        text = list(rev(group.labels),col = rev(group.colors)),
        points = list(pch = rev(group.symbols), col = rev(group.colors),
        fill = rev(group.fill))))
# attendance by opponent and day/night game
group.labels <- c("Day","Night")
group.symbols <- c(1,20)
group.symbols.size <- c(2,2.75)
bwplot(opponent ~ attend/1000, data = dodgers, groups = day_night,
    xlab = "Attendance (thousands)",
    panel = function(x, y, groups, subscripts, ...)
       {panel.grid(h = (length(levels(dodgers$opponent)) - 1), v = -1)
        panel.stripplot(x, y, groups = groups, subscripts = subscripts,
        cex = group.symbols.size, pch = group.symbols, col = "darkblue")
       },
    key = list(space = "top",
    text = list(group.labels,col = "black"),
    points = list(pch = group.symbols, cex = group.symbols.size,
    col = "darkblue")))

# employ training-and-test regimen for model validation
set.seed(1234) # set seed for repeatability of training-and-test split
training_test <- c(rep(1,length=trunc((2/3)*nrow(dodgers))),
rep(2,length=(nrow(dodgers) - trunc((2/3)*nrow(dodgers)))))
dodgers$training_test <- sample(training_test) # random permutation
dodgers$training_test <- factor(dodgers$training_test,
  levels=c(1,2), labels=c("TRAIN","TEST"))
dodgers.train <- subset(dodgers, training_test == "TRAIN")
print(str(dodgers.train)) # check training data frame
dodgers.test <- subset(dodgers, training_test == "TEST")
print(str(dodgers.test)) # check test data frame

# specify a simple model with bobblehead entered last
my.model <- {attend ~ ordered_month + ordered_day_of_week + bobblehead}
# fit the model to the training set
train.model.fit <- lm(my.model, data = dodgers.train)
# summary of model fit to the training set
print(summary(train.model.fit))
# training set predictions from the model fit to the training set
dodgers.train$predict_attend <- predict(train.model.fit)
# test set predictions from the model fit to the training set
dodgers.test$predict_attend <- predict(train.model.fit,
  newdata = dodgers.test)
```

```
# compute the proportion of response variance
# accounted for when predicting out-of-sample
cat("\n","Proportion of Test Set Variance Accounted for: ",
round((with(dodgers.test,cor(attend,predict_attend)^2)),
  digits=3),"\n",sep="")
# merge the training and test sets for plotting
dodgers.plotting.frame <- rbind(dodgers.train,dodgers.test)

# generate predictive modeling visual for management
group.labels <- c("No Bobbleheads","Bobbleheads")
group.symbols <- c(21,24)
group.colors <- c("black","black")
group.fill <- c("black","red")
xyplot(predict_attend/1000 ~ attend/1000 | training_test,
       data = dodgers.plotting.frame, groups = bobblehead, cex = 2,
       pch = group.symbols, col = group.colors, fill = group.fill,
       layout = c(2, 1), xlim = c(20,65), ylim = c(20,65),
       aspect=1, type = c("p","g"),
       panel=function(x,y, ...)
            {panel.xyplot(x,y,...)
             panel.segments(25,25,60,60,col="black",cex=2)
            },
       strip=function(...) strip.default(..., style=1),
       xlab = "Actual Attendance (thousands)",
       ylab = "Predicted Attendance (thousands)",
       key = list(space = "top",
              text = list(rev(group.labels),col = rev(group.colors)),
              points = list(pch = rev(group.symbols),
              col = rev(group.colors),
              fill = rev(group.fill))))
# use the full data set to obtain an estimate of the increase in
# attendance due to bobbleheads, controlling for other factors
my.model.fit <- lm(my.model, data = dodgers)  # use all available data
print(summary(my.model.fit))
# tests statistical significance of the bobblehead promotion
# type I anova computes sums of squares for sequential tests
print(anova(my.model.fit))
cat("\n","Estimated Effect of Bobblehead Promotion on Attendance: ",
round(my.model.fit$coefficients[length(my.model.fit$coefficients)],
digits = 0),"\n",sep="")
# standard graphics provide diagnostic plots
plot(my.model.fit)
# additional model diagnostics drawn from the car package
library(car)
residualPlots(my.model.fit)
marginalModelPlots(my.model.fit)
print(outlierTest(my.model.fit))

# Suggestions for the student:
# Examine regression diagnostics for the fitted model.
# Examine other linear predictors and other explanatory variables.
# See if you can improve upon the model with variable transformations.
```

**Exhibit 8.2.** *Shaking Our Bobbleheads Yes and No (Python)*

```
# Predictive Model for Los Angeles Dodgers Promotion and Attendance (Python)

# prepare for Python version 3x features and functions
from __future__ import division, print_function
from future_builtins import ascii, filter, hex, map, oct, zip

# import packages for analysis and modeling
import pandas as pd  # data frame operations
from pandas.tools.rplot import RPlot, TrellisGrid, GeomPoint,\
    ScaleRandomColour  # trellis/lattice plotting
import numpy as np  # arrays and math functions
from scipy.stats import uniform  # for training-and-test split
import statsmodels.api as sm  # statistical models (including regression)
import statsmodels.formula.api as smf  # R-like model specification
import matplotlib.pyplot as plt  # 2D plotting

# read in Dodgers bobbleheads data and create data frame
dodgers = pd.read_csv("dodgers.csv")
# examine the structure of the data frame
print("\nContents of dodgers data frame ---------------")
# attendance in thousands for plotting
dodgers['attend_000'] = dodgers['attend']/1000
# print the first five rows of the data frame
print(pd.DataFrame.head(dodgers))

mondays = dodgers[dodgers['day_of_week'] == 'Monday']
tuesdays = dodgers[dodgers['day_of_week'] == 'Tuesday']
wednesdays = dodgers[dodgers['day_of_week'] == 'Wednesday']
thursdays = dodgers[dodgers['day_of_week'] == 'Thursday']
fridays = dodgers[dodgers['day_of_week'] == 'Friday']
saturdays = dodgers[dodgers['day_of_week'] == 'Saturday']
sundays = dodgers[dodgers['day_of_week'] == 'Sunday']

# convert days' attendance into list of vectors for box plot
data = [mondays['attend_000'], tuesdays['attend_000'],
    wednesdays['attend_000'], thursdays['attend_000'],
    fridays['attend_000'], saturdays['attend_000'],
    sundays['attend_000']]
ordered_day_names = ['Mon', 'Tue', 'Wed', 'Thur', 'Fri', 'Sat', 'Sun']

# exploratory data analysis: box plot for day of the week
fig, axis = plt.subplots()
axis.set_xlabel('Day of Week')
axis.set_ylabel('Attendance (thousands)')
day_plot = plt.boxplot(data, sym='o', vert=1, whis=1.5)
plt.setp(day_plot['boxes'], color = 'black')
plt.setp(day_plot['whiskers'], color = 'black')
plt.setp(day_plot['fliers'], color = 'black', marker = 'o')
axis.set_xticklabels(ordered_day_names)
plt.show()
```

```
plt.savefig('fig_advert_promo_dodgers_eda_day_of_week_Python.pdf',
    bbox_inches = 'tight', dpi=None, facecolor='w', edgecolor='b',
    orientation='portrait', papertype=None, format=None,
    transparent=True, pad_inches=0.25, frameon=None)
april = dodgers[dodgers['month'] == 'APR']
may = dodgers[dodgers['month'] == 'MAY']
june = dodgers[dodgers['month'] == 'JUN']
july = dodgers[dodgers['month'] == 'JUL']
august = dodgers[dodgers['month'] == 'AUG']
september = dodgers[dodgers['month'] == 'SEP']
october = dodgers[dodgers['month'] == 'OCT']
data = [april['attend_000'], may['attend_000'],
    june['attend_000'], july['attend_000'],
    august['attend_000'], september['attend_000'],
    october['attend_000']]
ordered_month_names = ['April', 'May', 'June', 'July', 'Aug', 'Sept', 'Oct']

fig, axis = plt.subplots()
axis.set_xlabel('Month')
axis.set_ylabel('Attendance (thousands)')
day_plot = plt.boxplot(data, sym='o', vert=1, whis=1.5)
plt.setp(day_plot['boxes'], color = 'black')
plt.setp(day_plot['whiskers'], color = 'black')
plt.setp(day_plot['fliers'], color = 'black', marker = 'o')
axis.set_xticklabels(ordered_month_names)
plt.show()
plt.savefig('fig_advert_promo_dodgers_eda_month_Python.pdf',
    bbox_inches = 'tight', dpi=None, facecolor='w', edgecolor='b',
    orientation='portrait', papertype=None, format=None,
    transparent=True, pad_inches=0.25, frameon=None)

# trellis/lattice plot attendance by temp, conditioning on skies
# and day_night with bobblehead NO/YES shown in distinct colors
plt.figure()
plot = RPlot(dodgers,  x = 'temp', y = 'attend_000')
plot.add(TrellisGrid(['day_night', 'skies']))
plot.add(GeomPoint(colour = ScaleRandomColour('bobblehead')))
plot.render(plt.gcf())
plt.show()
plt.savefig('fig_advert_promo_dodgers_eda_many.pdf',
    bbox_inches = 'tight', dpi=None, facecolor='w', edgecolor='b',
    orientation='portrait', papertype=None, format=None,
    transparent=True, pad_inches=0.25, frameon=None)

# map day_of_week to ordered_day_of_week
day_to_ordered_day = {'Monday' : '1Monday',
     'Tuesday' : '2Tuesday',
     'Wednesday' : '3Wednesday',
     'Thursday' : '4Thursday',
     'Friday' : '5Friday',
     'Saturday' : '6Saturday',
     'Sunday' : '7Sunday'}
dodgers['ordered_day_of_week'] = dodgers['day_of_week'].map(day_to_ordered_day)
```

```
# map month to ordered_month
month_to_ordered_month = {'APR' : '1April',
     'MAY' : '2May',
     'JUN' : '3June',
     'JUL' : '4July',
     'AUG' : '5Aug',
     'SEP' : '6Sept',
     'OCT' : '7Oct'}
dodgers['ordered_month'] = dodgers['month'].map(month_to_ordered_month)

# employ training-and-test regimen for model validation
np.random.seed(1234)
dodgers['runiform'] = uniform.rvs(loc = 0, scale = 1, size = len(dodgers))
dodgers_train = dodgers[dodgers['runiform'] >= 0.33]
dodgers_test = dodgers[dodgers['runiform'] < 0.33]
# check training data frame
print('\ndodgers_train data frame (rows, columns): ',dodgers_train.shape)
print(dodgers_train.head())
# check test data frame
print('\ndodgers_test data frame (rows, columns): ',dodgers_test.shape)
print(dodgers_test.head())

# specify a simple model with bobblehead entered last
my_model = str('attend ~ ordered_month + ordered_day_of_week + bobblehead')

# fit the model to the training set
train_model_fit = smf.ols(my_model, data = dodgers_train).fit()
# summary of model fit to the training set
print(train_model_fit.summary())
# training set predictions from the model fit to the training set
dodgers_train['predict_attend'] = train_model_fit.fittedvalues

# test set predictions from the model fit to the training set
dodgers_test['predict_attend'] = train_model_fit.predict(dodgers_test)

# compute the proportion of response variance
# accounted for when predicting out-of-sample
print('\nProportion of Test Set Variance Accounted for: ',\
    round(np.power(dodgers_test['attend'].\
    corr(dodgers_test['predict_attend']),2),3))
# use the full data set to obtain an estimate of the increase in
# attendance due to bobbleheads, controlling for other factors
my_model_fit = smf.ols(my_model, data = dodgers).fit()
print(my_model_fit.summary())
print('\nEstimated Effect of Bobblehead Promotion on Attendance: ',\
    round(my_model_fit.params[13],0))
```

# 9 Recommending Products

Ira: "Barry can pick out the exact flavor of ice cream to follow any meal. Go ahead. Challenge him."

Mitch: "Challenge him?"

Ira: "Go on."

Mitch: "Franks and beans."

Barry: "Scoop of chocolate. Scoop of vanilla. Don't waste my time. Come on. Push me."

Ira: "Sea bass."

Barry: "Grilled?"

Mitch: "Sautéed"

Barry: "I'm with ya."

Mitch: "Potatoes au gratin. Asparagas."

Barry: "Rum Raisin."

Ira and Barry: [High-five hand slapping] "Woop!"

Mitch: "How do you know he's right?"

Ira: "How do we know? Fourteen hundred retail outlets across the country. That's how we know."

—David Paymer as Ira Shalowitz, Josh Mostel as Barry Shalowitz, and Billy Crystal as Mitch Robbins in *City Slickers* (1991)

Being a frequent customer of Amazon.com, I receive many suggestions for book purchases. In May 2013, before the release of the initial edition of my first *Modeling Techniques* book, Amazon sent me an e-mail list of ten recommended books, and the book I was writing at the time was at the top of the list. My publisher, being a good publisher, made the book available for pre-order, and Amazon was asking if I wanted to order it.

Who are those guys who seem to know so much about me? Amazon for books and other products, Netflix movies, Pandora music, among others—they seem to know a lot about what I like without me telling them very much.

Recommender systems build on *sparse matrices*. A well known recommender system problem, the Netflix $1 million prize competition (May 29, 2006 though September 21, 2009), presented Netflix data for 17,770 movies and 480,189 customers. Any given customer rents at most a couple hundred movies, so if we were to place the number one in each cell of a customers-by-movies matrix for rentals and a zero otherwise, the resulting matrix would have 99 percent zeroes—it would be 99 percent sparse.

Algorithms for recommender systems, or recommender engines as they are sometimes called, must be capable of dealing with sparse matrices. There are many possible algorithms. Content-based recommender systems are one class of recommender systems. These draw on customer personal characteristics, past orders, and revealed preferences. Content-based systems may also rely on product characteristics.

Another class of recommender systems is *collaborative filtering* (also called social or group filtering), which builds on the premise that customers with similar ratings for one set of products will have similar ratings for other sets of products. To predict what music Janiya is going to like, find Janiya's nearest neighbors and see what music they like.

Suppose Brit has chosen ten movies and a movie service provider wants to suggest a new movie for Brit. One way to find that movie is to identify other customers who have chosen many of the same movies as Brit. Then, searching across those nearest-neighbor customers, the service provider detects movies they have in common. Movies most in common across these customers provide a basis for recommendations to Brit.

Interestingly, a competition offered by Kaggle asked analysts to develop a recommendation engine for users of R packages. Conway and White (2012) show how to build a nearest-neighbor recommender system using data from this competition. We note that data from the Kaggle R competition could have been supplemented with information from the R programming environment itself, with its "depends" and "reverse depends" links. We can represent any programming environment as a network, with one package, function, or module calling another. And using the network topology, we can identify nearest neighbors or nodes in close proximity to any given node.

Suppose a video rental firm wants to develop a service for recommending movies to its customers. Much like Amazon.com recommends books to its customers based on prior book purchases, we can recommend movies to prospective renters based upon prior rentals.

How might we recommend movies to rental customers? There are difficulties with using customer membership data exclusively. Memberships are often associated with households rather than individual consumers. The movies preferred by the male head of household may be different from those preferred by the female head of household, which are different from the children.

One way of finding similar movies might be to match them by movie directors. If a person likes Woody Allen's *New York Stories*, she may like Allen's *Celebrity*. If she liked Altman's *Nashville*, she may like Altman's *Ready to Wear*. If she liked Antonioni's *Blow-Up*, she might go for *The Passenger* or *Zabriskie Point*. And if she liked DePalma's *Body Double*, there is a good chance she will like *Femme Fatale*. Leading actors would be another way to recommend movies. Or we could use movie genre. Customer choices, ratings, and reviews show preferences, providing data for recommending movies to that customer and to other customers.

Another way to learn about movie preferences is to do independent consumer research. There are ways to marry consumer research data with customer sales data, while protecting the interests of both sets of research participants. For example, a matching of current customers with survey research participants could be accomplished by common keys such as customer and consumer demographics or geography.

Suppose males between twenty-five and thirty-four who express a strong preference for *The Italian Job* also like *Ocean's Eleven*. A male customer is between twenty-five and thirty-four (from the member application form), and he just rented *The Italian Job* (from the last sales transaction). A recommender system might suggest that he rent *Ocean's Eleven*.

Association rules modeling is another way of building recommender systems. Association rules modeling asks, *What goes with what? What products are ordered or purchased together? What activities go together? What website areas are viewed together?* A good way of understanding association rules is to consider their application to market basket analysis.

Shopping on the weekend is a half-hour walk to the grocery store. Until a year ago, I was fairly predictable in what I bought—milk, juices, poultry items, granola, bread, a jar of peanut butter, and treats. The items for a shopping trip from last year are shown in table 9.1. These represented my market basket and would go into one row of the store's database along with thousands of shopping trips or market baskets from other customers.

Market baskets reveal consumer preferences and lifestyles in a way that no survey can fully capture. When I started a serious diet with protein shakes replacing popsicles and treats, my market baskets changed drastically. Anything with added sugar or corn starch was summarily excluded.

Each individual market basket is a unit of analysis in the market basket database. Thousands of columns in the database correspond to the full set of products available at a grocery store. Purchases of items across customers are part of the store's database, with system stocking units (SKUs), quantities, and prices duly noted. To prepare these data for market basket analysis, we would convert quantities to binary indicators. The purchase of granola bars would be represented by a one in the row of the input data for this shopping trip, as would the purchase of cereal and the nine other items in the market basket. The rest of the columns in this row of the input data would be set to zero. The resulting input data to market basket analysis would be a sparse binary matrix, thousands of rows and columns with ones and zeroes.[1]

---

[1] Note that we have simplified the market basket problem by using ones and zeroes in our data set or matrix. Three fruit bars in my market basket are represented by the number 1, not 3. Five frozen chicken dinners are represented by the number 1, not 5. The resulting binary matrix is easier to work with than a matrix with actual quantities in the cells.

***Table 9.1.*** *Market Basket for One Shopping Trip*

| | | | |
|---|---|---|---|
| | GROCERY | | |
| | | | 4.99 |
| BARS OATS N HONEY | | 5.49 | |
| Card Savings | | -0.50 | |
| | | | 2.49 |
| OATS & HONEY CEREAL | | 4.49 | |
| Card Savings | | -2.00 | |
| VEGGIE JCE | | | 3.29 |
| | | | 2.00 |
| OCEAN SPRAY | | 3.99 | |
| Card Savings | | -1.99 | |
| | | | 2.99 |
| SKIPPY RF CRMY | | 4.79 | |
| Card Savings | | -1.80 | |
| | REFRIG/FROZEN | | |
| | | | 3.00 |
| FLORIDAS NATURAL J | | 4.19 | |
| Card Savings | | -1.19 | |
| | | | 9.00 |
| 3 QTY DREYERS STRWBRY | | 13.47 | |
| Card Savings | | -2.97 | |
| Mfr Cpn | | -1.50 | |
| 5 QTY EATING RIGHT CHKN | | | 10.00 |
| PANTRY ESTNL MILK | | | 3.19 |
| | BAKED GOODS | | |
| PANTRY WHEAT BREAD | | | 0.99 |
| | DELI | | |
| DC TURKEY BREAST | | | 3.43 |
| | BAL | | 45.37 |

Market basket analysis (also called affinity or association analysis) asks, *What goes with what? What products are ordered or purchased together?* This is contingency table analysis on a very large scale, and our job is to determine which contingency tables to look at.

There are obvious examples of things that go together, such as hot dogs and hot dog buns, party favors and ice, peanut butter, jelly, and bread (or, in my case, just peanut butter and bread). There are reports, sometimes surprising, sometimes bogus, of less obvious things going together, such as diapers and beer. The job of market basket analysis is to find what things go together, providing information to guide product placement in stores, cross-category and co-marketing promotions, and product bundling plans.

To provide an example of market basket analysis, we draw upon a grocery data set first analyzed by Hahsler, Hornik, and Reutterer (2006) and available in Hahsler et al. (2014a). The data set consists of $N = 9,835$ market baskets across $K = 169$ generically-labeled grocery items. The data set, which represents one month of real transaction data from a grocery outlet, is small enough to be processed on a laptop computer and large enough to demonstrate methods of market basket analysis.

A key challenge in market basket analysis and association rule modeling in general is the sheer number of rules that are generated. An *item set* is a collection of items selected from all items in the store. The size of an item set is the number of items in that set. Item sets may be composed of two items, three items, and so on. The number of distinct item sets is very large, even for the grocery store data set.

An *association rule* is a division of each item set into two subsets with one subset, the *antecedent*, thought of as preceding the other subset, the *consequent*. There are more association rules than there are item sets.[2] The Apriori algorithm of Agraval et al. (1996) deals with the large-number-of-rules problem by using selection criteria that reflect the potential utility of association rules.

---

[2] For the grocery store data set, with its $K = 169$ items and its corresponding binary data matrix, the number of distinct item sets will be

$$2^K = 2^{169} \approx 7.482888 \times 10^{50}$$

The first criterion relates to the *support* or prevalence of an item set. Each item set is evaluated to determine the proportion of times it occurs in the store data set. If this proportion exceeds a minimum support threshold or criterion, then it is passed along to the next phase of analysis. A support criterion of 0.01 implies that one in every one hundred market baskets must contain the item set. A support criterion of 0.001 implies that one in every one thousand market baskets must contain the item set.

The second criterion relates to the *confidence* or predictability of an association rule. This is computed as the support of the item set divided by the support of the subset of items in the antecedent. This is an estimate of the conditional probability of the consequent, given the antecedent. In the selection of association rules, we set the confidence criterion much higher than the support criterion.[3] Support and confidence criteria are arbitrarily set by the researcher. These vary from one market basket problem to anther. For the groceries data set, we set the support criterion to 0.025 and make an initial plot showing item frequencies for individual items meeting this criterion. See figure 9.1.

We see that there are groups of similar items: (1) candy and specialty bar, (2) specialty chocolate and chocolate, (3) canned beer and bottled beer, and so on. We can combine similar items into categories to provide a more meaningful analysis, obtaining the reduced set of fifty-five item categories, thirty-two of which are shown in figure 9.2.

A set of 344 association rules may be obtained by setting thresholds for support and confidence of 0.025 and 0.05, respectively. Figure 9.3 provides a scatter plot of these rules with support on the horizontal axis and confidence on the vertical axis. Color coding of the points relates to *lift*, a measure of relative predictive confidence.[4]

---

[3] With item subsets identified as $A$ and $B$, there are two possible association rules: $(A \Rightarrow B)$ and $(B \Rightarrow A)$. We need to consider only one of these rules because we favor rules with higher confidence. Consider our confidence in the rule $(A \Rightarrow B)$. This is the conditional probability of $B$ given $A$. Similarly, confidence in rule $(B \Rightarrow A)$ is the conditional probability of $A$ given $B$:

$$P(B|A) = \frac{P(AB)}{P(A)} \qquad P(A|B) = \frac{P(AB)}{P(B)}$$

It follows that, if the item subset $A$ has more support than the item subset $B$, then $P(A) > P(B)$ and $P(A|B) > P(B|A)$. Given our preference for rules of higher confidence, then, it is clear that the item subset with higher support will take the role of the consequent.

[4] As with support and confidence, probability formulas define lift. Think of lift as the confidence we have in predicting the consequent $B$ with the rule $(A \Rightarrow B)$, divided by the confidence we would

*Figure 9.1. Market Basket Prevalence of Initial Grocery Items*

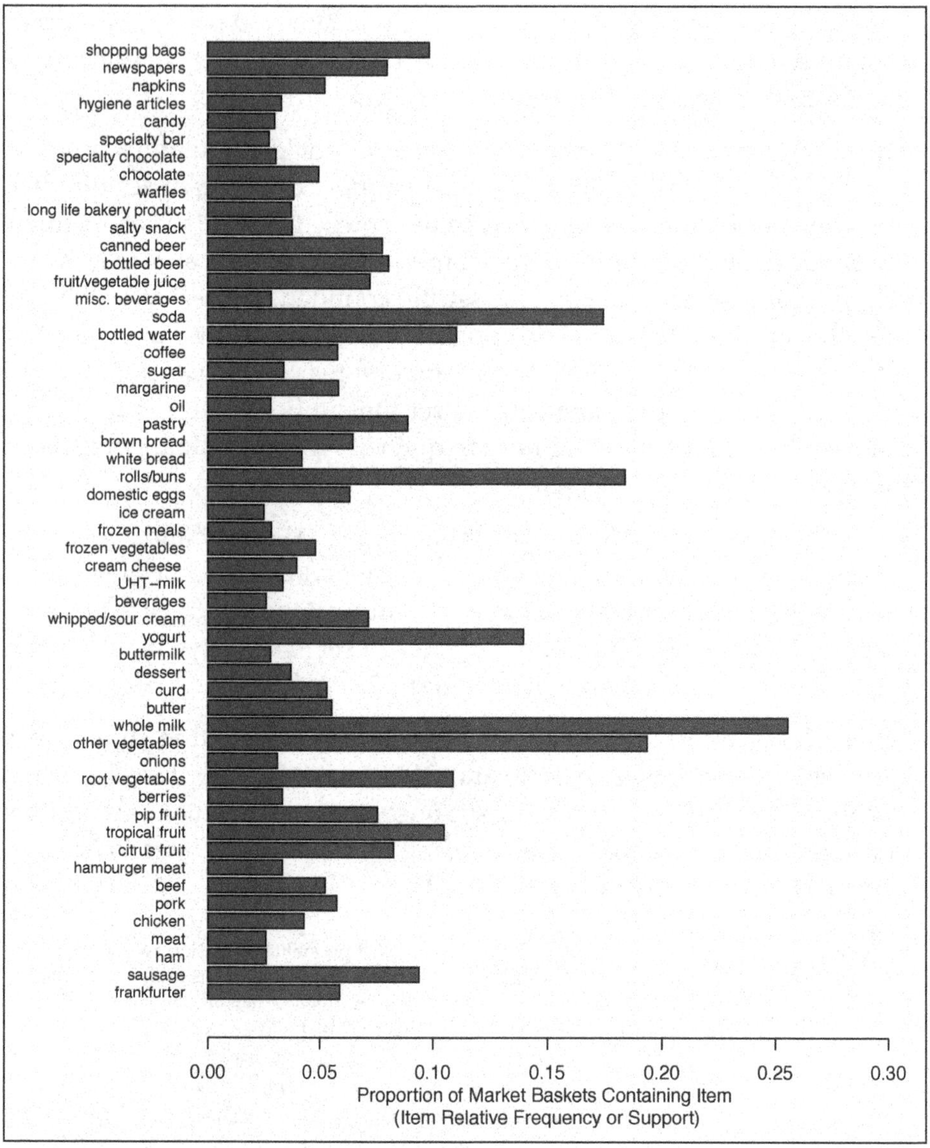

***Figure 9.2.*** *Market Basket Prevalence of Grocery Items by Category*

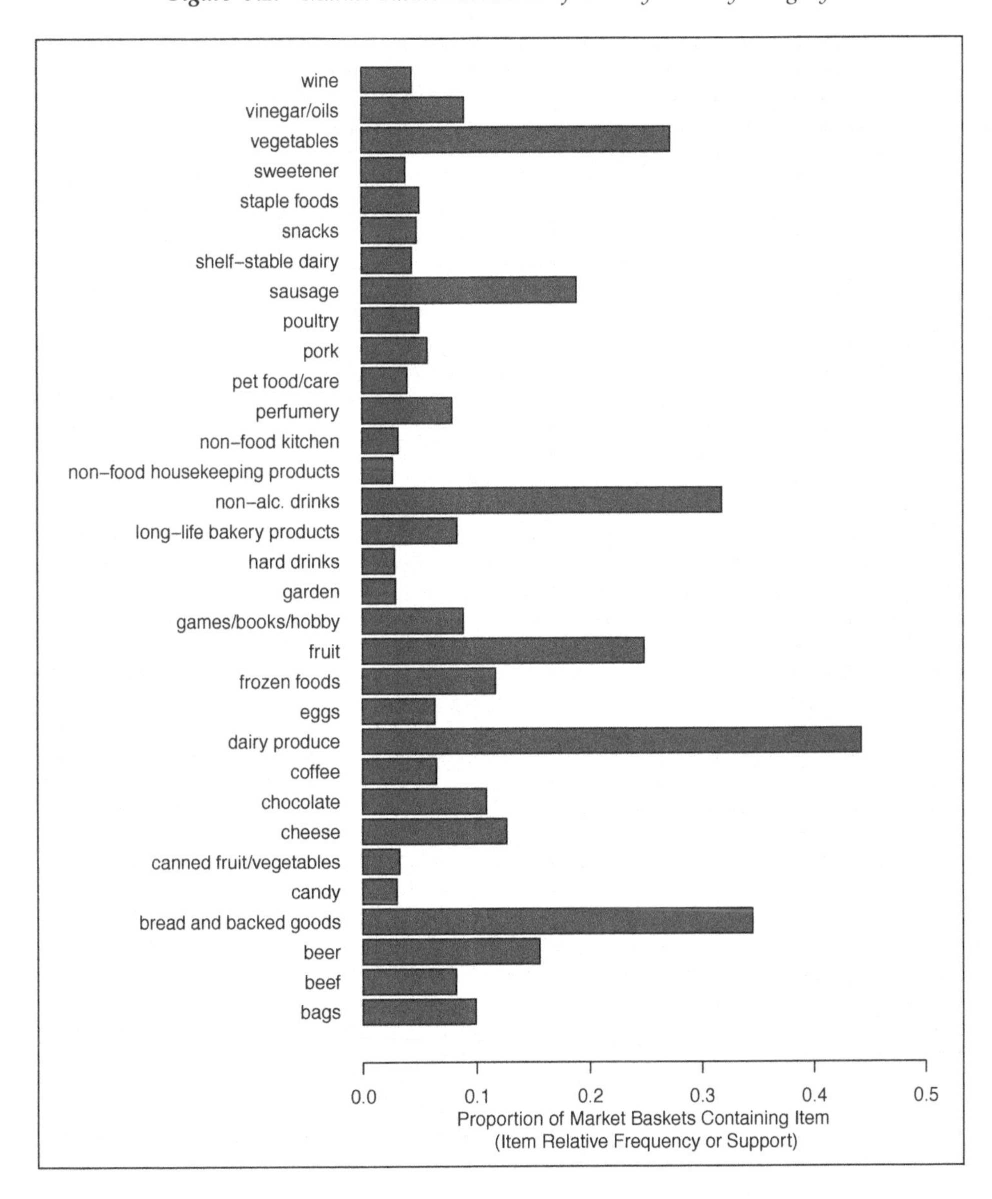

*Figure 9.3. Market Basket Association Rules: Scatter Plot*

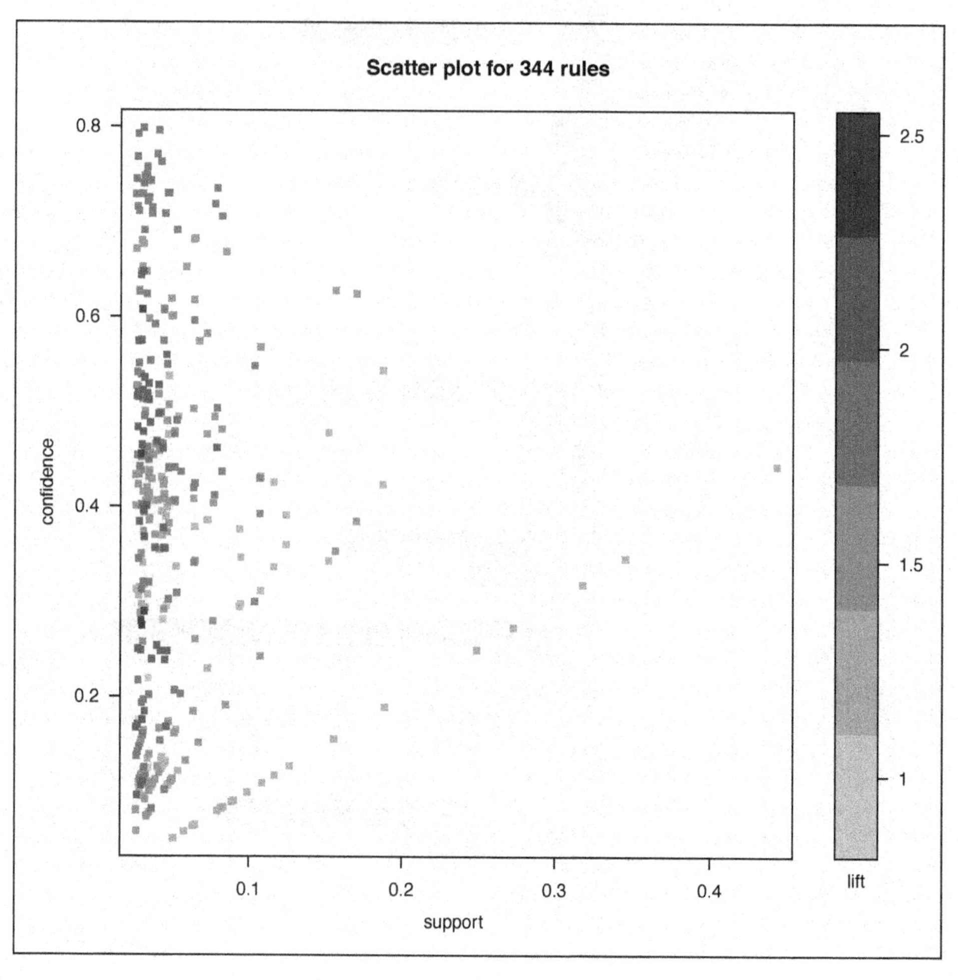

Figure 9.4 provides a clearer view of the identified association rules: a matrix bubble chart. An item in the antecedent subset or left-hand side (LHS) of an association rule provides the label at the top of the matrix, and an item in the consequent or right-hand side (RHS) of an association rule provides the label at the right of the matrix. Support relates to the size of each bubble and lift is reflected in its color intensity.

Suppose we are working for a local farmer, and we want to identify products that are commonly purchased along with vegetables. Selecting rules with consequent item subsets that include vegetables, we obtain forty-one rules from the set of 344. We rank these rules by lift and identify the top ten rules for display in table 9.2 and figure 9.5.

Reviewing the reported measures for association rules, support is prevalence. As a relative frequency or probability estimate it takes values from zero to one. Low values are tolerated, but values that are extremely low relative to other items in the store may indicate a lack of importance to shoppers.

Confidence relates to the predictability of the consequent, given the antecedent. As a conditional probability, confidence takes values from zero to one. Higher values are preferred. Lift, being a ratio of nonzero probabilities, takes positive values on the real number line. Lift needs to be above 1.00 to be of use to management. The higher the lift, the better.

Marketers use association rules to make decisions about store layout, product bundling, or cross-selling. Store managers run field experiments, observing how shoppers respond to new store layouts and cross-selling promotions. The ultimate test of market basket predictive models is provided by market response or sales.

---

have in predicting $B$ without the rule. Without knowledge of $A$ and the association rule $(A \Rightarrow B)$, our confidence in observing the item subset $B$ is $P(B)$. With knowledge of $A$ and the association rule $(A \Rightarrow B)$, our confidence in observing item subset $B$ is $P(B|A)$, as we have defined earlier. The ratio of these quantities is lift:

$$\frac{P(B|A)}{P(B)} = \left(\frac{P(AB)}{P(A)}\right)\left(\frac{1}{P(B)}\right) = \frac{P(AB)}{P(A)P(B)}$$

Looking at the numerator and denominator of the ratio on the far right-hand side of this equation, we note that these are equivalent under an independence assumption. That is, if there is no relationship between item subsets $A$ and $B$, then the joint probability of $A$ and $B$ is the product of their individual probabilities: $P(AB) = P(A)P(B)$. So lift is a measure of the degree to which item subsets in an association rule depart from independence.

*Figure 9.4. Market Basket Association Rules: Matrix Bubble Chart*

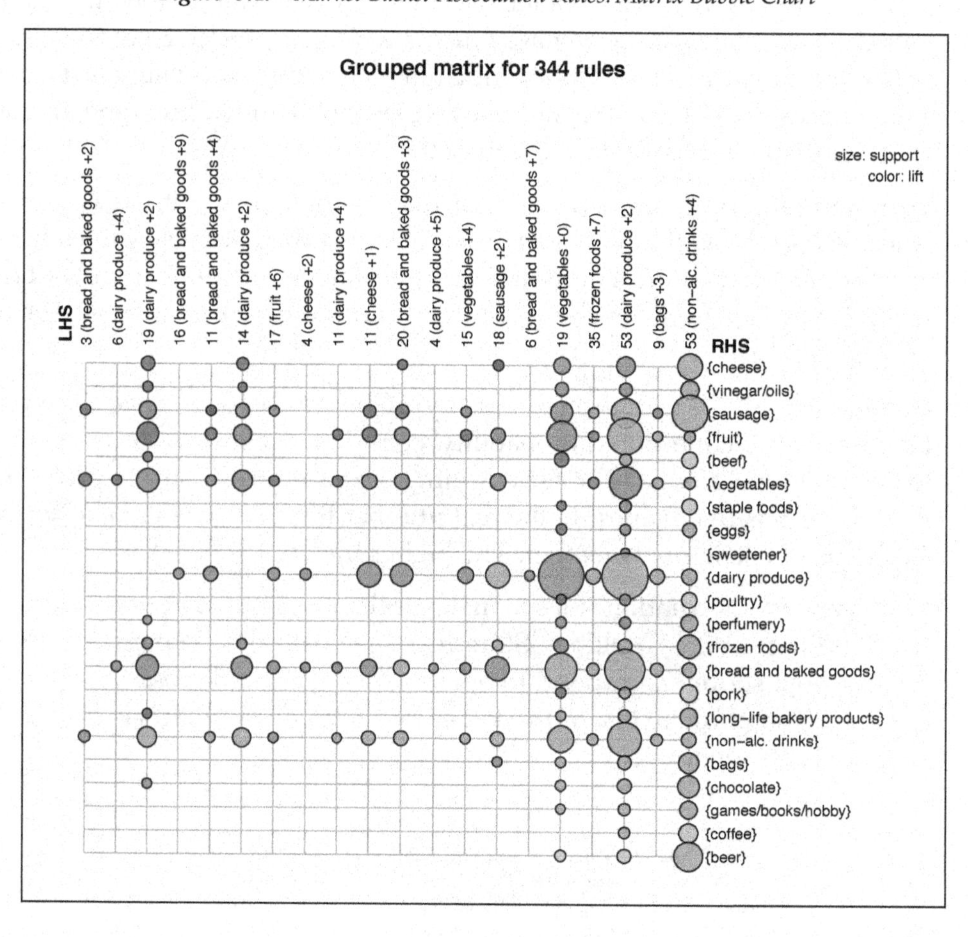

*Table 9.2. Association Rules for a Local Farmer*

| Rule No. | Left-Hand Side (Antecedent) | | Right-Hand Side (Consequent) | Support | Confidence | Lift |
|---|---|---|---|---|---|---|
| 1 | {beef, dairy produce} | => | {vegetables} | 0.030 | 0.607 | 2.225 |
| 2 | {poultry} | => | {vegetables} | 0.029 | 0.575 | 2.105 |
| 3 | {dairy produce, fruit, sausage} | => | {vegetables} | 0.027 | 0.574 | 2.103 |
| 4 | {beef} | => | {vegetables} | 0.046 | 0.560 | 2.050 |
| 5 | {dairy produce, vinegar/oils} | => | {vegetables} | 0.031 | 0.536 | 1.962 |
| 6 | {fruit, sausage} | => | {vegetables} | 0.034 | 0.529 | 1.938 |
| 7 | {bread and baked goods, dairy produce, fruit} | => | {vegetables} | 0.041 | 0.528 | 1.933 |
| 8 | {pork} | => | {vegetables} | 0.030 | 0.522 | 1.912 |
| 9 | {cheese, fruit} | => | {vegetables} | 0.027 | 0.520 | 1.904 |
| 10 | {dairy produce, fruit, non-alc. drinks} | => | {vegetables} | 0.033 | 0.518 | 1.899 |

*Figure 9.5. Association Rules for a Local Farmer: A Network Diagram*

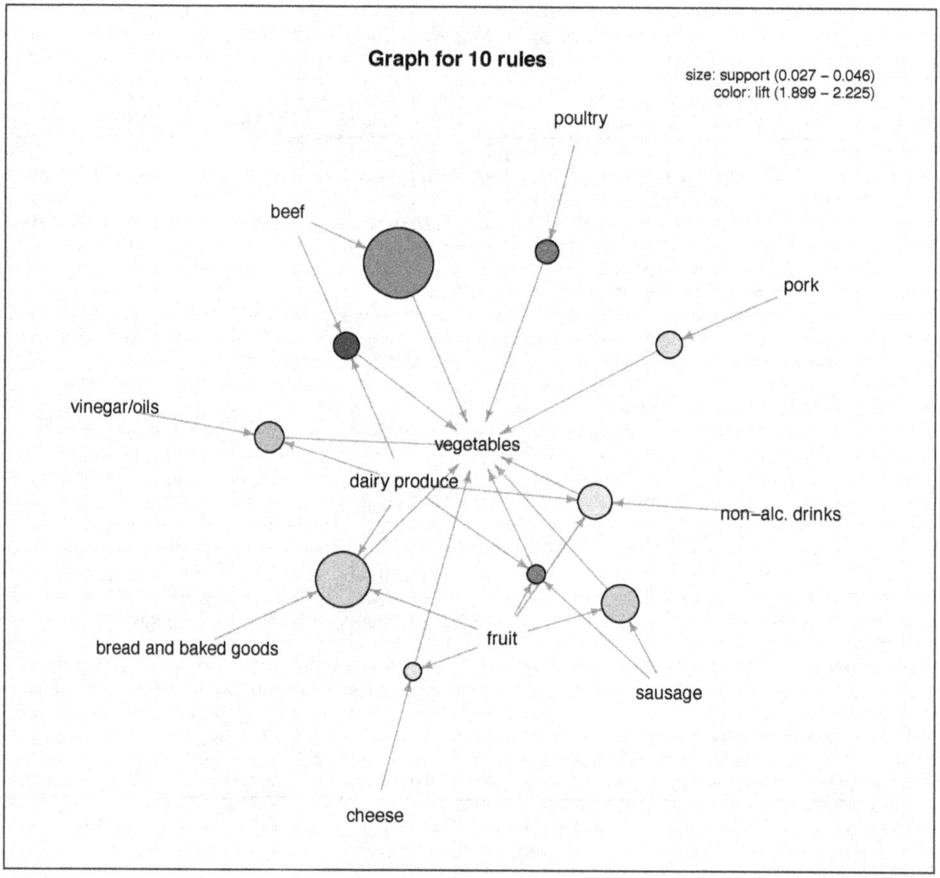

A data scientist may have to look through a lot of rules before finding one that is interesting. It is not surprising, then, that considerable attention has been paid to methods for selecting association rules (Hahsler, Buchta, and Hornik 2008), selecting variables in association rules modeling (Dippold and Hruschka 2013), and combining market segmentation and market basket analysis in the same analysis (Boztug and Reutterer 2008). Discussion of association rules may be found in the machine learning literature (Agrawal, Mannila, Srikant, Toivonen, and Verkamo 1996; Tan, Steinbach, and Kumar 2006; Hastie, Tibshirani, and Friedman 2009; Witten, Frank, and Hall 2011; Harrington 2012; Rajaraman and Ullman 2012). Bruzzese and Davino (2008) review data visualization of association rules.

Market basket and recommendation problems share an underlying directed graph structure, so that engines or algorithms appropriate for one area may be useful in the other. To build a recommender engine, we would organize the set of all purchases for a customer into a single record, and search for what goes with what. Association rules, frequent itemsets, nearest neighbors, and collaborative filtering—there are many ways to develop recommender engines. The challenge for the marketing data scientist is to find the method that works best.

Moving beyond association rules and market basket analysis, there are methods of biclustering, the joint clustering of consumers and products in the marketplace. To recommend new products for the consumer, we could conduct a simultaneous search across consumers and products. We could find segments of consumers that go with groups of products being purchased.

Recommender systems remain an active area of research and development. Reviews of recommender systems are provided in the edited volumes of Ricci et al. (2011) and Dehuri et al. (2012). Recommender system evaluation software is provided by Hahsler (2014a, 2014b).

Discussion of market basket analysis in R is provided by Hahsler, Grün, and Hornik (2005), Hahsler et al. (2011), and Hahsler et al. (2014a, 2014b), with data visualization support by Hahsler and Chelluboina (2014a, 2014b). Additional association rule algorithms are available through the R interface to Weka (Hornik 2014a, 2014b). Exhibit 9.1 shows the R program for the grocery store market basket analysis, and exhibit 9.2 shows a Python script calling R code for that same analysis.

*Exhibit 9.1. Market Basket Analysis of Grocery Store Data (R)*

```
# Association Rules for Market Basket Analysis (R)

library(arules)  # association rules
library(arulesViz)  # data visualization of association rules
library(RColorBrewer)  # color palettes for plots

data(Groceries)  # grocery transactions object from arules package

# show the dimensions of the transactions object
print(dim(Groceries))

print(dim(Groceries)[1])  # 9835 market baskets for shopping trips
print(dim(Groceries)[2])  # 169 initial store items

# examine frequency for each item with support greater than 0.025
pdf(file="fig_market_basket_initial_item_support.pdf",
  width = 8.5, height = 11)
itemFrequencyPlot(Groceries, support = 0.025, cex.names=0.8, xlim = c(0,0.3),
  type = "relative", horiz = TRUE, col = "dark red", las = 1,
  xlab = paste("Proportion of Market Baskets Containing Item",
    "\n(Item Relative Frequency or Support)"))
dev.off()

# explore possibilities for combining similar items
print(head(itemInfo(Groceries)))
print(levels(itemInfo(Groceries)[["level1"]]))  # 10 levels... too few
print(levels(itemInfo(Groceries)[["level2"]]))  # 55 distinct levels

# aggregate items using the 55 level2 levels for food categories
# to create a more meaningful set of items
groceries <- aggregate(Groceries, itemInfo(Groceries)[["level2"]])

print(dim(groceries)[1])  # 9835 market baskets for shopping trips
print(dim(groceries)[2])  # 55 final store items (categories)

pdf(file="fig_market_basket_final_item_support.pdf", width = 8.5, height = 11)
itemFrequencyPlot(groceries, support = 0.025, cex.names=1.0, xlim = c(0,0.5),
  type = "relative", horiz = TRUE, col = "blue", las = 1,
  xlab = paste("Proportion of Market Baskets Containing Item",
    "\n(Item Relative Frequency or Support)"))
dev.off()

# obtain large set of association rules for items by category and all shoppers
# this is done by setting very low criteria for support and confidence
first.rules <- apriori(groceries,
  parameter = list(support = 0.001, confidence = 0.05))
print(summary(first.rules))  # yields 69,921 rules... too many

# select association rules using thresholds for support and confidence
```

```
second.rules <- apriori(groceries,
  parameter = list(support = 0.025, confidence = 0.05))
print(summary(second.rules)) # yields 344 rules

# data visualization of association rules in scatter plot
pdf(file="fig_market_basket_rules.pdf", width = 8.5, height = 8.5)
plot(second.rules,
  control=list(jitter=2, col = rev(brewer.pal(9, "Greens")[4:9])),
  shading = "lift")
dev.off()

# grouped matrix of rules
pdf(file="fig_market_basket_rules_matrix.pdf", width = 8.5, height = 8.5)
plot(second.rules, method="grouped",
  control=list(col = rev(brewer.pal(9, "Greens")[4:9])))
dev.off()

# select rules with vegetables in consequent (right-hand-side) item subsets
vegie.rules <- subset(second.rules, subset = rhs %pin% "vegetables")
inspect(vegie.rules) # 41 rules

# sort by lift and identify the top 10 rules
top.vegie.rules <- head(sort(vegie.rules, decreasing = TRUE, by = "lift"), 10)
inspect(top.vegie.rules)

pdf(file="fig_market_basket_farmer_rules.pdf", width = 11, height = 8.5)
plot(top.vegie.rules, method="graph",
  control=list(type="items"),
  shading = "lift")
dev.off()

# Suggestions for the student:
# Suppose your client is someone other than the local farmer,
# a meat producer/butcher, dairy, or brewer perhaps.
# Determine association rules relevant to your client's products
# guided by the market basket model. What recommendations
# would you make about future marketplace actions?
# Use association rule modeling for another problem, such as
# the activities market baskets from the Wisconsin Dells case.
# Try other ways of making recommendations, such as nearest
# neighbors or frequent itemsets.
```

*Exhibit 9.2. Market Basket Analysis of Grocery Store Data (Python to R)*

```
# Association Rules for Market Basket Analysis (Python to R)

# import package for analysis and modeling
from rpy2.robjects import r   # interface from Python to R

r('library(arules)')  # association rules
r('library(arulesViz)')  # data visualization of association rules
r('library(RColorBrewer)')  # color palettes for plots

r('data(Groceries)')  # grocery transactions object from arules package

# show the dimensions of the transactions object
r('print(dim(Groceries))')

r('print(dim(Groceries)[1])')  # 9835 market baskets for shopping trips
r('print(dim(Groceries)[2])')  # 169 initial store items

# examine frequency for each item with support greater than 0.025
r('pdf(file="fig_market_basket_initial_item_support.pdf", \
    width = 8.5, height = 11)')
r('itemFrequencyPlot(Groceries, support = 0.025, \
    cex.names=0.8, xlim = c(0,0.3), \
    type = "relative", horiz = TRUE, col = "dark red", las = 1, \
    xlab = paste("Proportion of Market Baskets Containing Item", \
      "\n(Item Relative Frequency or Support)"))')
r('dev.off()')

# explore possibilities for combining similar items
r('print(head(itemInfo(Groceries)))')
r('print(levels(itemInfo(Groceries)[["level1"]]))')  # 10 levels... too few
r('print(levels(itemInfo(Groceries)[["level2"]]))')  # 55 distinct levels

# aggregate items using the 55 level2 levels for food categories
# to create a more meaningful set of items
r('groceries <- aggregate(Groceries, itemInfo(Groceries)[["level2"]])')

r('print(dim(groceries)[1])')  # 9835 market baskets for shopping trips
r('print(dim(groceries)[2])')  # 55 final store items (categories)

r('pdf(file="fig_market_basket_final_item_support.pdf", \
      width = 8.5, height = 11)')
r('itemFrequencyPlot(groceries, support = 0.025, \
      cex.names=1.0, xlim = c(0,0.5),\
      type = "relative", horiz = TRUE, col = "blue", las = 1,\
      xlab = paste("Proportion of Market Baskets Containing Item",\
    "\n(Item Relative Frequency or Support)"))')
r('dev.off()')

# obtain large set of association rules for items by category and all shoppers
# this is done by setting very low criteria for support and confidence
```

```
r('first.rules <- apriori(groceries, \
      parameter = list(support = 0.001, confidence = 0.05))')
r('print(summary(first.rules))')  # yields 69,921 rules... too many

# select association rules using thresholds for support and confidence
r('second.rules <- apriori(groceries, \
      parameter = list(support = 0.025, confidence = 0.05))')
r('print(summary(second.rules))')  # yields 344 rules

# data visualization of association rules in scatter plot
r('pdf(file="fig_market_basket_rules.pdf", width = 8.5, height = 8.5)')
r('plot(second.rules, \
      control=list(jitter=2, col = rev(brewer.pal(9, "Greens")[4:9])), \
  shading = "lift")')
r('dev.off()')

# grouped matrix of rules
r('pdf(file="fig_market_basket_rules_matrix.pdf", \
      width = 8.5, height = 8.5)')
r('plot(second.rules, method="grouped", \
      control=list(col = rev(brewer.pal(9, "Greens")[4:9])))')
r('dev.off()')

# select rules with vegetables in consequent (right-hand-side) item subsets
r('vegie.rules <- subset(second.rules, subset = rhs %pin% "vegetables")')
r('inspect(vegie.rules)')  # 41 rules

# sort by lift and identify the top 10 rules
r('top.vegie.rules <- head(sort(vegie.rules, \
      decreasing = TRUE, by = "lift"), 10)')
r('inspect(top.vegie.rules)')

r('pdf(file="fig_market_basket_farmer_rules.pdf", width = 11, height = 8.5)')
r('plot(top.vegie.rules, method="graph", \
      control=list(type="items"), \
      shading = "lift")')
r('dev.off()')
```

# 10 Assessing Brands and Prices

"Joker Products with a new secret ingredient—smiley. Now, let's go over our blind taste test. Uh-oh, he don't look happy. He's been usin' BRAND X. But with Joker brand, I get a grin again and again... That lucious tan, those ruby lips, and hair color only your undertaker knows for sure. I know what you're saying, 'Where can I get these fine new items?' Well, that's the gag. Chances are you bought 'em already. So remember, put on a happy face. "

—JACK NICHOLSON AS JOKER IN *Batman* (1989)

As a consumer, I am like two people. With almost everything—where I live and shop, how I travel, the clothes I buy, and the things I do for fun—I am extremely price-conscious. I walk rather than drive whenever I can, and I look for things on sale. As long as I am buying for myself, I rarely seek out particular brands. This is my economical side. I look for value.

There is another side to me, however. When it comes to computers and software, I am a person who buys only the best, highest quality, and most reliable products. Money is no object (until it runs out, of course). I read about fast computers and high-resolution displays, thinking, *Maybe that's what I should get next*. This most uneconomical side of me also applies to books. It is as though I have no control over my spending, no budget. When I see a book I like, I buy, and sometimes I pay for next-day delivery. I would be a good candidate for a "bibliophiles anonymous" if there were one.

The example for this chapter takes us back to 1998. Microsoft introduced a new operating system, and computer manufacturers were interested in making predictions about the personal computer marketplace. The study we consider, which we call the Computer Choice Study, is presented in appendix C (page 360). It involved eight computer brands, price, and four other attributes of interest: compatibility, performance, reliability, and learning time. Table C.4 shows attribute levels used in the study.

The Computer Choice Study used a survey consisting of sixteen pages, with each page showing a choice set of four product profiles. For each choice set, survey participants were asked first to select the computer they most preferred, and second, to indicate whether or not they would actually buy that computer. For our analysis we focus on the initial choice or most preferred computer in each set.

Each profile, defined in terms of its attributes, is associated with the consumer's binary response (0 = not chosen, 1 = chosen). Being diligent data scientists, we define an appropriate training-and-test regimen. In this context, we build predictive models on twelve choice sets and test on four.[1] The data for one individual are shown in table C.5.

To obtain useful conjoint measures for individuals, we employ hierarchical Bayes (HB) methods with constraints upon the signs of attribute coefficients other than brand. We like to specify the signs of attribute coefficients when the order of preference for levels within attributes is obvious . For the Computer Choice Study, we can specify the signs of coefficients for all attributes other than brand prior to fitting a model to the data. In particular, compatibility, performance, and reliability should have positive coefficients, whereas learning time and price should have negative coefficients.

As we have learned, the value of a model lies in the quality of its predictions. Here we are concerned with predictions for each individual respondent. We use the training set of twelve choice items to fit an HB model, estimating individual-level conjoint measures. Then we use the test set of

[1] When the prediction problem is to predict individual choice, the survey itself is divided into training and test. For this study, we split across the survey items or choice sets, arbitrarily selecting sets 3, 7, 11, and 15 as our hold-out choice sets. Items 1, 2, 4, 5, 6, 8, 9, 10, 12, 13, 14, and 16 serve as training data. With sixteen choice sets of four, we have 64 product profiles for each individual in the study. As a result of this training-and-test split, the training data include 48 rows of product profiles for each individual, and the test data include 16 rows of product profiles for each individual.

**Table 10.1.** *Contingency Table of Top-ranked Brands and Most Valued Attributes*

| *Top-Ranked* | *Most Valued Attribute Relative to Other Consumers* | | | | | | |
|---|---|---|---|---|---|---|---|
| *Brand* | Brand | Compatibility | Performance | Reliability | Learning | Price | Total |
| Apple | 9 | 0 | 4 | 6 | 12 | 7 | 38 |
| Compaq | 5 | 1 | 4 | 4 | 3 | 2 | 19 |
| Dell | 8 | 3 | 4 | 1 | 6 | 6 | 28 |
| Gateway | 10 | 4 | 10 | 5 | 9 | 4 | 42 |
| HP | 7 | 0 | 1 | 5 | 3 | 3 | 19 |
| IBM | 5 | 6 | 10 | 4 | 11 | 2 | 38 |
| Sony | 1 | 0 | 2 | 2 | 1 | 3 | 9 |
| Sun | 2 | 6 | 2 | 10 | 3 | 8 | 31 |
| Total | 47 | 20 | 37 | 37 | 48 | 35 | 224 |

four choice items to evaluate the fitted model. As a criterion of evaluation, we use the percentage of choices correctly predicted, noting that by chance alone we would expect to predict 25 percent of the choices correctly (because each choice item contains four product alternatives). The percentage of choices correctly predicted is the sensitivity or true positive rate in a binary classification problem (see figure A.1 in appendix A.).

How well does the HB model work? An HB model fit to the computer choice data provides training set sensitivity of 93.7 percent and test set sensitivity of 52.6 percent. As modeling techniques in predictive analytics, Bayesian methods hold considerable promise.

Returning to the original survey with sixteen choice sets, we estimate conjoint measures at the individual level with an HB model and place consumers into groups based upon their revealed preferences for computer products. Table 10.1 provides the cross-tabulation of top-ranked brands versus most valued attributes, where the top-ranked brands are determined within individuals and the most valued attributes are determined relative to all consumers in the study.[2] These data are rendered as a mosaic plot in figure 10.1.[3]

[2] The top-ranked brand is the brand with the highest estimated part-worth for the individual. Most favored attributes are determined relative to the entire study group. That is, we compute the value of an attribute to an individual as a standard score versus the entire study group. We do this because conventional attribute importance calculations are dependent upon relative ranges of sets of attribute levels utilized in surveys.

[3] Mosaic plots provide a convenient visualization of cross-tabulations (contingency tables). The relative heights of rows correspond to the relative row frequencies. The relative width of columns corresponds to the cell frequencies within rows. Mosaic plots are discussed in Hartigan and Kleiner (1984) and Friendly (2000), with R implementations reviewed in Meyer et al. (2006, 2014).

*Figure 10.1.* *Computer Choice Study: A Mosaic of Top Brands and Most Valued Attributes*

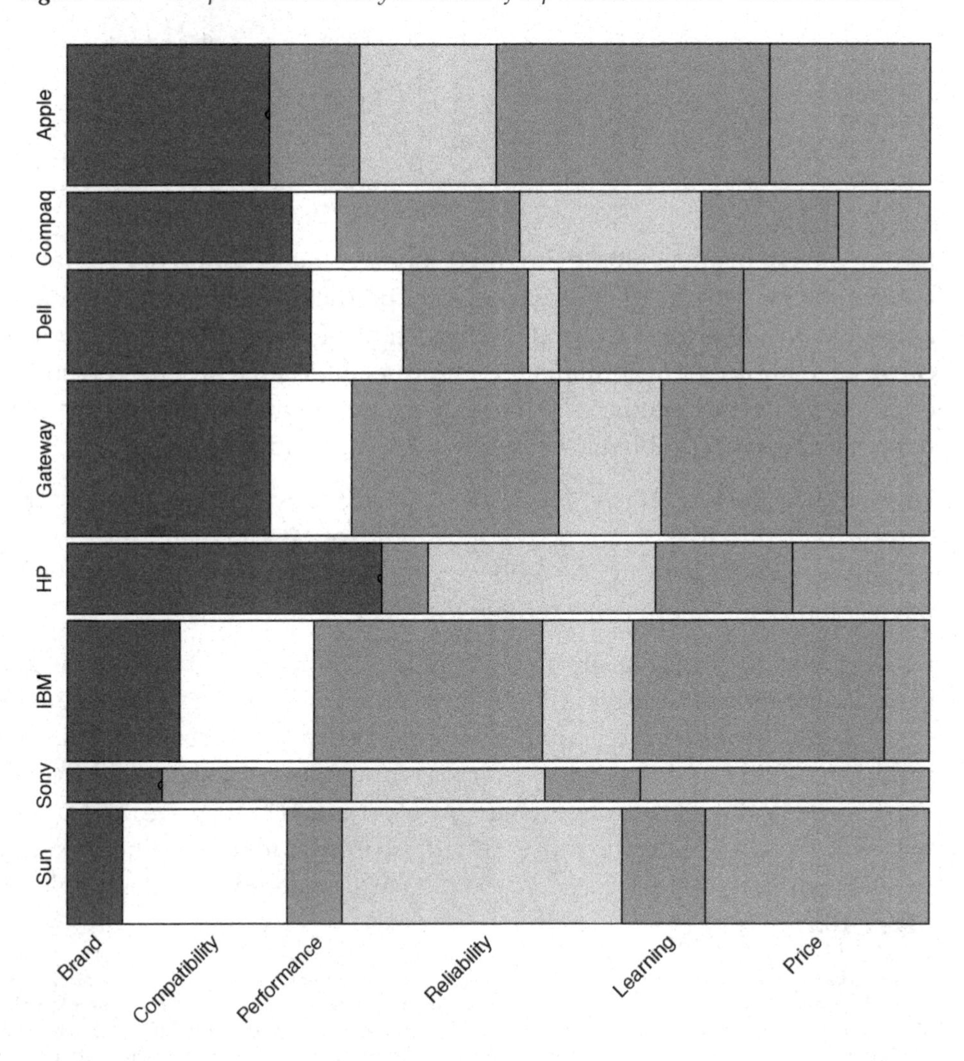

Brand and price are important. They help define how we behave in the marketplace—whether we buy, what we buy, when and how we buy. Product features are also important. Considered together, brand, price, and product features are key inputs to models of consumer preference and market response. Marketing managers benefit by having data and models presented in a clear and consistent manner. The model illustrated in figure 10.2, provides a framework for communicating with management. The plot in figure 10.3, referred to as a ternary plot or triplot, provides a visual of consumer preference and choice. Brand-loyal consumers fall to the bottom-left vertex, price-sensitive consumers fall to the bottom-right vertex, and feature-focused consumers are closest to the top vertex.

From the distribution of points across the ternary plot for the Computer Choice Study, we can see wide variability or heterogeneity in consumer preferences. What does this mean for computer suppliers?

Suppose we focus on three brands, Apple, Dell, and HP, and select the subset of consumers for whom one of these brands is the top-ranked brand. We use density plots to examine the distributions of values for brand loyalty, price sensitivity, and feature focus across this subset of consumers. A density plot provides a smoothed picture of a distribution of measures. Showing densities for consumer groups on the same plot provides a picture of the degree to which there is overlap in these distributions.

The densities in figure 10.4 suggest that consumers rating Dell highest tend to be less price-sensitive and more feature-focused than consumers who rate Apple and HP highest. We can see, as well, that consumers rating HP highest have higher brand-loyalty than consumers rating Dell or Apple highest. A general finding that emerges from a review of these densities is that, in terms of the three measures from our ternary model, there is considerable overlap across consumers rating Apple, Dell, and HP highest.

A concern of marketers in many product categories is the extent to which consumers are open to switching from one brand to another. Parallel coordinate plots may be used to explore the potential for brand switching.

*Figure 10.2. Framework for Describing Consumer Preference and Choice*

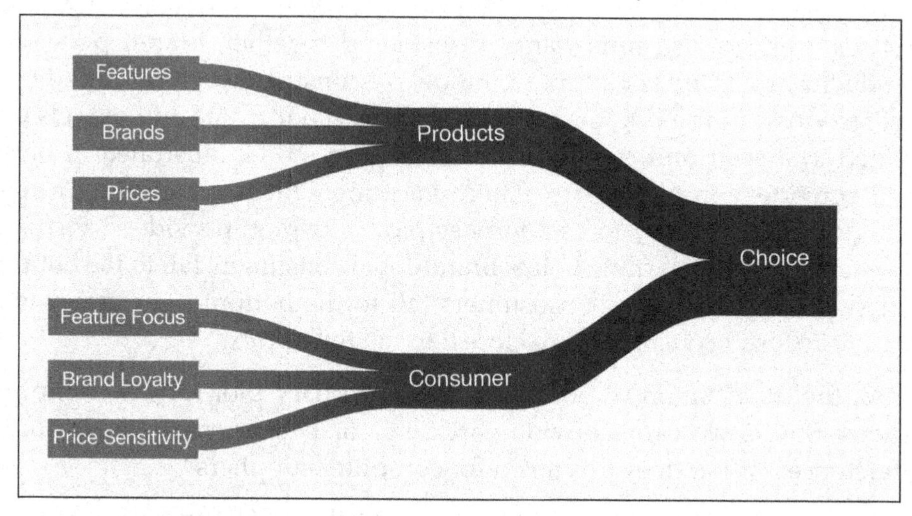


*Figure 10.3. Ternary Plot of Consumer Preference and Choice*

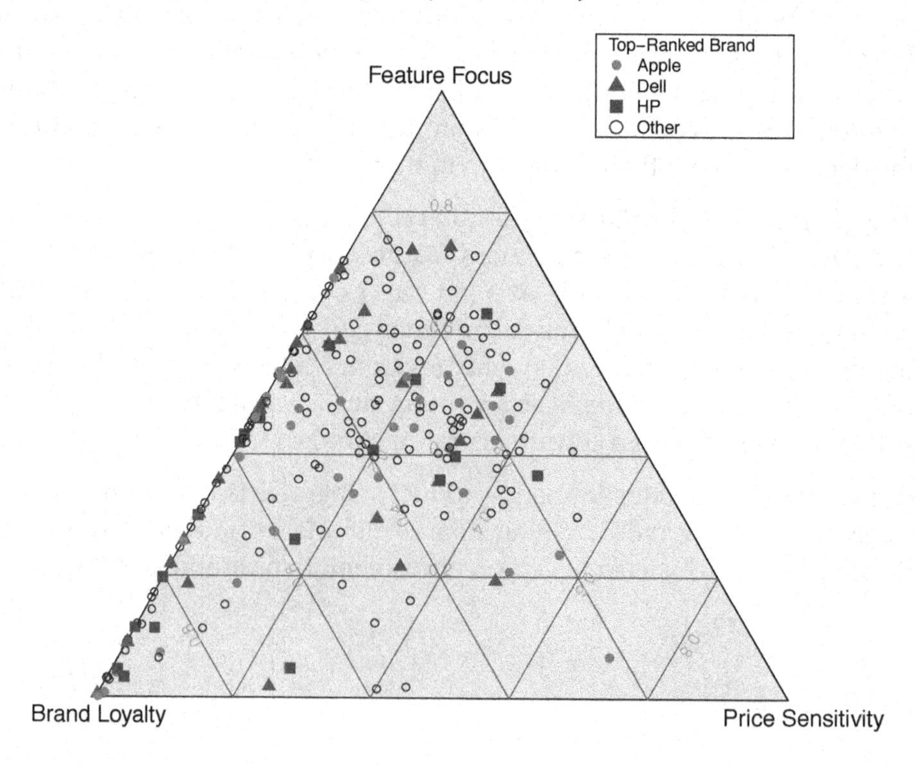

*Figure 10.4.* *Comparing Consumers with Differing Brand Preferences*

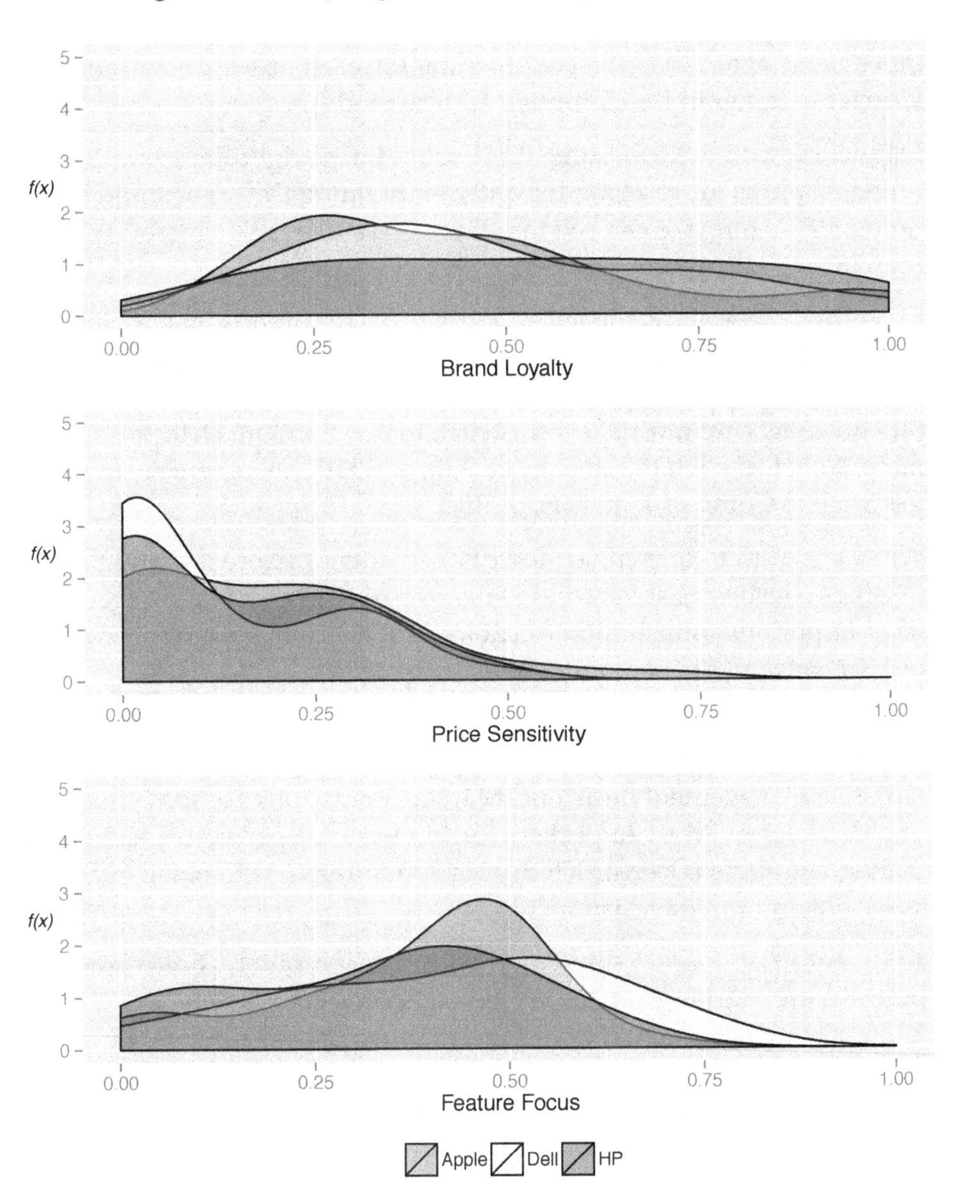

A parallel coordinates plot shows relationships among many variables. It is like a univariate scatter plot of all displayed variables standardized and stacked parallel to one another (Inselberg 1985; Inselberg 1985; Wegman 1990; Moustafa and Wegman 2006). Parallel coordinates show common movements across variables with a line for each observational unit (individual consumer in the Computer Choice Study).

Parallel coordinates in figure 10.5 display thirty-eight lines for consumers rating Apple as the top brand, twenty-eight rating Dell as top, and nineteen rating HP as top. We can see that there is considerable variability in these individuals' part-worth profiles.

In figure 10.6 we show the mean part-worths for brands on parallel coordinates. Lines further to the right show stronger preference for a brand and stronger likelihood of switching to that brand. The figure shows that Apple consumers are most likely to switch to Sony or Sun, Dell consumers are most likely to switch to Gateway or Sun, and HP consumers are most likely to switch to Compaq, Gateway, or IBM.

While it is most interesting to describe consumer preferences and the extent to which consumers are brand-loyal, price-sensitive, or feature-focused, the most important contribution of choice models comes from their ability to predict consumer behavior in the marketplace.

We use models of consumer preference and choice to develop market simulations, exploring a variety of marketplace conditions and evaluating alternative management decisions. Market simulations are sometimes called *what-if analyses*. Most important for the work of predictive analytics are market simulations constructed from individual-level conjoint measures, as we obtain from Bayesian methods.

To demonstrate market simulation with the Computer Choice Study, suppose we are working for the Apple computer company, and we want to know what price to charge for our computer, given three other competitors in the market: Dell, Gateway, and HP. In addition, suppose that we have an objective of commanding a 25 percent share in the market. This is a hypothetical example but one that demonstrates market simulation as a modeling technique.

*Figure 10.5.* *Potential for Brand Switching: Parallel Coordinates for Individual Consumers*

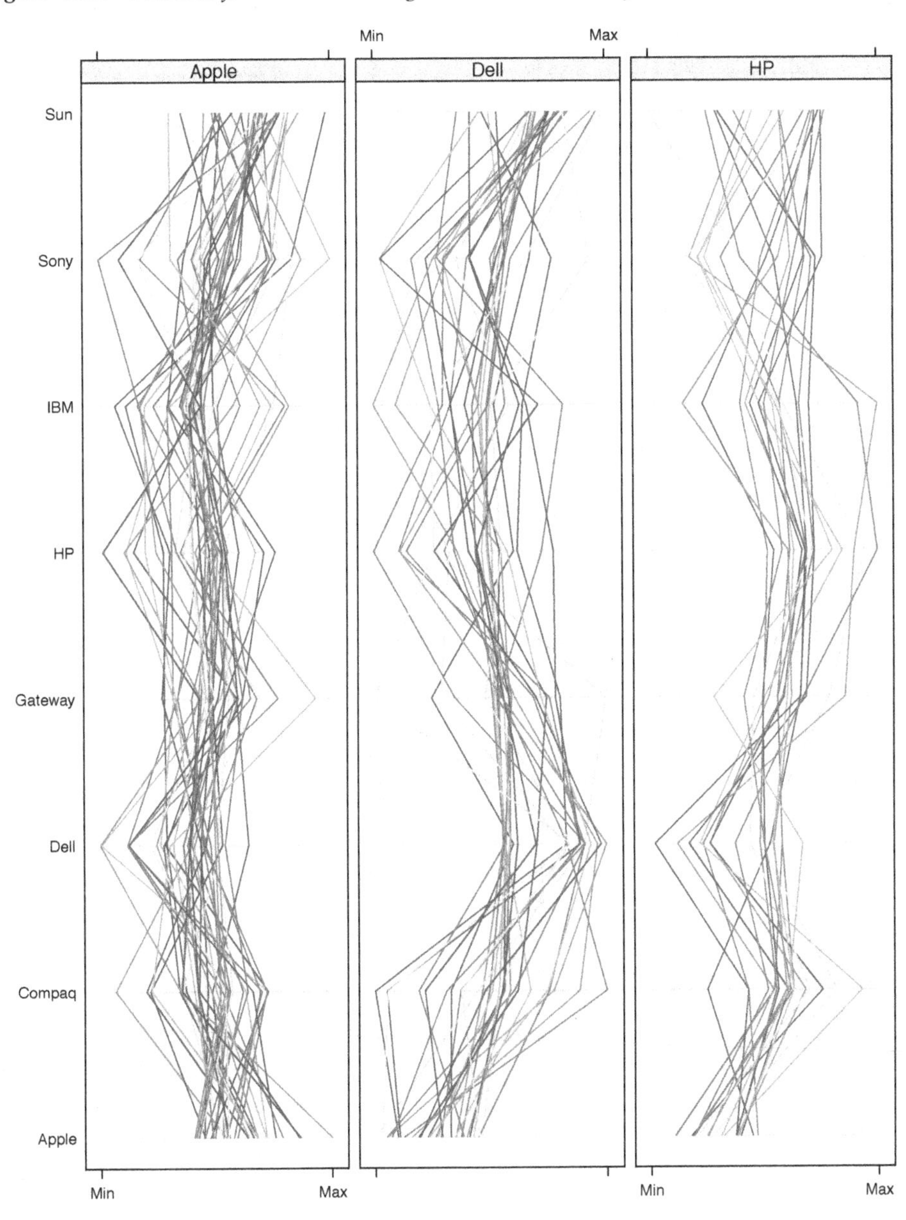

***Figure 10.6.*** *Potential for Brand Switching: Parallel Coordinates for Consumer Groups*

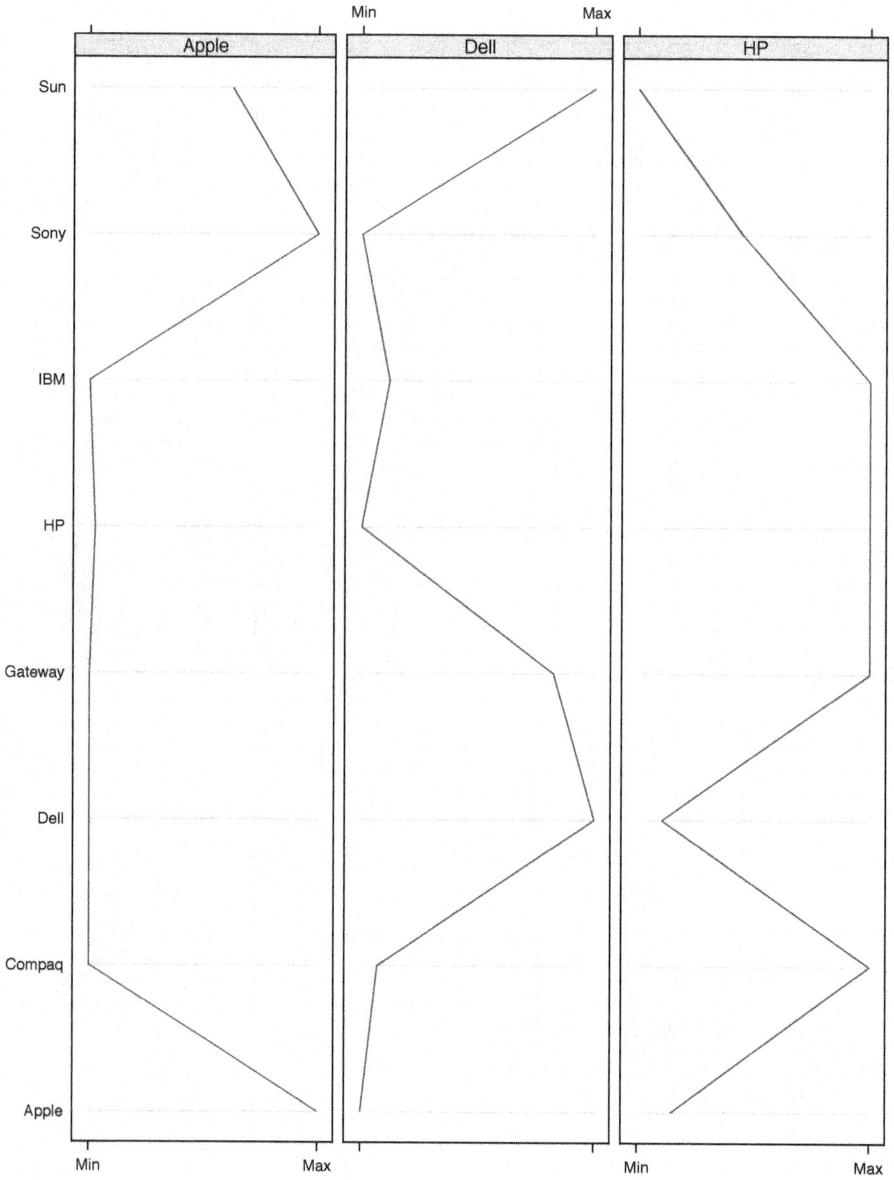

We describe the competitive products in terms of attributes, creating simulated choice sets for input to the market simulation. Let us imagine that the Dell system is a high-end system, 100 percent compatible with earlier systems, four times as fast, and less likely to fail. It takes sixteen hours to learn and costs $1,750. The Gateway offering is 90 percent compatible, twice as fast, and just as likely to fail as earlier systems. The Gateway takes eight hours to learn and costs only $1,250. Finally, among the marketplace competitors, we have an HP system, 90 percent compatible, three times as fast, and less likely to fail. It takes eight hours to learn and costs $1,500. Suppose that Apple is entering this market with a system that, like the Dell system, is four times as fast as earlier systems and less likely to fail. The Apple system is only 50 percent compatible with prior systems and takes eight hours to learn. We allow Apple prices to vary across the full range of prices from the Computer Choice Study, defining eight choice sets for the market simulation.

Choice sets for the market simulation are displayed in table 10.2 with coding of profile cells as required for use with the HB-estimated individual-level part-worths. Looking across the simulation choice sets, we can see that they are identical except for the one factor that is being studied—the price of the Apple computer. The variation in Apple prices is represented by the highlighted prices for Apple in the far right-hand-side column of the table.

Data visualization of market simulation results is provided by painting a mosaic of preference shares,[4] as shown in figure 10.7. The length/width of tiles in each row of the mosaic reflects brand market share. There is a row for each Apple price or simulation choice set. Obeying the law of demand, higher prices translate into lower market shares for Apple. The precise results of the simulation, shown in table 10.3, suggest that Apple would need to set its price below $1,750 to capture a 25 percent share.

We can take market simulations further, considering the actions and reactions of competitive firms in the marketplace. Suppose the battle lines had been drawn and the world of laptop computers had come down to two

[4] The term *market share* is used with caution in this context. Many conjoint researchers, including Orme (2013), prefer to use the term *preference share,* highlighting the fact that estimates are based on choice-based surveys or revealed preferences rather than on actual behavior or choices in the marketplace. Of course, managers want estimates of market share, regardless of what we call it.

*Table 10.2. Market Simulation: Choice Set Input*

| *profile* | setid | brand | compat | perform | reliab | learn | price |
|---|---|---|---|---|---|---|---|
| *1* | 1 | Dell | 8 | 4 | 2 | 4 | 4 |
| *2* | 1 | Gateway | 6 | 2 | 1 | 2 | 2 |
| *3* | 1 | HP | 6 | 3 | 2 | 2 | 3 |
| *4* | 1 | Apple | 5 | 4 | 2 | 1 | 1 |
| *5* | 2 | Dell | 8 | 4 | 2 | 4 | 4 |
| *6* | 2 | Gateway | 6 | 2 | 1 | 2 | 2 |
| *7* | 2 | HP | 6 | 3 | 2 | 2 | 3 |
| *8* | 2 | Apple | 5 | 4 | 2 | 1 | 2 |
| *9* | 3 | Dell | 8 | 4 | 2 | 4 | 4 |
| *10* | 3 | Gateway | 6 | 2 | 1 | 2 | 2 |
| *11* | 3 | HP | 6 | 3 | 2 | 2 | 3 |
| *12* | 3 | Apple | 5 | 4 | 2 | 1 | 3 |
| *13* | 4 | Dell | 8 | 4 | 2 | 4 | 4 |
| *14* | 4 | Gateway | 6 | 2 | 1 | 2 | 2 |
| *15* | 4 | HP | 6 | 3 | 2 | 2 | 3 |
| *16* | 4 | Apple | 5 | 4 | 2 | 1 | 4 |
| *17* | 5 | Dell | 8 | 4 | 2 | 4 | 4 |
| *18* | 5 | Gateway | 6 | 2 | 1 | 2 | 2 |
| *19* | 5 | HP | 6 | 3 | 2 | 2 | 3 |
| *20* | 5 | Apple | 5 | 4 | 2 | 1 | 5 |
| *21* | 6 | Dell | 8 | 4 | 2 | 4 | 4 |
| *22* | 6 | Gateway | 6 | 2 | 1 | 2 | 2 |
| *23* | 6 | HP | 6 | 3 | 2 | 2 | 3 |
| *24* | 6 | Apple | 5 | 4 | 2 | 1 | 6 |
| *25* | 7 | Dell | 8 | 4 | 2 | 4 | 4 |
| *26* | 7 | Gateway | 6 | 2 | 1 | 2 | 2 |
| *27* | 7 | HP | 6 | 3 | 2 | 2 | 3 |
| *28* | 7 | Apple | 5 | 4 | 2 | 1 | 7 |
| *29* | 8 | Dell | 8 | 4 | 2 | 4 | 4 |
| *30* | 8 | Gateway | 6 | 2 | 1 | 2 | 2 |
| *31* | 8 | HP | 6 | 3 | 2 | 2 | 3 |
| *32* | 8 | Apple | 5 | 4 | 2 | 1 | 8 |

*Figure 10.7. Market Simulation: A Mosaic of Preference Shares*

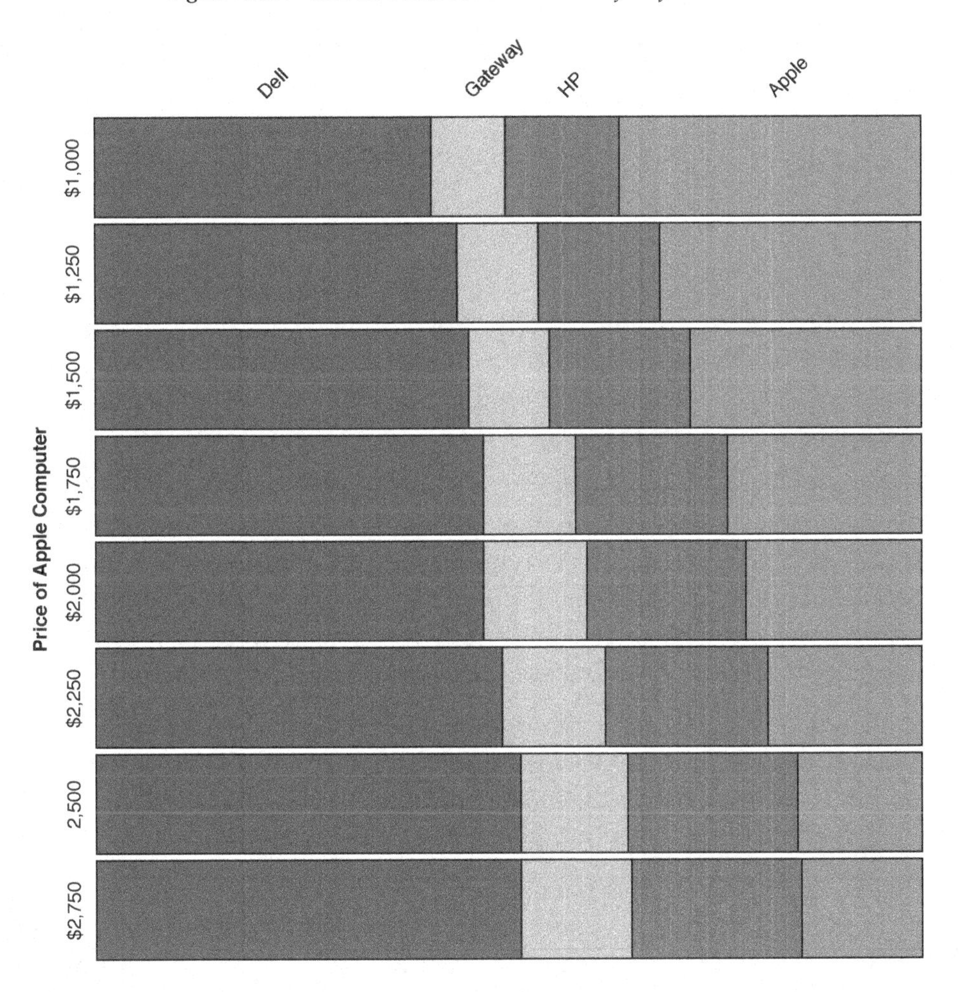

*Table 10.3. Market Simulation: Preference Shares in a Hypothetical Four-brand Market*

| | *Brand Preference Share (percentage)* | | | |
|---|---|---|---|---|
| *Apple Price* | Apple | Dell | Gateway | HP |
| $1,000 | 36.6 | 40.6 | 8.9 | 13.8 |
| 1,250 | 31.7 | 43.8 | 9.8 | 14.7 |
| 1,500 | 28.1 | 45.1 | 9.8 | 17.0 |
| 1,750 | 23.7 | 46.9 | 11.2 | 18.3 |
| 2,000 | 21.4 | 46.9 | 12.5 | 19.2 |
| 2,250 | 18.8 | 49.1 | 12.5 | 19.6 |
| 2,500 | 15.2 | 51.3 | 12.9 | 20.5 |
| 2,750 | 14.7 | 51.3 | 13.4 | 20.5 |

manufacturers: Apple and HP. The Apple system is selling for $2,000 and the HP for $1,750. Suppose further that both Apple and HP have the option of lowering their prices by $100, which each firm might do in order to increase its market share.

Those with training in economics may recognize this Apple-HP scenario as a two-player competitive game. We could run market simulations with Apple and HP in the simulation choice sets, obtaining what we think of as market shares under each of four outcomes defined by no-price-change and price-change strategies. Knowing unit costs, we would have everything needed to make entries in the game outcome or payoff table.

Business managers look for profit-improving strategies, and modeling techniques in predictive analytics may be used to evaluate alternative strategies and how they play out in the marketplace. Consumer preferences translate into choices. Choices turn into sales, sales into market shares.

Recognize that all models have limitations. In a choice simulation we attempt to predict what consumers will do when faced with sets of potential (not actual) products with hypothetical (not actual) prices within a marketplace of possible (not actual) competitors. We would be most surprised if all such predictions were correct. Our uncertainty about prediction is something that we readily admit and estimate within the context of Bayesian analysis. Error bands can be provided around all conjoint measures, drawing on posterior distributions obtained from an HB estimation procedure.

To understand customers is to predict what they might do when the firm introduces a new product or changes the price of a current product. A good model, when paired with a market simulation, provides predictions to guide management action.

The term *brand equity* is used to describe the special value that a brand name has for products. Brand equity is built over time. Hollis (2005) talks about brand *presence,* which is awareness of what a brand stands for, *relevance* to the consumer, *performance* and *advantage* within the category of products being considered, and *bonding,* which is associated with a consumer's loyalty to or relationship with a brand. Brand equity and loyalty translate into long-term customer value and product profitability.

Choice-based conjoint studies with product profiles defined in terms of brand and price provide a way of measuring what a brand is worth relative to other brands. In fact, with HB-estimated part-worths, brand equity can be assessed for each individual in a conjoint study.[5]

An introduction to pricing theory and the economics of demand may be found in books about microeconomics (Stigler 1987; Varian 2005; Pindyck and Rubinfeld 2012) and econometrics (Greene 2012). For pricing research methods, see Lyon (2000, 2002), and Feldman (2002a). Marder (1997) reviews pricing research with monadic methods (each consumer seeing one and only one price). Nagle and Hogan (2005) and Krishnamurthi (2001) review pricing strategy and price planning for business managers.

Special considerations in the construction of choice studies have been discussed by Kuhfeld, Tobias, and Garratt (1994), Huber and Zwerina (1996), Zwerina (1997), Louviere, Hensher, and Swait (2000), and Hensher, Rose,

---

[5] To compute brand equity for an individual, we begin by observing the range in part-worths for price. Suppose that the price part-worths for an individual extend from -0.55 to 0.55. Then one unit of part-worth would be equivalent to $1,501, as shown in the following calculation:

$$\frac{|\$2,750 - \$1,000|}{|0.55 - (-0.55)|} = \frac{\$1,750}{1.10} = \$1,501$$

This is because the range of prices displayed to consumers in the Computer Choice Study extends from $1,000 to $2,750. The computed conversion factor $1,501 is called a *dollar-metric* and can be applied to part-worth differences across any attribute in the study, including brand. Suppose this individual's part-worth for Sony is 0.44 and part-worth for Gateway is 0.36. Then we would say that Sony has a brand equity of $120 over Gateway for this individual. That is, a difference of 0.08 in part-worth units is worth $120:

$$(0.44 - 0.36) \times \$1,501 = 0.08 \times \$1,501 = \$120.$$

and Greene (2005). Reviews of discrete choice methods may be found in the work of Ben-Akiva and Lerman (1985), McFadden (2001), Train (1985, 2003), and Greene (2012). Many discrete choice models fall within the general class of generalized linear models.

Orme (2013) provides a review of conjoint and choice methods for product design and pricing research, including discussion of brand equity estimation and market simulations. Louviere, Hensher, and Swait (2000) provide extensive discussion of choice study research design and analysis. In addition, specialized methods are available for market share estimation (Cooper and Nakanishi 1988; Cooper 1999).

Multinomial logit and conditional logit modeling methods implemented in R are especially useful in the analysis of choice study data (Elff 2014; Croissant 2014; Therneau 2014; Train and Croissant 2014). R programming for Bayesian analysis of choice studies is reviewed by Rossi, Allenby, and McCulloch (2005) and Rossi (2014).

Exhibit 10.1 shows an R program to obtain hierarchical Bayes estimates for the Computer Choice Study. This code shows how to implement training and test choice sets from the conjoint survey. Market simulation utilities are provided in exhibit D.2. This program and the next call upon an packages developed by Sarkar (2014), Meyer, Zeileis, Hornik, and Friendly (2014), Wickham and Chang (2014), Sermas (2014), and Kuhn (2014). Comparable packages are not yet available in Python.

Exhibit 10.2 shows how to use individual-level part-worths in the modeling of consumer preferences. The program provides an analysis of the complete data from consumer choice study, following the logic of the ternary model introduced in this chapter. The program shows how to generate mosaic, ternary, and density visualizations. The program code also shows how to conduct a market simulation using individual-level part worths.

***Exhibit 10.1.*** *Training and Testing a Hierarchical Bayes Model (R)*

```
# Hierarchical Bayes Part-Worth Estimation: Training and Test

# load market simulation utilities
load(file="mspa_market_simulation_utilities.RData")

library(ChoiceModelR)  # for Hierarchical Bayes Estimation

library(caret)  # for confusion matrix function

# read in the data from a case study in computer choice.
complete.data.frame <- read.csv("computer_choice_study.csv")

print.digits <- 2
# user-defined function for printing conjoint measures
if (print.digits == 2)
  pretty.print <- function(x) {sprintf("%1.2f",round(x,digits = 2))}
if (print.digits == 3)
  pretty.print <- function(x) {sprintf("%1.3f",round(x,digits = 3))}

# set up sum contrasts for effects coding
options(contrasts=c("contr.sum","contr.poly"))

# employ a training-and-test regimen across survey sets/items
test.set.ids <- c("3","7","11","15")  # select four sets/items
training.set.ids <- setdiff(unique(complete.data.frame$setid),test.set.ids)
training.data.frame <-
  subset(complete.data.frame,subset=(setid %in% training.set.ids))
test.data.frame <-
  subset(complete.data.frame,subset=(setid %in% test.set.ids))

UniqueID <- unique(training.data.frame$id)
# set up zero priors
cc.priors <- matrix(0,nrow=length(UniqueID),ncol=13)

# we could use coefficients from aggregate model as starting values
# here we comment out the code needed to do that
# aggregate.cc.betas <- c(as.numeric(conjoint.results$coefficients)[2:7],
#  -sum(as.numeric(conjoint.results$coefficients)[2:7]),
#  as.numeric(conjoint.results$coefficients)[8:13])
# clone aggregate part-worths across the individuals in the study
# set up Bayesian priors
# cc.priors <- matrix(0,nrow=length(UniqueID),ncol=length(aggregate.cc.betas))
# for(index.for.ID in seq(along=UniqueID))
# cc.priors[index.for.ID,] <- aggregate.cc.betas

colnames(cc.priors) <- c("A1B1","A1B2","A1B3","A1B4","A1B5","A1B6","A1B7",
  "A1B8","A2B1","A3B1","A4B1","A5B1","A6B1")
# note that the actual names are as follows:
AB.names <- c("Apple","Compaq","Dell","Gateway","HP","IBM","Sony","Sun",
  "Compatibility","Performance","Reliability","Learning","Price")
```

```
# set up run parameters for the MCMC
# using aggregate beta estimates to get started
truebetas <- cc.priors
cc.xcoding <- c(0,1,1,1,1,1)  # first variable categorical others continuous
cc.attlevels <- c(8,8,4,2,8,8) # test run with brand price and performance
# no constraint for order on brand so 8x8 matrix of zeroes
c1 <- matrix(0,ncol=8,nrow=8)
# compatibility is ordered higher numbers are better
# continuous attributes have 1x1 matrix representation
c2 <- matrix(1, ncol = 1, nrow = 1, byrow = TRUE)
# performance is ordered higher numbers are better
# continuous attributes have 1x1 matrix representation
c3 <- matrix(1, ncol = 1, nrow = 1, byrow = TRUE)
# reliability is ordered higher numbers are better
# continuous attributes have 1x1 matrix representation
c4 <- matrix(1, ncol = 1, nrow = 1, byrow = TRUE)
# learning has expected order... higher prices less valued
# continuous attributes have 1x1 matrix representation
c5 <- matrix(-1, ncol = 1, nrow = 1, byrow = TRUE)
# price has expected order... higher prices less valued
# continuous attributes have 1x1 matrix representation
c6 <- matrix(-1, ncol = 1, nrow = 1, byrow = TRUE)
cc.constraints <- list(c1,c2,c3,c4,c5,c6)
# controls for length of run and sampling from end of run
# cc.mcmc <- list(R = 10, use = 10) # fast trial run
# set run parameters 10000 total iterations with estimates based on last 2000
cc.mcmc <- list(R = 10000, use = 2000) # run parameters
# run options
cc.options <- list(none=FALSE, save=TRUE, keep=1)

# set up the data frame for analysis
# redefine set ids so they are a complete set 1-12 as needed for HB functions
training.data.frame$newsetid <- training.data.frame$setid
training.data.frame$newsetid <- ifelse((training.data.frame$newsetid == 16),
  3,training.data.frame$newsetid)
training.data.frame$newsetid <- ifelse((training.data.frame$newsetid == 14),
  7,training.data.frame$newsetid)
training.data.frame$newsetid <- ifelse((training.data.frame$newsetid == 13),
  11,training.data.frame$newsetid)

UnitID <- training.data.frame$id
Set <- as.integer(training.data.frame$newsetid)
Alt <- as.integer(training.data.frame$position)
X_1 <- as.integer(training.data.frame$brand) # categories by brand
X_2 <- as.integer(training.data.frame$compat)  # integer values 1 to 8
X_3 <- as.integer(training.data.frame$perform)  # integer values 1 to 4
X_4 <- as.integer(training.data.frame$reliab)  # integer values 1 to 2
X_5 <- as.integer(training.data.frame$learn)  # integer values 1 to 8
X_6 <- as.integer(training.data.frame$price)  # integer values 1 to 8
y <- as.numeric(training.data.frame$choice)  # using special response coding

cc.data <- data.frame(UnitID,Set,Alt,X_1,X_2,X_3,X_4,X_5,X_6,y)
```

```
# now for the estimation... be patient
set.seed(9999)  # for reproducible results
out <- choicemodelr(data=cc.data, xcoding = cc.xcoding,
  mcmc = cc.mcmc, options = cc.options, constraints = cc.constraints)

# out provides a list for the posterior parameter estimates
# for the runs sampled (use = 2000)

# the MCMC beta parameter estimates are traced on the screen as it runs

# individual part-worth estimates are provided in the output file RBetas.csv
# the final estimates are printed to RBetas.csv with columns labeled as
#  A1B1 = first attribute first level
#  A1B2 = first attribute second level
#  ....
#  A2B1 = second attribute first level
#  ....
# gather data from HB posterior parameter distributions
# we imposed constraints on all continuous parameters so we use betadraw.c
posterior.mean <- matrix(0, nrow = dim(out$betadraw.c)[1],
  ncol = dim(out$betadraw.c)[2])
posterior.sd <- matrix(0, nrow = dim(out$betadraw.c)[1],
  ncol = dim(out$betadraw.c)[2])
for(index.row in 1:dim(out$betadraw.c)[1])
for(index.col in 1:dim(out$betadraw.c)[2]) {
  posterior.mean[index.row,index.col] <-
    mean(out$betadraw.c[index.row,index.col,])
  posterior.sd[index.row,index.col] <-
    sd(out$betadraw.c[index.row,index.col,])
  }
# HB program uses effects coding for categorical variables and
# mean-centers continuous variables across the levels appearing in the data
# working with data for one respondent at a time we compute predicted choices
# for both the training and test choice sets
create.design.matrix <- function(input.data.frame.row) {
  xdesign.row <- numeric(12)
  if (input.data.frame.row$brand == "Apple")
    xdesign.row[1:7] <- c(1,0,0,0,0,0,0)
  if (input.data.frame.row$brand == "Compaq")
    xdesign.row[1:7] <- c(0,1,0,0,0,0,0)
  if (input.data.frame.row$brand == "Dell")
    xdesign.row[1:7] <- c(0,0,1,0,0,0,0)
  if (input.data.frame.row$brand == "Gateway")
    xdesign.row[1:7] <- c(0,0,0,1,0,0,0)
  if (input.data.frame.row$brand == "HP")
    xdesign.row[1:7] <- c(0,0,0,0,1,0,0)
  if (input.data.frame.row$brand == "IBM")
    xdesign.row[1:7] <- c(0,0,0,0,0,1,0)
  if (input.data.frame.row$brand == "Sony")
    xdesign.row[1:7] <- c(0,0,0,0,0,0,1)
  if (input.data.frame.row$brand == "Sun")
    xdesign.row[1:7] <- c(-1,-1,-1,-1,-1,-1,-1)
```

```
  xdesign.row[8] <- input.data.frame.row$compat -4.5
  xdesign.row[9] <- input.data.frame.row$perform -2.5
  xdesign.row[10] <- input.data.frame.row$reliab -1.5
  xdesign.row[11] <- input.data.frame.row$learn -4.5
  xdesign.row[12] <- input.data.frame.row$price -4.5
  t(as.matrix(xdesign.row))  # return row of design matrix
  }

# evaluate performance in the training set
training.choice.utility <- NULL  # initialize utility vector
# work with one row of respondent training data frame at a time
# create choice predictions using the individual part-worths
list.of.ids <- unique(training.data.frame$id)
for (index.for.id in seq(along=list.of.ids)) {
  this.id.part.worths <- posterior.mean[index.for.id,]
  this.id.data.frame <- subset(training.data.frame,
    subset=(id == list.of.ids[index.for.id]))
  for (index.for.profile in 1:nrow(this.id.data.frame)) {
    training.choice.utility <- c(training.choice.utility,
      create.design.matrix(this.id.data.frame[index.for.profile,]) %*%
      this.id.part.worths)
    }
  }

training.predicted.choice <-
  choice.set.predictor(training.choice.utility)
training.actual.choice <- factor(training.data.frame$choice, levels = c(0,1),
  labels = c("NO","YES"))
# look for sensitivity > 0.25 for four-profile choice sets
training.set.performance <- confusionMatrix(data = training.predicted.choice,
  reference = training.actual.choice, positive = "YES")
# report choice prediction sensitivity for training data
cat("\n\nTraining choice set sensitivity = ",
  sprintf("%1.1f",training.set.performance$byClass[1]*100)," Percent",sep="")

# evaluate performance in the test set
test.choice.utility <- NULL  # initialize utility vector
# work with one row of respondent test data frame at a time
# create choice prediction using the individual part-worths
list.of.ids <- unique(test.data.frame$id)
for (index.for.id in seq(along=list.of.ids)) {
  this.id.part.worths <- posterior.mean[index.for.id,]
  this.id.data.frame <- subset(test.data.frame,
    subset=(id == list.of.ids[index.for.id]))
  for (index.for.profile in 1:nrow(this.id.data.frame)) {
    test.choice.utility <- c(test.choice.utility,
      create.design.matrix(this.id.data.frame[index.for.profile,]) %*%
      this.id.part.worths)
    }
  }
```

```
test.predicted.choice <-
  choice.set.predictor(test.choice.utility)
test.actual.choice <- factor(test.data.frame$choice, levels = c(0,1),
  labels = c("NO","YES"))
# look for sensitivity > 0.25 for four-profile choice sets
test.set.performance <- confusionMatrix(data = test.predicted.choice,
  reference = test.actual.choice, positive = "YES")
# report choice prediction sensitivity for test data
cat("\n\nTest choice set sensitivity = ",
  sprintf("%1.1f",test.set.performance$byClass[1]*100)," Percent",sep="")

# Suggestions for the student:
# Having demonstrated the predictive power of the HB model...
# return to the complete set of 16 choice sets to obtain
# part-worths for individuals based upon the complete survey
# (the next program will provide guidance on how to do this).
# After estimating part-worths for individuals, average across
# individuals to obtain an aggregate profile of conjoint measures.
# Standardize the aggregate part-worths and display them
# on a spine chart using the spine chart plotting utility
# provided in the appendix of code and utilities.
# Interpret the spine chart, compare attribute importance values,
# compare the brands, compute brand equity for each brand
# relative to each of the other brands in the study.
```

***Exhibit 10.2.*** *Analyzing Consumer Preferences and Building a Market Simulation (R)*

```
# Analyzing Consumer Preferences and Building a Market Simulation (R)

# having demonstrated the predictive power of the HB model...
# we now return to the complete set of 16 choice sets to obtain
# individual-level part-worths for further analysis
# analysis guided by ternary model of consumer preference and market response
# brand loyalty... price sensitivity... and feature focus... are key aspects
# to consider in determining pricing policy

library(lattice)  # package for lattice graphics
library(vcd)  # graphics package with mosaic plots for mosaic and ternary plots
library(ggplot2)  # package ggplot implements Grammar of Graphics approach
library(ChoiceModelR)  # for Hierarchical Bayes Estimation
library(caret)  # for confusion matrix... evaluation of choice set predictions

# load split-plotting utilities for work with ggplot
load("mtpa_split_plotting_utilities.Rdata")
# load market simulation utilities
load(file="mspa_market_simulation_utilities.RData")

# read in the data from a case study in computer choice.
complete.data.frame <- read.csv("computer_choice_study.csv")
# we employed a training-and-test regimen in previous research work
# here we will be using the complete data from the computer choice study
working.data.frame <- complete.data.frame

# user-defined function for plotting descriptive attribute names
effect.name.map <- function(effect.name) {
  if(effect.name=="brand") return("Manufacturer/Brand")
  if(effect.name=="compat") return("Compatibility with Windows 95")
  if(effect.name=="perform") return("Performance")
  if(effect.name=="reliab") return("Reliability")
  if(effect.name=="learn") return("Learning Time (4 to 32 hours)")
  if(effect.name=="price") return("Price ($1,000 to $2,750)")
  }
print.digits <- 2
# user-defined function for printing conjoint measures
if (print.digits == 2)
  pretty.print <- function(x) {sprintf("%1.2f",round(x,digits = 2))}
if (print.digits == 3)
  pretty.print <- function(x) {sprintf("%1.3f",round(x,digits = 3))}

# set up sum contrasts for effects coding
options(contrasts=c("contr.sum","contr.poly"))

UniqueID <- unique(working.data.frame$id)
# set up zero priors
cc.priors <- matrix(0,nrow=length(UniqueID),ncol=13)
colnames(cc.priors) <- c("A1B1","A1B2","A1B3","A1B4","A1B5","A1B6","A1B7",
  "A1B8","A2B1","A3B1","A4B1","A5B1","A6B1")
```

```
# note that the actual names are as follows:
AB.names <- c("Apple","Compaq","Dell","Gateway","HP","IBM","Sony","Sun",
  "Compatibility","Performance","Reliability","Learning","Price")

# set up run parameters for the MCMC
# using aggregate beta estimates to get started
truebetas <- cc.priors
cc.xcoding <- c(0,1,1,1,1,1)  # first variable categorical others continuous
cc.attlevels <- c(8,8,4,2,8,8) # test run with brand price and performance
# no constraint for order on brand so 8x8 matrix of zeroes
c1 <- matrix(0,ncol=8,nrow=8)
# compatibility is ordered higher numbers are better
# continuous attributes have 1x1 matrix representation
c2 <- matrix(1, ncol = 1, nrow = 1, byrow = TRUE)
# performance is ordered higher numbers are better
# continuous attributes have 1x1 matrix representation
c3 <- matrix(1, ncol = 1, nrow = 1, byrow = TRUE)
# reliability is ordered higher numbers are better
# continuous attributes have 1x1 matrix representation
c4 <- matrix(1, ncol = 1, nrow = 1, byrow = TRUE)
# learning has expected order... higher prices less valued
# continuous attributes have 1x1 matrix representation
c5 <- matrix(-1, ncol = 1, nrow = 1, byrow = TRUE)
# price has expected order... higher prices less valued
# continuous attributes have 1x1 matrix representation
c6 <- matrix(-1, ncol = 1, nrow = 1, byrow = TRUE)
cc.constraints <- list(c1,c2,c3,c4,c5,c6)
# controls for length of run and sampling from end of run
# cc.mcmc <- list(R = 10, use = 10) # fast trial run
# set run parameters 10000 total iterations with estimates based on last 2000
cc.mcmc <- list(R = 10000, use = 2000) # run parameters
# run options
cc.options <- list(none=FALSE, save=TRUE, keep=1)

# set up the data frame for analysis
UnitID <- working.data.frame$id
Set <- as.integer(working.data.frame$setid)
Alt <- as.integer(working.data.frame$position)
X_1 <- as.integer(working.data.frame$brand) # categories by brand
X_2 <- as.integer(working.data.frame$compat)  # integer values 1 to 8
X_3 <- as.integer(working.data.frame$perform)  # integer values 1 to 4
X_4 <- as.integer(working.data.frame$reliab)  # integer values 1 to 2
X_5 <- as.integer(working.data.frame$learn)  # integer values 1 to 8
X_6 <- as.integer(working.data.frame$price)  # integer values 1 to 8
y <- as.numeric(working.data.frame$choice)  # using special response coding

cc.data <- data.frame(UnitID,Set,Alt,X_1,X_2,X_3,X_4,X_5,X_6,y)

# the estimation begins here... be patient
set.seed(9999)  # for reproducible results
out <- choicemodelr(data=cc.data, xcoding = cc.xcoding,
  mcmc = cc.mcmc, options = cc.options, constraints = cc.constraints)
```

```
# out provides a list for the posterior parameter estimates
# for the runs sampled (use = 2000)
# the MCMC beta parameter estimates are traced on the screen as it runs
# individual part-worth estimates are provided in the output file RBetas.csv
# the final estimates are printed to RBetas.csv with columns labeled as
#  A1B1 = first attribute first level
#  A1B2 = first attribute second level
#  ....
#  A2B1 = second attribute first level
#  ....
# gather data from HB posterior parameter distributions
# we imposed constraints on all continuous parameters so we use betadraw.c
posterior.mean <- matrix(0, nrow = dim(out$betadraw.c)[1],
  ncol = dim(out$betadraw.c)[2])
posterior.sd <- matrix(0, nrow = dim(out$betadraw.c)[1],
  ncol = dim(out$betadraw.c)[2])
for(index.row in 1:dim(out$betadraw.c)[1])
for(index.col in 1:dim(out$betadraw.c)[2]) {
  posterior.mean[index.row,index.col] <-
    mean(out$betadraw.c[index.row,index.col,])
  posterior.sd[index.row,index.col] <-
    sd(out$betadraw.c[index.row,index.col,])
  }
# HB program uses effects coding for categorical variables and
# mean-centers continuous variables across the levels appearing in the data
# working with data for one respondent at a time we compute predicted choices
# for the full set of consumer responses
create.design.matrix <- function(input.data.frame.row) {
  xdesign.row <- numeric(12)
  if (input.data.frame.row$brand == "Apple")
    xdesign.row[1:7] <- c(1,0,0,0,0,0,0)
  if (input.data.frame.row$brand == "Compaq")
    xdesign.row[1:7] <- c(0,1,0,0,0,0,0)
  if (input.data.frame.row$brand == "Dell")
    xdesign.row[1:7] <- c(0,0,1,0,0,0,0)
  if (input.data.frame.row$brand == "Gateway")
    xdesign.row[1:7] <- c(0,0,0,1,0,0,0)
  if (input.data.frame.row$brand == "HP")
    xdesign.row[1:7] <- c(0,0,0,0,1,0,0)
  if (input.data.frame.row$brand == "IBM")
    xdesign.row[1:7] <- c(0,0,0,0,0,1,0)
  if (input.data.frame.row$brand == "Sony")
    xdesign.row[1:7] <- c(0,0,0,0,0,0,1)
  if (input.data.frame.row$brand == "Sun")
    xdesign.row[1:7] <- c(-1,-1,-1,-1,-1,-1,-1)
  xdesign.row[8] <- input.data.frame.row$compat -4.5
  xdesign.row[9] <- input.data.frame.row$perform -2.5
  xdesign.row[10] <- input.data.frame.row$reliab -1.5
  xdesign.row[11] <- input.data.frame.row$learn -4.5
  xdesign.row[12] <- input.data.frame.row$price -4.5
  t(as.matrix(xdesign.row))  # return row of design matrix
  }
```

```
# evaluate performance in the full set of consumer responses
working.choice.utility <- NULL  # initialize utility vector
# work with one row of respondent training data frame at a time
# create choice prediction using the individual part-worths
list.of.ids <- unique(working.data.frame$id)
for (index.for.id in seq(along=list.of.ids)) {
  this.id.part.worths <- posterior.mean[index.for.id,]
  this.id.data.frame <- subset(working.data.frame,
    subset=(id == list.of.ids[index.for.id]))
  for (index.for.profile in 1:nrow(this.id.data.frame)) {
    working.choice.utility <- c(working.choice.utility,
      create.design.matrix(this.id.data.frame[index.for.profile,]) %*%
      this.id.part.worths)
    }
  }

working.predicted.choice <-
  choice.set.predictor(working.choice.utility)
working.actual.choice <- factor(working.data.frame$choice, levels = c(0,1),
  labels = c("NO","YES"))
# look for sensitivity > 0.25 for four-profile choice sets
working.set.performance <- confusionMatrix(data = working.predicted.choice,
  reference = working.actual.choice, positive = "YES")
# report choice prediction sensitivity for the full data
cat("\n\nFull data set choice set sensitivity = ",
  sprintf("%1.1f",working.set.performance$byClass[1]*100)," Percent",sep="")
#
# results: Full data set choice set sensitivity = 89.1 Percent
#

# to continue with our analysis of consumer preferences...
# we build a data frame for the consumers with the full set of eight brands
ID <- unique(working.data.frame$id)
Apple <- posterior.mean[,1]
Compaq <- posterior.mean[,2]
Dell <- posterior.mean[,3]
Gateway <- posterior.mean[,4]
HP <- posterior.mean[,5]
IBM <- posterior.mean[,6]
Sony <- posterior.mean[,7]
Sun <- -1 * (Apple + Compaq + Dell + Gateway + HP + IBM + Sony)
Compatibility <- posterior.mean[,8]
Performance <- posterior.mean[,9]
Reliability <- posterior.mean[,10]
Learning <- posterior.mean[,11]
Price <- posterior.mean[,12]

# creation of data frame for analysis of consumer preferences and choice
# starting with individual-level part-worths... more to be added shortly
id.data <- data.frame(ID,Apple,Compaq,Dell,Gateway,HP,IBM,Sony,Sun,
  Compatibility,Performance,Reliability,Learning,Price)
```

```
# compute attribute importance values for each attribute
id.data$brand.range <- numeric(nrow(id.data))
id.data$compatibility.range <- numeric(nrow(id.data))
id.data$performance.range <- numeric(nrow(id.data))
id.data$reliability.range <- numeric(nrow(id.data))
id.data$learning.range <- numeric(nrow(id.data))
id.data$price.range <- numeric(nrow(id.data))
id.data$sum.range <- numeric(nrow(id.data))
id.data$brand.importance <- numeric(nrow(id.data))
id.data$compatibility.importance <- numeric(nrow(id.data))
id.data$performance.importance <- numeric(nrow(id.data))
id.data$reliability.importance <- numeric(nrow(id.data))
id.data$learning.importance <- numeric(nrow(id.data))
id.data$price.importance <- numeric(nrow(id.data))

for(id in seq(along=id.data$ID)) {
  id.data$brand.range[id] <- max(id.data$Apple[id],
    id.data$Compaq[id],id.data$Dell[id],
    id.data$Gateway[id],id.data$HP[id],
    id.data$IBM[id],id.data$Sony[id],
    id.data$Sun[id]) -
    min(id.data$Apple[id],
    id.data$Compaq[id],id.data$Dell[id],
    id.data$Gateway[id],id.data$HP[id],
    id.data$IBM[id],id.data$Sony[id],
    id.data$Sun[id])

  id.data$compatibility.range[id] <- abs(8*id.data$Compatibility[id])
  id.data$performance.range[id] <- abs(4*id.data$Performance[id])
  id.data$reliability.range[id] <- abs(2*id.data$Reliability[id])
  id.data$learning.range[id] <- abs(8*id.data$Learning[id])
  id.data$price.range[id] <-  abs(8*id.data$Price[id])

  id.data$sum.range[id] <- id.data$brand.range[id] +
    id.data$compatibility.range[id] +
    id.data$performance.range[id] +
    id.data$reliability.range[id] +
    id.data$learning.range[id] +
    id.data$price.range[id]

  id.data$brand.importance[id] <-
    id.data$brand.range[id]/id.data$sum.range[id]
  id.data$compatibility.importance[id] <-
    id.data$compatibility.range[id]/id.data$sum.range[id]
  id.data$performance.importance[id] <-
    id.data$performance.range[id]/id.data$sum.range[id]
  id.data$reliability.importance[id] <-
    id.data$reliability.range[id]/id.data$sum.range[id]
  id.data$learning.importance[id] <-
    id.data$learning.range[id]/id.data$sum.range[id]
  id.data$price.importance[id] <-
    id.data$price.range[id]/id.data$sum.range[id]
```

```
# feature importance relates to the most important product feature
# considering product features as not brand and not price
  id.data$feature.importance[id] <- max(id.data$compatibility.importance[id],
    id.data$performance.importance[id],
    id.data$reliability.importance[id],
    id.data$learning.importance[id])
  }

# identify each individual's top brand defining top.brand factor variable
id.data$top.brand <- integer(nrow(id.data))
for(id in seq(along=id.data$ID)) {
  brand.index <- 1:8
  brand.part.worth <- c(id.data$Apple[id],id.data$Compaq[id],
    id.data$Dell[id],id.data$Gateway[id],id.data$HP[id],id.data$IBM[id],
    id.data$Sony[id],id.data$Sun[id])
  temp.data <- data.frame(brand.index,brand.part.worth)
  temp.data <- temp.data[sort.list(temp.data$brand.part.worth, decreasing = TRUE),]
  id.data$top.brand[id] <- temp.data$brand.index[1]
  }
id.data$top.brand <- factor(id.data$top.brand, levels = 1:8,
  labels = c("Apple","Compaq","Dell","Gateway",
  "HP","IBM","Sony","Sun"))

# note that the standard importance measures from conjoint methods are
# ipsative... their sum is always 1 for proportions or 100 for percentages
# this has advantages for triplots (ternary plots) but because importance
# is so dependent upon the levels of attributes, it has significant
# disadvantages as well... so we consider a relative-value-based measure
# lets us define an alternative to importance called "attribute value"

# compute "attribute value" relative to the consumer group
# it is a standardized measure... let "attribute value" be mean 50 sd 10
# here are user-defined functions to use to obtain "value"

standardize <- function(x) {
# standardize x so it has mean zero and standard deviation 1
  (x - mean(x))/sd(x)
  }
compute.value <- function(x) {
# rescale x so it has the same mean and standard deviation as y
  standardize(x) * 10 + 50
  }

id.data$brand.value <- compute.value(id.data$brand.range)
id.data$compatibility.value <- compute.value(id.data$compatibility.range)
id.data$performance.value <- compute.value(id.data$performance.range)
id.data$reliability.value <- compute.value(id.data$reliability.range)
id.data$learning.value <- compute.value(id.data$learning.range)
id.data$price.value <- compute.value(id.data$price.range)
```

```
# identify each individual's top value using computed relative attribute values
id.data$top.attribute <- integer(nrow(id.data))
for(id in seq(along=id.data$ID)) {
  attribute.index <- 1:6
  attribute.value <- c(id.data$brand.value[id],id.data$compatibility.value[id],
    id.data$performance.value[id],id.data$reliability.value[id],
    id.data$learning.value[id],id.data$price.value[id])
  temp.data <- data.frame(attribute.index,attribute.value)
  temp.data <-
    temp.data[sort.list(temp.data$attribute.value, decreasing = TRUE),]
  id.data$top.attribute[id] <- temp.data$attribute.index[1]
  }
id.data$top.attribute <- factor(id.data$top.attribute, levels = 1:6,
  labels = c("Brand","Compatibility","Performance","Reliability",
  "Learning","Price"))


# mosaic plot of joint frequencies top ranked brand by top value
pdf(file="fig_price_top_top_mosaic_plot.pdf", width = 8.5, height = 11)
  mosaic( ~ top.brand + top.attribute, data = id.data,
  highlighting = "top.attribute",
  highlighting_fill =
    c("blue", "white", "green","lightgray","magenta","black"),
  labeling_args =
  list(set_varnames = c(top.brand = "", top.attribute = ""),
  rot_labels = c(left = 90, top = 45),
  pos_labels = c("center","center"),
  just_labels = c("left","center"),
  offset_labels = c(0.0,0.0)))
dev.off()


# an alternative representation that is often quite useful in pricing studies
# is a triplot/ternary plot with three features identified for each consumer
# using the idea from importance caluclations we now use price, brand, and
# feature importance measures to obtain data for three-way plots
# as the basis for three relative measures, which we call brand.loyalty,
# price.sensitivity, and feature_focus...


id.data$brand.loyalty <- numeric(nrow(id.data))
id.data$price.sensitivity <- numeric(nrow(id.data))
id.data$feature.focus <- numeric(nrow(id.data))
for(id in seq(along=id.data$ID)) {
  sum.importances <- id.data$brand.importance[id] +
  id.data$price.importance[id] +
  id.data$feature.importance[id]  # less than 1.00 feature is an average
  id.data$brand.loyalty[id] <- id.data$brand.importance[id]/sum.importances
  id.data$price.sensitivity[id] <- id.data$price.importance[id]/sum.importances
  id.data$feature.focus[id] <- id.data$feature.importance[id]/sum.importances
  }
```

```
# ternary model of consumer response... the plot
pdf("fig_price_ternary_three_brands.pdf", width = 11, height = 8.5)
ternaryplot(id.data[,c("brand.loyalty","price.sensitivity","feature.focus")],
dimnames = c("Brand Loyalty","Price Sensitivity","Feature Focus"),
prop_size = ifelse((id.data$top.brand == "Apple"), 0.8,
            ifelse((id.data$top.brand == "Dell"),0.7,
            ifelse((id.data$top.brand == "HP"),0.7,0.5))),
pch = ifelse((id.data$top.brand == "Apple"), 20,
      ifelse((id.data$top.brand == "Dell"),17,
      ifelse((id.data$top.brand == "HP"),15,1))),
col = ifelse((id.data$top.brand == "Apple"), "red",
      ifelse((id.data$top.brand == "Dell"),"mediumorchid4",
      ifelse((id.data$top.brand == "HP"),"blue","darkblue"))),
grid_color = "#626262",
bg = "#E6E6E6",
dimnames_position = "corner", main = ""
)
grid_legend(0.725, 0.8, pch = c(20, 17, 15, 1),
col = c("red", "mediumorchid4", "blue", "darkblue"),
c("Apple", "Dell", "HP", "Other"), title = "Top-Ranked Brand")
dev.off()
# another way of looking at these data is to employ comparative densities
# for the three selected brands: Apple, Dell, and HP
# using those individual how selected these as the top brand
selected.brands <- c("Apple","Dell","HP")
selected.data <- subset(id.data, subset = (top.brand %in% selected.brands))
# plotting objects for brand.loyalty, price.sensitivity, and feature.focus
# create these three objects and then plot them together on one page
pdf("fig_price_density_three_brands.pdf", width = 8.5, height = 11)
first.object <- ggplot(selected.data,
  aes(x = brand.loyalty, fill = top.brand))  +
  labs(x = "Brand Loyalty",
       y = "f(x)") +
  theme(axis.title.y = element_text(angle = 0, face = "italic", size = 10)) +
  geom_density(alpha = 0.4) +
  coord_fixed(ratio = 1/15) +
  theme(legend.position = "none") +
  scale_fill_manual(values = c("red","white","blue"),
    guide = guide_legend(title = NULL)) +
  scale_x_continuous(limits = c(0,1)) +
  scale_y_continuous(limits = c(0,5))
second.object <- ggplot(selected.data,
  aes(x = price.sensitivity, fill = top.brand))  +
  labs(x = "Price Sensitivity",
       y = "f(x)") +
  theme(axis.title.y = element_text(angle = 0, face = "italic", size = 10)) +
  geom_density(alpha = 0.4) +
  coord_fixed(ratio = 1/15) +
  theme(legend.position = "none") +
  scale_fill_manual(values = c("red","white","blue"),
    guide = guide_legend(title = NULL)) +
  scale_x_continuous(limits = c(0,1))  +
  scale_y_continuous(limits = c(0,5))
```

```
third.object <- ggplot(selected.data,
  aes(x = feature.focus, fill = top.brand))  +
  labs(x = "Feature Focus",
       y = "f(x)") +
  theme(axis.title.y = element_text(angle = 0, face = "italic", size = 10)) +
  geom_density(alpha = 0.4) +
  coord_fixed(ratio = 1/15) +
  theme(legend.position = "bottom") +
  scale_fill_manual(values = c("red","white","blue"),
    guide = guide_legend(title = NULL)) +
  scale_x_continuous(limits = c(0,1)) +
  scale_y_continuous(limits = c(0,5))

three.part.ggplot.print.with.margins(ggfirstplot.object.name = first.object,
  ggsecondplot.object.name = second.object,
  ggthirdplot.object.name = third.object,
  left.margin.pct=5,right.margin.pct=5,
  top.margin.pct=10,bottom.margin.pct=9,
  first.plot.pct=25,second.plot.pct=25,
  third.plot.pct=31)
dev.off()

# to what extent are consumers open to switching from one brand to another
# can see this trough parallel coordinates plots for the brand part-worths
pdf(file = "fig_price_parallel_coordinates_individuals.pdf",
  width = 8.5, height = 11)
parallelplot(~selected.data[,c("Apple","Compaq","Dell","Gateway",
  "HP","IBM","Sony","Sun")] | top.brand, selected.data, layout = c (3,1))
dev.off()

# these get a little messy or cluttered...
# more easily interpreted are parallel coordinate plots of mean part-worths
# for brand part-worth columns and aggreate by top brand (Apple, Dell, or HP)
brands.data <- aggregate(x = selected.data[,2:9],
  by = selected.data[29], mean)

pdf(file = "fig_price_parallel_coordinates_groups.pdf",
  width = 8.5, height = 11)
parallelplot(~brands.data[,c("Apple","Compaq","Dell","Gateway",
  "HP","IBM","Sony","Sun")] | top.brand, brands.data, layout = c (3,1),
   lwd = 3, col = "mediumorchid4")
dev.off()

# market simulation for hypothetical set of products in the marketplace
# suppose we work for Apple and we focus upon a market with three
# competitors: Dell, Gateway, and HP.... we define the products in the
# market using values from the computer choice study just as we did
# in fitting the HB model... we create the simuation input data frame
# and use the previously designed function create.design.matrix
# along with simulation utility functions
```

```
# first product in market is Dell Computer defined as follows:
brand <- "Dell"
compat <- 8  # 100 percent compatibility
perform <- 4 # four times as fast as earlier generation system
reliab <- 2  # Less likely to fail
learn <- 4  # 16 hours to learn
price <- 4  # $1750
dell.competitor <- data.frame(brand,compat,perform,reliab,learn,price)

# second product in market is Gateway defined as follows:
brand <- "Gateway"
compat <- 6  # 90 percent compatibility
perform <- 2 # twice as fast as earlier generation system
reliab <- 1  # just as likely to fail
learn <- 2  # 8 hours to learn
price <- 2  # $1250
gateway.competitor <- data.frame(brand,compat,perform,reliab,learn,price)

# third product in market is HP defined as follows:
brand <- "HP"
compat <- 6  # 90 percent compatibility
perform <- 3 # three times as fast as earlier generation system
reliab <- 2  # less likely to fail
learn <- 2  # 8 hours to learn
price <- 3  # $1500
hp.competitor <- data.frame(brand,compat,perform,reliab,learn,price)

# Apple product has price varying across many choice sets:
brand <- "Apple"
compat <- 5  # 85 percent compatibility
perform <- 4 # four times as fast as earlier generation system
reliab <- 2  # less likely to fail
learn <- 1  # 4 hours to learn
price <- 1  # $1000 Apple price in first choice set
apple1000 <- data.frame(brand,compat,perform,reliab,learn,price)
price <- 2  # $1250 Apple price in second choice set
apple1250 <- data.frame(brand,compat,perform,reliab,learn,price)
price <- 3  # $1500 Apple price in third choice set
apple1500 <- data.frame(brand,compat,perform,reliab,learn,price)
price <- 4  # $1750 Apple price in fourth choice set
apple1750 <- data.frame(brand,compat,perform,reliab,learn,price)
price <- 5  # $2000 Apple price in fifth choice set
apple2000 <- data.frame(brand,compat,perform,reliab,learn,price)
price <- 6  # $2250 Apple price in sixth choice set
apple2250 <- data.frame(brand,compat,perform,reliab,learn,price)
price <- 7  # $2500 Apple price in seventh choice set
apple2500 <- data.frame(brand,compat,perform,reliab,learn,price)
price <- 8  # $2750 Apple price in eighth choice set
apple2750 <- data.frame(brand,compat,perform,reliab,learn,price)

# the competitive products are fixed from one choice set to the next
competition <- rbind(dell.competitor,gateway.competitor,hp.competitor)
```

```
# build the simulation choice sets with Apple varying across choice sets
simulation.choice.sets <-
  rbind(competition, apple1000, competition, apple1250,
  competition, apple1500, competition, apple1750, competition, apple2000,
  competition, apple2250, competition, apple2500, competition, apple2750)

# add set id to the simuation.choice sets for ease of analysis
setid <- NULL
for(index.for.set in 1:8) setid <- c(setid,rep(index.for.set, times = 4))
simulation.choice.sets <- cbind(setid,simulation.choice.sets)

# list the simulation data frame to check it out
print(simulation.choice.sets)

# create the simulation data frame for all individuals in the study
# by cloning the simulation choice sets for each individual
simulation.data.frame <- NULL  # initialize
list.of.ids <- unique(working.data.frame$id)  # ids from original study
for (index.for.id in seq(along=list.of.ids)) {
  id <- rep(list.of.ids[index.for.id], times = nrow(simulation.choice.sets))
  this.id.data <- cbind(data.frame(id),simulation.choice.sets)
  simulation.data.frame <- rbind(simulation.data.frame, this.id.data)
  }

# check structure of simulation data frame
print(str(simulation.data.frame))
print(head(simulation.data.frame))
print(tail(simulation.data.frame))

# using create.design.matrix function we evalutate the utility
# of each product profile in each choice set for each individual
# in the study... HP part-worths are used for individuals
# this code is similar to that used previously for original data
# from the computer choice study... except now we have simulation data
simulation.choice.utility <- NULL  # initialize utility vector
# work with one row of respondent training data frame at a time
# create choice prediction using the individual part-worths
list.of.ids <- unique(simulation.data.frame$id)
simulation.choice.utility <- NULL # intitialize
for (index.for.id in seq(along=list.of.ids)) {
  this.id.part.worths <- posterior.mean[index.for.id,]
  this.id.data.frame <- subset(simulation.data.frame,
    subset=(id == list.of.ids[index.for.id]))
  for (index.for.profile in 1:nrow(this.id.data.frame)) {
    simulation.choice.utility <- c(simulation.choice.utility,
      create.design.matrix(this.id.data.frame[index.for.profile,]) %*%
      this.id.part.worths)
    }
  }
# use choice.set.predictor function to predict choices in market simulation
simulation.predicted.choice <-
  choice.set.predictor(simulation.choice.utility)
```

```
# add simulation predictions to simulation data frame for analysis
# of the results from the market simulation
simulation.analysis.data.frame <-
  cbind(simulation.data.frame,simulation.predicted.choice)

# contingency table shows results of market simulation
with(simulation.analysis.data.frame,
  table(setid,brand,simulation.predicted.choice))

# summary table of preference shares
YES.data.frame <- subset(simulation.analysis.data.frame,
  subset = (simulation.predicted.choice == "YES"), select = c("setid","brand"))

# check YES.data.frame to see that it reproduces the information
# from the contingency table
print(with(YES.data.frame,table(setid,brand)))

# create market share estimates by dividing by number of individuals
# no need for a spreasheet program to work with tables
table.work <- with(YES.data.frame,as.matrix(table(setid,brand)))
table.work <- table.work[,c("Apple","Dell","Gateway","HP")] # order columns
table.work <- round(100 *table.work/length(list.of.ids), digits = 1)  # percent
Apple.Price <- c(1000,1250,1500,1750,2000,2250,2500,2750)  # new column
table.work <- cbind(Apple.Price,table.work) # add price column to table
print(table.work)  # print the market/preference share table

# data visualization of market/preference share estimates from the simulation
mosaic.data.frame <- YES.data.frame
mosaic.data.frame$setid <- factor(mosaic.data.frame$setid, levels = 1:8,
  labels = c("$1,000","$1,250","$1,500","$1,750",
  "$2,000","$2,250","2,500","$2,750"))
# mosaic plot of joint frequencies from the market simulation
# length/width of the tiles in each row reflects market share
# rows relate to Apple prices... simulation choice sets
pdf(file="fig_price_market_simulation_results.pdf", width = 8.5, height = 11)
  mosaic( ~ setid + brand, data = mosaic.data.frame,
  highlighting = "brand",
  highlighting_fill =
    c("mediumorchid4", "green", "blue","red"),
  labeling_args =
  list(set_varnames = c(brand = "", setid = "Price of Apple Computer"),
  rot_labels = c(left = 90, top = 45),
  pos_labels = c("center","center"),
  just_labels = c("left","center"),
  offset_labels = c(0.0,0.0)))
dev.off()

# Suggestions for the student:
# Try setting up your own market simulation study with a hypothetical client
# and competitive products. Vary prices on your client's product and see
# what happens to preference shares. Try alternative model specifications.
```

# 11 Utilizing Social Networks

Economics Teacher: "Bueller? Bueller? Bueller? Bueller?"

Simone Adamley: "He's sick. My best friend's sister's boyfriend's brother's girlfriend heard from this guy who knows this kid who's going with a girl who saw Ferris pass-out at 31 Flavors last night. I guess it's pretty serious."

—BEN STEIN AS ECONOMICS TEACHER AND KRISTY SWANSON AS SIMONE ADAMLEY IN *Ferris Bueller's Day Off* (1986)

Part of my exercise regimen is to walk between five and six in the morning. Except for an old man I often see, I have the sidewalks to myself. As far as I can tell, he lives in a nearby park, or at least that is where he spends most of his time. When I pass him on the sidewalk, I wave, and he waves back. That is the extent of our social interaction. It is as though we are saying, "We are here sharing this space in the city, and that is okay."

Couples marry, and we trace genealogy. Contracts bind the party of the first part to the party of the second part, whatever those parties and parts happen to be. There are mergers and acquisitions, service providers and clients, students and teachers, friends and lovers. There are groups of people attending events, congregations, and conventions. Some connections are strong, others less so. It is not always easy to understand why people do what they do, but we can at the very least observe their relationships with one another.

People's motives play out in communities of association, which we represent as social networks. And it is through social networks that we can understand product adoption across communities.

To review social network analysis methods relevant to marketing data science, we begin with mathematical models of networks, making predictions about network phenomena and studying relationships across network measures. Three types of models are of special interest: random graph, preferential attachment, and small-world network models.

The first mathematical model of interest, developed by Paul Erdös and Alfred Rényi (1959, 1960, 1961), lays the foundation for all models to follow. A *random graph* is a set of nodes connected by links in a purely random fashion. Figure 11.1 uses a circular layout to show a random graph with fifty nodes and one hundred links. No discernable pattern is detected in this network. Links are randomly associated with nodes, which is to say that all pairs of nodes are equally likely to be linked.

A second mathematical model for networks is the *preferential attachment* model and follows from the work of Barabási and Albert (1999). This model begins as a random graph and adds new links in a manner that gives preference to nodes that are already well connected (nodes with higher degree centrality).

Figure 11.2 shows a preferential attachment network in a circular layout. Note that certain nodes have many more links than other nodes. In describing preferential attachment models, we can think of the saying, "The rich get richer, and the poor get poorer."

Networks arising from preferential attachment are sometimes called *scale-free* or *long-tail* networks. In the literature of network science, there have been many variations on the preferential attachment model (Albert and Barabási 2002). These models are especially useful for representing online networks.

A third mathematical model is especially important in representing social networks. This is the *small-world network* (Watts and Strogatz 1998). The small-world model defines a structure in which many nodes or actors are connected to nearby neighbors and some nodes are linked to nodes that are not nearby.

*Figure 11.1.* A Random Graph

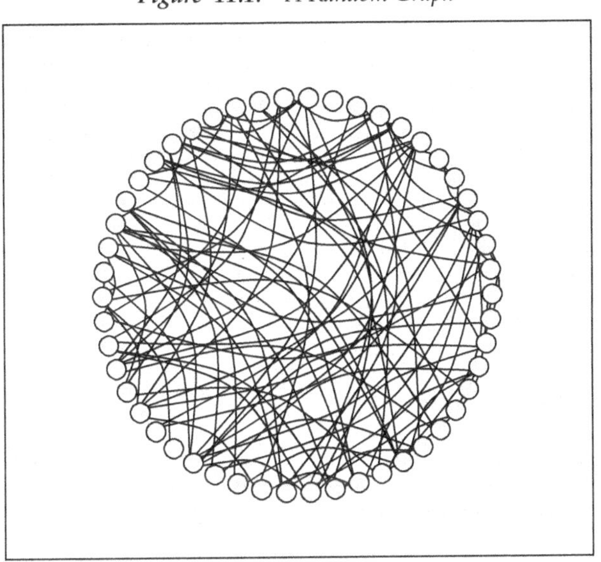

*Figure 11.2.* Network Resulting from Preferential Attachment

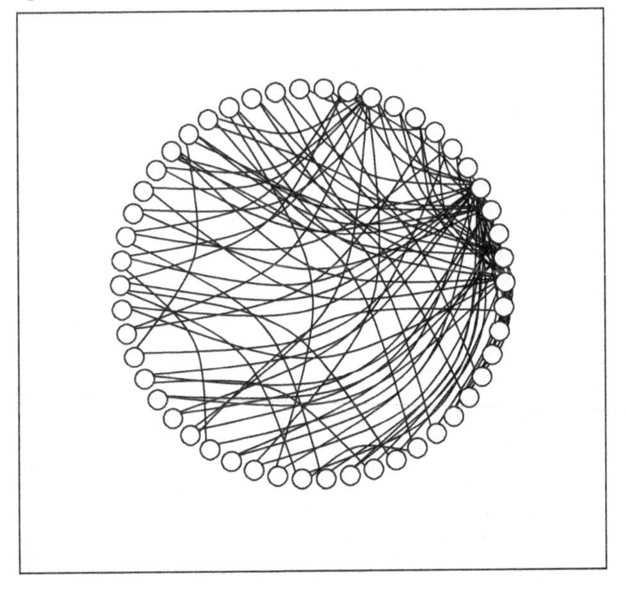

*Figure 11.3. Building the Baseline for a Small World Network*

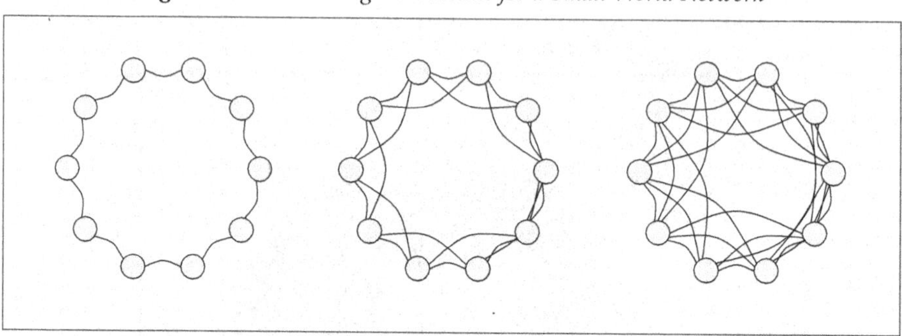

Small-world networks have been studied extensively in social psychology, sociology, and network science (Milgram 1967; Travers and Milgram 1969; Watts 1999; Schnettler 2009). To generate an undirected small-world network, we first specify the number of nodes and the size of the small-world neighborhood. Then we perturb the small-world by rewiring a selected proportion of the nodes. The probability of random attachment or rewiring, like the size of the neighborhood, is a parameter defining the network.

Figure 11.3 shows baseline small-world networks prior to the rewiring of nodes. Using a circular layout, we see a ten-node network at the left with nodes associated only with adjacent nodes. Each node has two links. As neighborhoods grow in size, moving from left to right in the figure, the number of links between nodes grows. The network at the center has nodes linking to four other nodes, and the network at the right has nodes linking to six other nodes. The defining property of a small-world network is that many nodes are associated with nearby nodes.

Moving from a baseline small-world network to an actual small-world network involves rewiring. We select links to rewire, detaching them from one of the nodes and reattaching to another node. The selection of links is conducted at random.

Again using a circular layout, we show a small-world network with fifty nodes and one hundred links in figure 11.4. Half of the links in this network are associated with nearest neighbors, the other half are associated with randomly selected nodes.

*Figure 11.4.* *A Small-World Network*

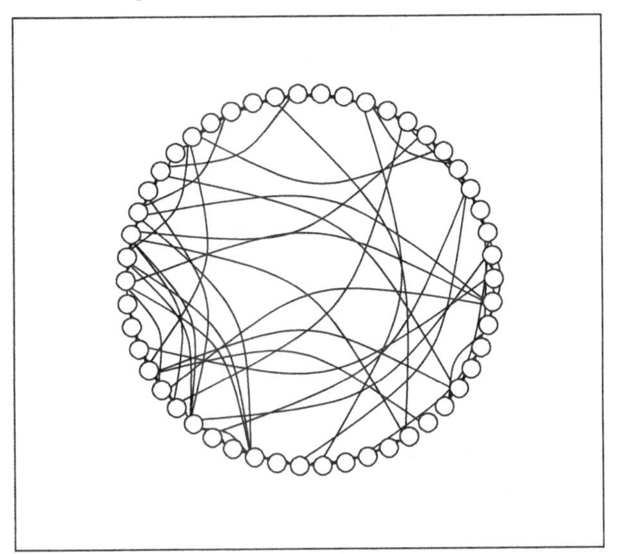

Figure 11.5 shows the degree distributions of networks generated from random graph, small-world, and preferential attachment models. Each of the models used to generate the figure contained about fifty nodes and one hundred links. Notice how small-world networks have a preponderance of low-degree nodes, compared with random graphs. Notice how the the preferential attachment model has a highly skewed or long-tail degree distribution.

We can use a mathematical model as a starting point in the statistical simulation of networks. They also have value in making statistical inferences about network parameters and predictions about the future state of networks.

Before using a particular network model for inference or prediction, however, we may want to pose initial questions about the suitability of the model for the problem under investigation. Is it reasonable to believe that sample data come from a random graph, preferential attachment, small-world, or other mathematical model?

*Figure 11.5.* *Degree Distributions for Network Models*

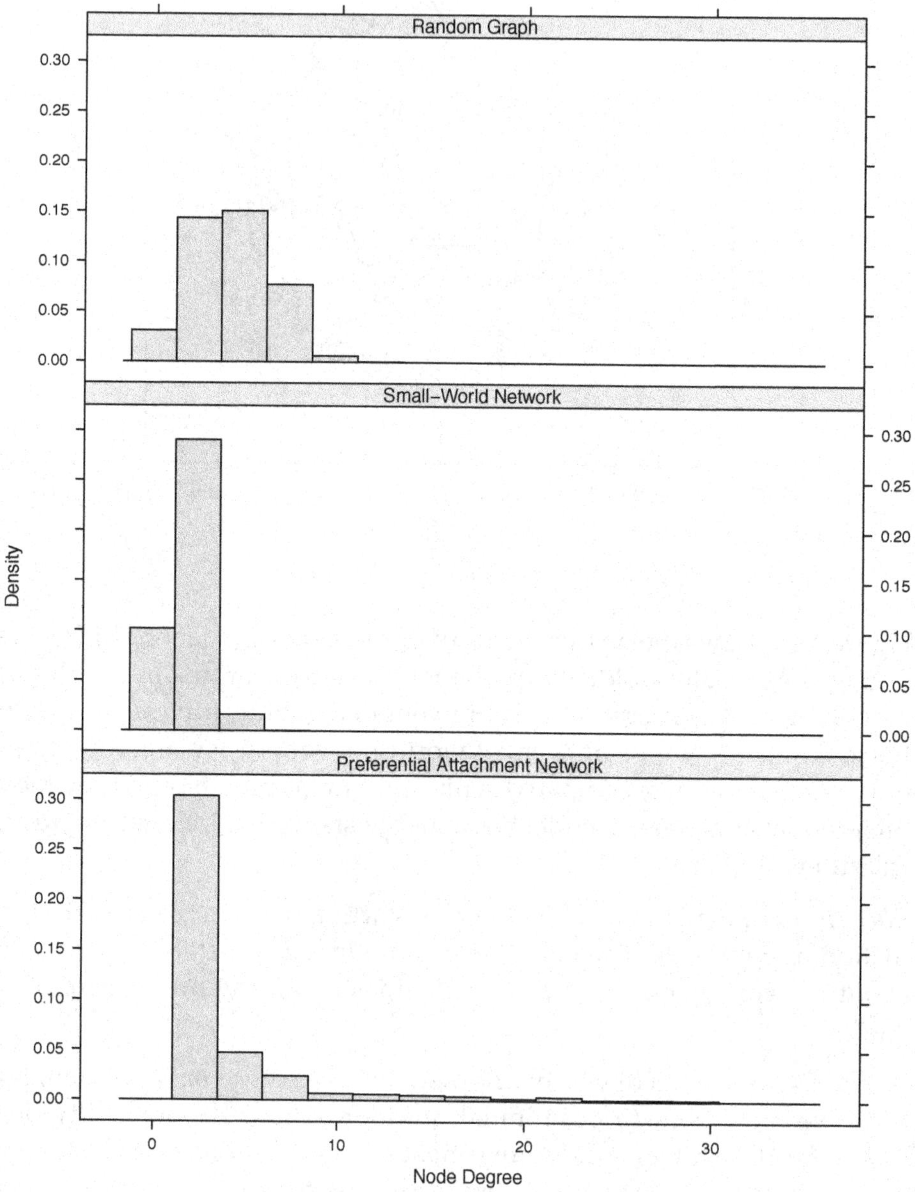

Random Graph
Small-World Network
Preferential Attachment Network
Density
Node Degree
0 10 20 30
0.00 0.05 0.10 0.15 0.20 0.25 0.30

Measures of node importance are useful for finding sources of power, influence, or importance. We might want to characterize the importance of a politician to his party, thinking of measures of centrality as indicators of political power. We consider the importance of a page to a website, or the degree to which a journal article influences a scientific discipline.

We can begin by counting the links connected to the node. And for a directed network, we can count the number of links in and out—we measure the *degree* of each node, both *in-degree* and *out-degree*. The *degree distribution* of a network is one way to characterize the structure of a network, and the average degree or *degree centrality* is a way to summarize the connectedness of an entire network.

While the degree centrality of a node may be computed by looking only at that single node, other measures of centrality depend on the structure of the entire network. One nodal measure is *closeness centrality*, which characterizes how close a node is to all other nodes in the network, where closeness is the number of hops or links between nodes. Computing closeness centrality can be problematic in disconnected networks, so it is less often used as a measure of centrality than other measures (Borgatti, Everett, and Johnson 2013).

When we measure *betweenness centrality*, we consider the proportion of times a node lies on the shortest path between other pairs of nodes. Betweenness centrality reflects the degree to which a node is a broker or go-between node, affecting the traffic or flow of information across the network.

What if we consider a node as important if it is close to other nodes of importance, which are in turn close to other nodes of importance, and so on. This thinking moves us in the direction of the notion of *eigenvector centrality*. Much as the first principle component characterizes common variability in a set of variables, eigenvector centrality characterizes the degree to which a node is central to the set of nodes comprising the network.

Eigenvalue centrality is a summary measure of overall connectedness, where a node acquires higher scores by being connected to other well connected nodes. In this regard eigenvector centrality is a good summary measure of importance.

It should come as no surprise that various measures of node importance correlate highly. We can see this by randomly generating data from mathematical models for networks, computing alternative indices of importance or centrality for each node in each network, and then correlating the indices.

The importance of a person follows from the importance of relatives, friends, and associates. With measures of centrality, we have ways of finding the sources of credibility, power, or influence. These are associated with individual actors, the nodes of a social network. What about the links?

When looking at flow across a network or critical paths through a network, we focus on the links connecting nodes. To distinguish among links, we typically use betweenness. Links, like nodes, can lie on the path between pairs of nodes. And the more pairs of nodes potentially communicating across a link, the higher the betweenness of that link.

Various network-wide measures consider both nodes and links. One such measure is *density*. A completely unconnected or disconnected set of nodes has zero links, while a completely connected network of $n$ nodes or clique has $\frac{n(n-1)}{2}$ links. If $l$ is the number of links in a network, then network density is given by the formula $\frac{l}{n(n-1)/2}$.

Network density is easily calculated and takes values between zero and one. It is also easy to interpret as the proportion of actual links out of the set of all possible links. But density has limited value as a descriptive network measure across networks differing in size.

Social networks concern people, and people are constrained by time, distance, and communication media. There is a limit to the number of social connections or links people can maintain. As networks grow large, observed densities tend to decline.

Another network-wide measure is *transitivity*, which is the proportion of closed triads. A triad or triple is a set of three nodes, and a closed triad is a set of three nodes with links between each pair of nodes. We might think of a closed triad as a mini-clique. Transitivity, like other network measures, can be computed from the adjacency matrix.

There are many other measures of networks that we might consider. A comprehensive review of measures obtained from graphs and matrices is provided by Iacobucci in the volume by Wasserman and Faust (1994).

The measures we have been discussing relate to network topology or structure. They describe nodes relative to other nodes, links relative to other links, and with density, links relative to the number of possible links. These measures help to identify cliques and core communities within a larger network.

Moving beyond topology or structure alone, we have what are known as metadata. There are node characteristics relating to player roles, demographics, attitudes, and behavior, and there are link characteristics relating to width, bandwidth, length/distance, and cost of traversing a network. When we think about making predictions from social networks, we often think in terms of metadata.

Will a consumer buy a product? Will the citizen vote for a candidate? How much will a person spend on sporting events next year? Knowing the social network of the consumer, voter, or sports fan, can help to answer these questions.

Nearest-neighbor methods for predicting people's attitudes and behavior make no assumption about underlying processes or statistical distributions. These methods are data-driven rather than theory-dependent, and because of this, they may be used for all kinds of data structures.

There are temporal, geographic, demographic, and psychographic nearest neighbors. Consider temporal nearest neighbors. Observations closer to one another in time are usually more highly correlated than observations distant in time. Nearest neighbors play out in autoregressive and moving average models for time series.

All other things being equal, observations closer together in space are more highly correlated than observations distant in space. We have geographic nearest neighbors. Someone living four miles from Dodger Stadium would be more likely a Dodgers fan than someone living in Madison, Wisconsin.

We can compute multivariate distance measures using sets of demographic or psychographic variables. We determine the distance between each pair of individuals in a study and use those distances to identify each person's five, ten, or fifteen nearest neighbors, for example. Who will get Mary's vote? Look at the votes of Mary's nearest neighbors to answer the question. See for whom they voted.

To employ nearest-neighbor methods, we need distances in time, physical space, demographics, psychographics, or network topology. For networks, we identify the neighborhood around each person and then identify the people in that neighborhood. Is Joe going to buy a multipurpose watch that displays physiological functions as well as time? Who are the people closest to Joe in his personal or ego-centric network? Are they buying the multipurpose watch?

Alternative versions of closeness may offer disparate nearest-neighbor predictions. The old man I see on my morning walks is close to me in geography and (I must acknowledge) close in age, but on other dimensions we have little in common. We do not frequent the same social circles.

Figure 11.6 provides an overview of network modeling techniques, beginning with mathematical models and ending with agent-based models. With mathematical models, known solutions follow from assumptions—they are proven to be correct. The Erdős-Rényi random graph model (Erdős and Rényi, 1959, 1960, 1961) is an example of a mathematical model for networks. Mathematical models of networks provide the foundation of network science (Lewis 2009; Kolaczyk 2009; Newman 2010).

Discrete event simulation of fixed networks builds on mathematical models and provides a flexible structure for modeling. With discrete event simulation, we demonstrate solutions rather than prove solutions. We define a known, fixed network structure in advance and study performance within that fixed network structure.

Discrete event simulation of fixed networks is most useful in the study of websites and physical communication networks, transportation and production problems. The nodes and links are fixed. Resource limitations are known. And traffic or demand for resources varies across simulated time.

Agent-based modeling of networks can build on discrete event simulation of fixed networks. We acknowledge the fact that individual nodes may differ from one another in roles, motives, behavior, or interactions. Agent-based techniques provide a facility for modeling networks that change with time (dynamic networks).

***Figure 11.6.*** *Network Modeling Techniques*

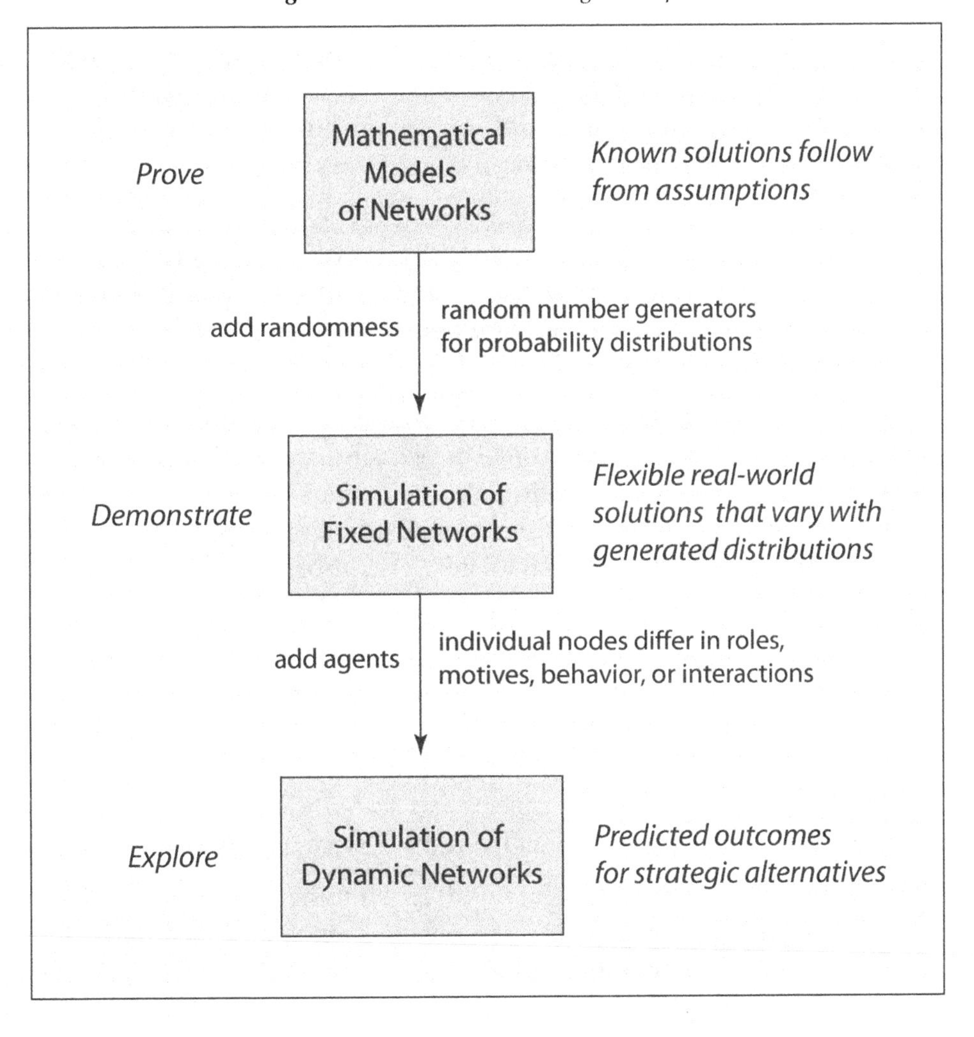

With agent-based models, we can define the initial structure of a network (the nature of the links and the characteristics of the nodes), and then we can use that information to build a behavioral model of the network. We can simulate the behavior of individual nodes or agents, including their interactions with one another.

With agent-based models, we create a microworld intended to simulate real world phenomena. This presents a rich platform for modeling complexity, most useful in the domain of network data science. We can set the stage for agent-based models and for models of dynamic networks by first working with discrete event simulation and fixed network models.

Many questions of interest to management involve agents of various types and networks that change with time. By introducing agents into a simulation model, we can explore alternative agent inputs and management actions, and in the process predict outcomes associated with strategic alternatives (North and Macal 2007).

Franchi (2014) discusses requirements for agent-based simulation of social networks. Distributed and parallel processing may be needed for large networks. The system must simulate the actions of thousands of agents operating independently and simultaneously, suggesting the necessity for multithreading or callback mechanisms. Agent-based simulation requires a robust object-oriented framework as provided by Python.

Agents are autonomous software objects—they operate without human intervention. Agents have the ability to react to events, including events generated by other agents. Agents, like people, have goals and exhibit goal-driven behavior, which is to say that they are proactive, not just reactive. Agents embodied in software objects have identifiable boundaries. They maintain an internal state distinct from the system or network they inhabit.

Software agents can communicate with one another by passing messages. Communication can be asynchronous or synchronous. We utilize data structures or databases to maintain information about agents, messages among agents, and the structure of the network. Information about the network itself can be made available to all agents. And we can give agents the power to create and destroy links in the network.

Implementing agent-based simulation requires software to control the simulation clock, agent creation, and agent destruction. The simulation runs until certain events occur or until some number of ticks have expired on the simulation clock.

An important marketing phenomenon to be explored with agent-based network simulation is diffusion of innovations. The new product development cycle has stages identified as opportunity identification, design, testing, introduction, and product management. The last stage assumes success in the marketplace. What makes some products successful and others not? Network data science may hold the answer to this question.

Members of a social network share information with one another. They talk about new products or see new products in action. The more connections members have with one another (higher degree or betweeness), the more likely a new idea will travel quickly through the network. And more well connected, respected, or influential members of a network can be of special value in the sharing of information. Diffusion of innovations is a social network process, not unlike a biological model of infection.

Franchi (2014) implements a biological model of infection in Python. This is the SIR model, with each agent having one of three states: susceptible (S), infected (I), or recovered (R). System-wide parameters include the proportion of initially infected nodes, the probability that a susceptible agent will become infected, and the probability that an infected node will recover. We specify the initial network structure, the total number of nodes, and the number of initially infected nodes. During each simulation trial, the exact topology and infected nodes are selected at random. A random graph or small-world network could be used for this purpose.

We can put a marketing spin on the infection/information diffusion model by thinking of susceptible nodes/agents as being open to new product adoption. An infected node is a consumer who has purchased the new product, and a recovered node is a former consumer of the product. We compute market share as the proportion of infected nodes.

Suppose we consider a market penetration experiment with consumer electronics. A manufacturing firm plans to offer a new product in the wearable technology space. The firm is interested in testing alternative methods for promoting the product. Mass media offers wide coverage of the population with limited contact with each individual consumer. Direct marketing to target groups entails narrow coverage. But with a narrow, more focused target group, it is possible to tailor a message for that group, resulting in more extensive contact with individual consumers.

The success of mass versus targeted marketing will depend on the coverage of the population, the levels of exposure or contact with individuals, and the responsiveness of individuals to the message. Just as the spread of disease in a population depends on the structure of the network and the susceptibility of individuals to the disease, the adoption of a new product depends on the willingness of individual consumers to purchase and the structure of the network of consumers. Some consumers are more willing to accept new ideas, information, and products than others. And we can use agent-based models to explore individual differences among consumers and their effect on the network.

The response of consumers to an initial marketing message begins the process. Some purchase the wearable technology product and wear it. They interact with other consumers, influencing their purchases. The process plays out in simulation time. As the simulation progresses, we note the numbers of consumers exposed to the idea and their reaction to the idea (purchase or not). An agent-based framework allows us to fine-tune consumer characteristics to match our understanding of individual characteristics in the marketplace. We can test alternative marketing actions (mass versus direct), and their effect on consumer adoption of the product.

Agent-based studies, like many simulation studies, are set up as factorial experiments. We look at alternative network configurations and agent characteristics and see how they play out in simulation time. For a market penetration experiment, we can begin with small-world representations of the network. Worlds vary in size (number of nodes) across experimental conditions. Networks also vary in degree of connectedness. Individual nodes have levels of susceptibility or openness to new product adoption. We can fix the initial market share and recovery rates for the simulation, although these, too, could be experimental factors.

We imagine that each tick of the simulation clock is one week of marketing activity. For each trial of the experiment under each treatment condition, we record market share or the proportion of infected nodes at the end of

104 weeks or two years. Distributions of market shares across one hundred trials of each treatment combination could be computed and analyzed as part of the simulation study.

We analyze the results of agent-based simulations with statistical models, much as we have with actual marketing data throughout this book. Analysis of variance and linear models work for continuous response variables. Analysis of deviance and generalized linear models work for binary response variables, proportions, and counts. Market share is a proportion, so we can employ analysis of deviance in our evaluation of factors.

Discussions of agent-based models and complexity go hand-in-hand. This is because agent-based models are most useful when the entire system is less well understood than its individual components. So rather than build a complex systems model, we build many small models or what might be called "micro-worlds" or "micro-models."

Validation of agent-based models presents challenges. Is it sufficient to see the simulation acting like what we see in the real world? And how shall we judge the degree of similarity?

Agent-based models remain in an experimental state, not widely employed in business research. To move from the academic world into business—to be viewed as more than a modeler's playground—agent-based models need to do things that are useful and understood by people outside the modeling community.

We learn about agent-based simulation as we learn about many things—by doing, by programming. For marketing data scientists who are up to the challenge, agent-based simulation may hold the key to developing marketplace predictions that go beyond those obtained with traditional statistical models or machine learning algorithms.

Social network analysis has been an active area of research in psychology, sociology, anthropology, and political science for many years. The invention of the sociogram and concepts of social structure may be traced back more than eighty years (Moreno 1934; Radcliffe-Brown 1940). Research topics include isolation and popularity, prestige, power, and influence, social cohesion, subgroups and cliques, status and roles within organizations, balance and reciprocity, marketplace relationships, and measures of centrality and connectedness.

An introduction to social networks is provided by Kadushin (2012). Ackland (2013) applies social network concepts to web and social media networks. Reviews of network science are provided by Watts (2003), Lewis (2009), and Newman (2010). Works by Jackson (2008) and Easley and Kleinberg (2010) attempt to integrate network science and economic game theory.

Detection of clusters, information communities, core groups, and cliques is an important area of research in social network analysis. A recent review is provided by Zubcsek, Chowdhury, and Katona (2014).

Social network analyses often involve multidimensional scaling, hierarchical cluster analysis, log linear modeling, and a variety of specialized methods. Kolaczyk (2009) and Kolaczyk and Csárdi (2014) review statistical methods in network science. Wasserman and Faust (1994) and contributors to volume edited by Brandes and Erlebach (2005) review methods of social network analysis. Log-linear models have been used extensively in this area of research. For an overview of log linear models, see Bishop, Fienberg, and Holland (1975), Christensen (1997), and Fienberg (2007). Application of log-linear and logistic regression models to social networks is discussed in Wasserman and Iacobucci (1986) and Wasserman and Pattison (1996).

Stochastic blockmodels can be especially useful in the analysis of large networks. Individual nodes and their connections may be combined to form blocks within a network, and then these blocks may themselves be analyzed as networks. Holland et al. (1983) and Wasserman and Anderson (1987) review blockmodels.

Network science and social networks provide interesting fodder for the popular press, as we can see with books by Buchanan (2002), Barabási (2003, 2010), Christakis and Fowler (2009), and Pentland (2014).

Law (2014) discusses the domain of simulation modeling, including agent-based modeling. For discussion of complexity theory and agent-based modeling, see North and Macal (2007), Miller and Page (2007), and Šalamon (2011). Much of the work in this area is biologically inspired, modeling the behavior of individual organisms to see what happens in the community (Resnick 1998; Mitchell 2009).

For discussion of nearest-neighbor methods, see Duda, Hart, and Stork (2001), Shakhnarovich, Darrell, and Indyk (2006), and Hastie, Tibshirani, and Friedman (2009). Kolaczyk (2009) reviews nearest neighbors and other predictive modeling methods, including Markov random fields and kernel regression.

Traditional models for diffusion of innovations build on early work by Bass (1969). Discussion of mathematical models is available from various sources (Mahajan, Muller, and Bass 1995; Rogers 2003; Ehrenberg and Bound 2000; Mahajan, Muller, and Wind 2000). Network models for the diffusion of innovations, infection, and information cascades have been reviewed by Valente (1995) and Watts (2003), among others.

Exhibit 11.1 shows an R program for generating networks from random graph, small-world, and preferential attachment models. This program draws on R packages developed by Csardi (2014b), Kolaczyk and Csárdi (2014), and Sarkar (2014). Reviews of random network models have been provided by Albert and Barabási (2002), Cami and Deo (2008), Kolaczyk (2009), and Toivonen et al. (2009).

Exhibit 11.2 shows an example of analysis of deviance using R. With factorial designs, interaction plots provide a useful summary of experimental results. The program shows how to generate a two-way or two-factor interaction plot using software developed by Wickham and Chang (2014).

Exhibit 11.3 shows a Python program for generating a small-world network, a good starting point in agent-based models for diffusion of innovations. The program builds on work by Hagbert and Schult (2014) and the NetworkX development team.

Exhibit 11.4 shows Python code for conducting analysis of deviance in an agent-based modeling experiment.

**Exhibit 11.1.** *Network Models and Measures (R)*

```
# Network Models and Measures (R)

# install necessary packages

# bring packages into the workspace
library(igraph)  # network/graph models and methods
library(lattice)  # statistical graphics

# load correlation heat map utility
load(file = "correlation_heat_map.RData")

# note. evcent() often produces warning about lack of convergence
# we will ignore these warnings in this demonstration program
options(warn = -1)
# number of iterations for the statistical simulations
NITER <- 100

# user-defined function to compute centrality index correlations
get_centrality_matrix <- function(graph_object) {
    adjacency_mat <- as.matrix(get.adjacency(graph_object))
    node_degree <- degree(graph_object)
    node_betweenness <- betweenness(graph_object)
    node_closeness <- closeness.estimate(graph_object,
        mode = "all", cutoff = 0)
    node_evcent <- evcent(graph_object)$vector
    centrality <- cbind(node_degree, node_betweenness,
    node_closeness, node_evcent)
        colnames(centrality) <-  c("Degree", "Betweenness",
            "Closeness", "Eigenvector")
    return(cor(centrality))
    }
# --------------------------------
# Random Graphs
# --------------------------------
# show the plot of the first random graph model
set.seed(1)
# generate random graph with 50 nodes and 100 links/edges
random_graph <- erdos.renyi.game(n = 50, type = "gnm", p.or.m = 100)
pdf(file = "fig_network_random_graph.pdf", width = 5.5, height = 5.5)
plot(random_graph, vertex.size = 10, vertex.color = "yellow",
    vertex.label = NA, edge.arrow.size = 0.25, edge.color = "black",
    layout = layout.circle, edge.curved = TRUE)
dev.off()

# express adjacency matrix in standard matrix form
random_graph_mat <- as.matrix(get.adjacency(random_graph))
# verify that the network has one hundred links/edges
print(sum(degree(random_graph))/2)

aggregate_degree <- NULL  # initialize collection of node degree values
correlation_array <- array(NA, dim = c(4, 4, NITER))  # initialize array
```

```
for (i in 1:NITER) {
    set.seed(i)
    random_graph <- erdos.renyi.game(n = 50, type = "gnm", p.or.m = 100)
    aggregate_degree <- c(aggregate_degree,
        degree(random_graph))
    correlation_array[,,i] <- get_centrality_matrix(random_graph)
    }
average_correlation <- matrix(NA, nrow = 4, ncol = 4,
    dimnames = list(c("Degree", "Betweenness", "Closeness", "Eigenvector"),
        c("Degree", "Betweenness", "Closeness", "Eigenvector")))
for (i in 1:4)
    for(j in 1:4)
        average_correlation[i, j] <- mean(correlation_array[i, j, ])

pdf(file = "fig_network_random_graph_heat_map.pdf", width = 11,
  height = 8.5)
correlation_heat_map(cormat = average_correlation)
dev.off()

# create data frame for node degree distribution
math_model <- rep("Random Graph", rep = length(aggregate_degree))
random_graph_degree_data_frame <- data.frame(math_model, aggregate_degree)

# --------------------------------
# Small-World Networks
# --------------------------------
# example of a small-world network (no random links)
set.seed(1)
# one-dimensional small-world model with 10 nodes,
# links to additional adjacent nodes in a lattice
# (nei = 1 implies degree = 2 for all nodes prior to rewiring)
# rewiring probability of 0.00... no rewiring
small_world_network_prelim <- watts.strogatz.game(dim = 1, size = 10,
    nei = 1, p = 0.00, loops = FALSE, multiple = FALSE)
# remove any multiple links/edges
small_world_network_prelim <- simplify(small_world_network_prelim)
# express adjacency matrix in standard matrix form
# show that each node has four links
print(degree(small_world_network_prelim))
# verify that the network has one hundred links/edges
print(sum(degree(small_world_network_prelim))/2)
pdf(file = "fig_network_small_world_nei_1.pdf", width = 5.5, height = 5.5)
plot(small_world_network_prelim, vertex.size = 25, vertex.color = "yellow",
    vertex.label = NA, edge.arrow.size = 0.25, edge.color = "black",
    layout = layout.circle, edge.curved = TRUE)
dev.off()

# another of a small-world network (no random links)
set.seed(1)
# one-dimensional small-world model with 10 nodes,
# links to additional adjacent nodes in a lattice
# (nei = 2 implies degree = 2 for all nodes prior to rewiring)
# rewiring probability of 0.00... no rewiring
```

```
small_world_network_prelim <- watts.strogatz.game(dim = 1, size = 10,
    nei = 2, p = 0.00, loops = FALSE, multiple = FALSE)
# remove any multiple links/edges
small_world_network_prelim <- simplify(small_world_network_prelim)
# express adjacency matrix in standard matrix form
# show that each node has four links
print(degree(small_world_network_prelim))
# verify that the network has one hundred links/edges
print(sum(degree(small_world_network_prelim))/2)

pdf(file = "fig_network_small_world_nei_2.pdf", width = 5.5, height = 5.5)
plot(small_world_network_prelim, vertex.size = 25, vertex.color = "yellow",
    vertex.label = NA, edge.arrow.size = 0.25, edge.color = "black",
    layout = layout.circle, edge.curved = TRUE)
dev.off()

# yet another small-world network (no random links)
set.seed(1)
# one-dimensional small-world model with 10 nodes,
# links to additional adjacent nodes in a lattice
# (nei = 3 implies degree = 6 for all nodes prior to rewiring)
# rewiring probability of 0.00... no rewiring
small_world_network_prelim <- watts.strogatz.game(dim = 1, size = 10,
    nei = 3, p = 0.00, loops = FALSE, multiple = FALSE)
# remove any multiple links/edges
small_world_network_prelim <- simplify(small_world_network_prelim)
# express adjacency matrix in standard matrix form
# show that each node has four links
print(degree(small_world_network_prelim))
# verify that the network has one hundred links/edges
print(sum(degree(small_world_network_prelim))/2)
pdf(file = "fig_network_small_world_nei_3.pdf", width = 5.5, height = 5.5)
plot(small_world_network_prelim, vertex.size = 25, vertex.color = "yellow",
    vertex.label = NA, edge.arrow.size = 0.25, edge.color = "black",
    layout = layout.circle, edge.curved = TRUE)
dev.off()

# rewire a selected proportion of the links to get small world model
set.seed(1)
small_world_network <- watts.strogatz.game(dim = 1, size = 50, nei = 2,
    p = 0.2, loops = FALSE, multiple = FALSE)
# remove any multiple links/edges
small_world_network <- simplify(small_world_network)
# express adjacency matrix in standard matrix form
# show that each node has four links
print(degree(small_world_network))
# verify that the network has one hundred links/edges
print(sum(degree(small_world_network))/2)
pdf(file = "fig_network_small_world.pdf", width = 5.5, height = 5.5)
plot(small_world_network, vertex.size = 10, vertex.color = "yellow",
    vertex.label = NA, edge.arrow.size = 0.25, edge.color = "black",
    layout = layout.circle, edge.curved = TRUE)
dev.off()
```

```
aggregate_degree <- NULL  # initialize collection of node degree values
correlation_array <- array(NA, dim = c(4, 4, NITER))  # initialize array
for (i in 1:NITER) {
    set.seed(i)
    small_world_network <- watts.strogatz.game(dim = 1, size = 50, nei = 1,
         p = 0.2, loops = FALSE, multiple = FALSE)
    aggregate_degree <- c(aggregate_degree,
        degree(small_world_network))
    correlation_array[,,i] <- get_centrality_matrix(small_world_network)
    }
average_correlation <- matrix(NA, nrow = 4, ncol = 4,
    dimnames = list(c("Degree", "Betweenness", "Closeness", "Eigenvector"),
        c("Degree", "Betweenness", "Closeness", "Eigenvector")))
for (i in 1:4)
    for(j in 1:4)
        average_correlation[i, j] <- mean(correlation_array[i, j, ])

pdf(file = "fig_network_small_world_correlation_heat_map.pdf", width = 11,
  height = 8.5)
correlation_heat_map(cormat = average_correlation)
dev.off()

# create data frame for node degree distribution
math_model <- rep("Small-World Network", rep = length(aggregate_degree))
small_world_degree_data_frame <- data.frame(math_model, aggregate_degree)

# ----------------------------------
# Scale-Free Networks
# ----------------------------------
# show the plot of the first scale-free network model
set.seed(1)
# directed = FALSE to generate an undirected graph
# fifty nodes to be consistent with the models above
scale_free_network <- barabasi.game(n = 50, m = 2, directed = FALSE)

# remove any multiple links/edges
scale_free_network <- simplify(scale_free_network)

pdf(file = "fig_network_scale_free.pdf", width = 5.5, height = 5.5)
plot(scale_free_network, vertex.size = 10, vertex.color = "yellow",
    vertex.label = NA, edge.arrow.size = 0.25, edge.color = "black",
    layout = layout.circle, edge.curved = TRUE)
dev.off()

# express adjacency matrix in standard matrix form
scale_free_network_mat <- as.matrix(get.adjacency(scale_free_network))

# note that this model yields a graph with almost 100 links/edges
print(sum(degree(scale_free_network))/2)

aggregate_degree <- NULL  # initialize collection of node degree values
correlation_array <- array(NA, dim = c(4, 4, NITER))  # initialize array
```

```
for (i in 1:NITER) {
    set.seed(i)
    scale_free_network <- barabasi.game(n = 50, m = 2, directed = FALSE)
    # remove any multiple links/edges
     scale_free_network <- simplify(scale_free_network)
    aggregate_degree <- c(aggregate_degree,
        degree(scale_free_network))
    correlation_array[,,i] <- get_centrality_matrix(scale_free_network)
    }
average_correlation <- matrix(NA, nrow = 4, ncol = 4,
    dimnames = list(c("Degree", "Betweenness", "Closeness", "Eigenvector"),
        c("Degree", "Betweenness", "Closeness", "Eigenvector")))
for (i in 1:4)
    for(j in 1:4)
        average_correlation[i, j] <- mean(correlation_array[i, j, ])

pdf(file = "fig_network_scale_free_correlation_heat_map.pdf", width = 11,
  height = 8.5)
correlation_heat_map(cormat = average_correlation)
dev.off()

# create data frame for node degree distribution
math_model <- rep("Preferential Attachment Network",
    rep = length(aggregate_degree))
scale_free_degree_data_frame <- data.frame(math_model, aggregate_degree)

# ---------------------------------
# Compare Degree Distributions
# ---------------------------------
plotting_data_frame <- rbind(scale_free_degree_data_frame,
    small_world_degree_data_frame,
    random_graph_degree_data_frame)

# use lattice graphics to compare degree distributions

pdf(file = "fig_network_model_degree_distributions.pdf", width = 8.5,
  height = 11)
lattice_object <- histogram(~aggregate_degree | math_model,
    plotting_data_frame, type = "density",
    xlab = "Node Degree", layout = c(1,3))
print(lattice_object)
dev.off()

# Suggestions for the student.
# Experiment with the three models, varying the numbers of nodes
# and methods for constructing links between nodes. Try additional
# measures of centrality to see how they relate to the four measures
# we explored in this program.  Explore summary network measures
# of centrality and connectedness to see how they relate across
# networks generated from random graph, preferential attachment,
# and small-world models.
```

***Exhibit 11.2.*** *Analysis of Agent-Based Simulation (R)*

```
# Analysis of Agent-Based Simulation (R)

# install necessary packages

library(ggplot2)  # for pretty plotting

# -------------------------------
# Simulation study background
# (same as under Python exhibit)
# -------------------------------
# note. it is possible to run NetLogo simulations within R
# using the RNetLogo package from Thiele (2014)

# -----------------------------
# Analysis of Deviance
# -----------------------------
options(warn = -1)  # drop glm() warnings about non-integer responses

# read in summary results and code the experimental factors
virus <- read.csv("virus_results.csv")

# define factor variables
virus$Connectivity <- factor(virus$degree,
  levels = c(3, 5), labels = c("LOW", "HIGH"))
virus$Susceptibility <- factor(virus$spread,
  levels = c(5, 10), labels = c("LOW", "HIGH"))

virus$Market_Share <- virus$infected

# show the mean proportions by cell in the 2x2 design
with(virus, print(by(Market_Share, Connectivity * Susceptibility, mean)))

# generalized linear model for response variable that is a proportion
virus_fit <- glm(Market_Share ~ Connectivity + Susceptibility +
  Connectivity:Susceptibility,
  data = virus, family = binomial(link = "logit"))
print(summary(virus_fit))

# analysis of deviance
print(anova(virus_fit, test="Chisq"))

# compute market share cell means for use in interaction plot
virus_means <- aggregate(Market_Share ~ Connectivity * Susceptibility,
  data = virus, mean)

# -----------------------------
# Interaction Plotting
# -----------------------------
# generate an interaction plot for market share as a percentage
```

```
interaction_plot <- ggplot(virus_means,
  aes(x = Connectivity, y = 100*Market_Share,
    group = Susceptibility, fill = Susceptibility)) +
  geom_line(linetype = "solid", size = 1, colour = "black") +
  geom_point(size = 4, shape = 21) +
  ylab("Market Share (Percentage)") +
  ggtitle("Network Interaction Effects in Innovation") +
  theme(plot.title = element_text(lineheight=.8, face="bold"))
print(interaction_plot)
```

***Exhibit 11.3.*** *Defining and Visualizing a Small-World Network (Python)*

```
# Defining and Visualizing a Small World Network (Python)

# prepare for Python version 3x features and functions
from __future__ import division, print_function

# load package into the workspace for this program
import networkx as nx
import matplotlib.pyplot as plt  # 2D plotting
import numpy as np

# generate small-world network with n nodes
# k is number of nearby nodes to which each node is connected
# and p probability of rewiring from nearby node to random node
small_world = nx.watts_strogatz_graph(n = 100, k = 3, p = 0.25, seed = None)

# create an adjacency matrix object for the line network
# use nodelist argument to order the rows and columns
small_world_mat = nx.adjacency_matrix(small_world)
# print(small_world_mat)  # undirected networks are symmetric

# examine alternative layouts for plotting the small_world
# plot the network/graph with default layout
fig = plt.figure()
nx.draw_networkx(small_world, node_size = 200, node_color = 'yellow')
plt.show()

# spring layout
fig = plt.figure()
nx.draw_networkx(small_world, node_size = 200, node_color = 'yellow',\
    pos = nx.spring_layout(small_world))
plt.show()

# circlular layout
fig = plt.figure()
nx.draw_networkx(small_world, node_size = 200, node_color = 'yellow',\
    pos = nx.circular_layout(small_world))
plt.show()

# shell/concentric circles layout
fig = plt.figure()
nx.draw_networkx(small_world, node_size = 200, node_color = 'yellow',\
    pos = nx.shell_layout(small_world))
plt.show()

# Gephi provides interactive network plots
# dump the graph object in GraphML format for input to Gephi
# and for initial network structure in pynetsim
nx.write_graphml(small_world,'small_world_network.graphml')
```

*Exhibit 11.4. Analysis of Agent-Based Simulation (Python)*

```
# Analysis of Agent-Based Simulation (Python)

# install necessary packages

# import packages into the workspace for this program
import numpy as np
import pandas as pd
import statsmodels.api as sm

# -----------------------------
# Simulation study background
# -----------------------------
# an agent-based simulation was run with NetLogo, a public-domain
# program available from Northwestern University (Wilensky 1999)

# added one line of code to the Virus on a Network program:

if ticks = 200 [stop]

# this line was added to stop the simulation at exactly 200 ticks
# the line was added to the <to go> code block as shown here:
#
# to go
#   if all? turtles [not infected?]
#     [ stop ]
#   ask turtles
#   [
#      set virus-check-timer virus-check-timer + 1
#      if virus-check-timer >= virus-check-frequency
#        [ set virus-check-timer 0 ]
#   ]
#   if ticks = 200 [stop]
#   spread-virus
#   do-virus-checks
#   tick
# end

# the simulation stops when no nodes/turtles are infected
# or when the simulation reaches 200 ticks

# To see the results of the simulation at 200 ticks, we route the simulation
# world to a file using the GUI File/Export/Export World
# this gives an a comma-delimited text file of the status of the network
# at 200 ticks. Specifically, we enter the following Excel command into
# cell D1 of the results spreadsheet to compute the proportion of nodes
# infected:   = COUNTIF(N14:N163, TRUE)/M10

# NetLogo turtle infected status values were given in cells N14 through N163.
# The detailed results of the simulation runs or trials are shown in the files
# <trial01.csv> through <trial20.csv> under the directory NetLogo_Results
```

```
# this particular experiment, has average connectivity or node degree
# at 3 or 5 and the susceptibility or virus spread chance to 5 or 10 percent.
# we have a completely crossed 2 x 2 design with 5 replications of each cell
# that is, we run each treatment combination 5 times, obtaining 20 independent
# observations or trials. for each trial, we note the percentage of infected
# nodes after 200 ticks---this is the response variable
# results are summarized in the comma-delimited file <virus_results.csv>.

# -----------------------------
# Analysis of Deviance
# -----------------------------

# read in summary results and code the experimental factors
virus = pd.read_csv("virus_results.csv")

# check input DataFrame
print(virus)

Intercept = np.array([1] * 20)

# use dictionary object for mapping to 0/1 binary codes
degree_to_binary = {3 : 0, 5 : 1}
Connectivity = np.array(virus['degree'].map(degree_to_binary))

# use dictionary object for mapping to 0/1 binary codes
spread_to_binary = {5 : 0, 10 : 1}
Susceptibility = np.array(virus['spread'].map(spread_to_binary))

Connectivity_Susceptibility = Connectivity * Susceptibility

Design_Matrix = np.array([Intercept, Connectivity, Susceptibility,\
     Connectivity_Susceptibility]).T

print(Design_Matrix)

Market_Share = np.array(virus['infected'])

# generalized linear model for a response variable that is a proportion
glm_binom = sm.GLM(Market_Share, Design_Matrix, family=sm.families.Binomial())
res = glm_binom.fit()
print res.summary()
```

# 12 Watching Competitors

Del: "I never did introduce myself. Del Griffith. American Light and Fixture—Director of Sales, Shower Curtain Ring Division. I sell shower curtain rings. Best in the world.... And you are?"

Neal: "Uh, Neal Page."

Del: "Neal Page. Pleased to meet you, Neal Page.
So what do you do for a living, Neal Page?"

Neal: "Marketing."

Del: "Marketing? Super. Super. Fabulous. Isn't that nice?"

Neal: "Uh, look, I don't want to be rude, but... I'm not much of a conversationalist. I'd like to finish this article. A friend wrote it, so..."

Del: "Don't let me stand in your way. Please, don't let me stand in your way. Last thing I want to be remembered as is an annoying blabbermouth. You know, nothing grinds my gears worse than some chowder head who doesn't know when to keep his trap shut. If you catch me running off at the mouth, just give me a poke in the chops."

—JOHN CANDY AS DEL GRIFFITH AND STEVE MARTIN AS NEAL PAGE
IN *Planes, Trains, and Automobiles* (1987)

Last year I flew Spirit Airlines from Los Angeles to Chicago. Known for its low fares, Spirit charges a base fare and then adds amenities. My first and only amenity was a seat. I bought seat 10F in each direction: $2 \times \$14 = \$28$. I could have purchased seats further forward at $25 each, but that seemed a bit excessive. Ordering online was a snap except for a barrage of pop-up ads. I was asked if I wanted to pay extra to board the plane early. I did not. And I accepted no luggage charges.

At the airport I spent twenty minutes rearranging my bags, converting two canvas bags of books and clothing into one. This benefitted me because, according to the gate agent, I would be charged one hundred dollars at the gate if I had two carry-on bags. I put my jacket on and stuffed its pockets with a computer mouse, extra glasses, Python pocket guide, and such.

Shortly after takeoff, an infant began his crying. This lasted for much of the duration of the flight, three hours and twenty minutes. This was not Spirit's fault. But I half expected the flight attendant to announce, "And for your flying enjoyment today, we offer ear plugs for twenty dollars and noise-canceling headphones for one hundred."

I got to Chicago safely and on time, and two days later the return trip was equally efficient, sans the screaming infant. But having to buy my seat and the inconvenience with the bags made me wonder about flying Spirit again. There are many carriers serving the LA–Chicago route. I wonder about Spirit's competitive position.

We could start the research by crawling and scraping websites relating to Spirit Airlines. But suppose we step back from that obvious activity for a moment to consider the work of competitive intelligence in general.

Competition is evident in most areas of business, whether we characterize it in economic terms as perfect competition, monopolistic competition, or rivalry among a few leading firms (oligopoly). Whatever form competition takes, most managers would agree that there are decided advantages to knowing the competition.

Compared to other types of research, competitive intelligence is little understood, misunderstood, often maligned, and sometimes feared. Stories of unethical business practices, corporate espionage, and spying make for good reading in the popular press (Penenberg and Barry 2000), but the real work of competitive intelligence is mundane.

Recall the whimsical quotation attributed to Yogi Berra: *You can observe a lot just by watching*. Such is the essence of competitive intelligence. We study, we observe, we learn.

It is fine to begin a competitive intelligence project with little understanding of the competitors. In fact, it may be an advantage to have few preconceptions, to enter the marketplace with open eyes.

Competitive intelligence is forward looking. It is not enough to know what competitors have done in the past. We must anticipate the future. We make make predictions about the future actions of firms, their products, services, and prices. Done right, competitive intelligence guides business strategy.

The pace of business is fast, and firms must respond quickly to changes in the competitive landscape, the political and regulatory environment, and developments in technology. Increasingly, firms face global competition, so we must study, look, and learn worldwide.

Many game theory models assume that players have common knowledge and equal access to information. In practice, asymmetric information is common. A player in a competitive game with less information than others has a competitive disadvantage.

Sources of competitive information are many, and much of this information is online. We attend to trade publications and look for lists of association members. We interview business insiders. It may be surprising to see how much information is in the public domain, available to anyone with access to a web browser and a good search tool. Many firms reveal more than they should about themselves on their corporate web sites. There will be information about current and future products, organizational structure, biographies of key employees, and mission statements.

Woodward and Bernstein could "follow the money," and so can we. Business transactions reveal information. When a corporate competitor decides to build a new manufacturing facility, it engages in numerous business transactions. Many people are involved, including corporate managers, sellers of land, contractors, realtors, lawyers, and bankers. There are public records, including permits and filings with local and state governments, bank filings, and environmental impact statements. There could be news reports and community gossip. Information revealed in public sources will point to the exact location of the facility and its size. Through environmental impact statements, it may be possible to infer the type of materials being produced and the quantity of production.

Registered business process patents are public records; these provide information about current and future products. The federal government provides access to patent information, as do many corporate websites.

Bids to governmental projects are open for public inspection. These provide information about products and pricing. Additional information about competitors may be gleaned from published price guides, press releases, newspaper and magazine articles, conference proceedings, white papers, technical reports, annual reports, and financial analyst reports.

News media provide extensive information about competitors. There are recorded radio and television interviews with key employees and industry consultants. Competitive intelligence professionals know to look locally, searching small news sources, as well as the national media.

Additional sources of information include syndicated data sources with restricted access to subscribers. Extensive legal and business databases are searchable within the syndicated offerings.These fee-based information services can be quite useful in competitive intelligence.

In competitive intelligence, professionals distinguish between primary and secondary sources (not to be confused with primary and secondary research). A primary source is a direct source of information, a representative of the competitive firm being studied. A secondary source is someone who gets information from a primary source or another secondary source, and may have altered that information, either intentionally or unintentionally. In journalism and law it is normal procedure to view primary sources as being more trustworthy and accurate than secondary sources.

Moving beyond published sources of information, competitive intelligence professionals turn to one of their favorite activities—talking with people. Employees and former employees, salespeople, and customers—these represent potential sources of competitive intelligence.

We assume, as we do for all types of research, that the work of competitive intelligence will be done in a legal and ethical manner. Ethical questions often arise when working in the area of competitive intelligence. It is entirely legal and ethical to ask questions as long as the researcher identifies herself properly. She uses the skills of a journalist while explaining that she works for a particular firm. Thompson (2000) and Houston, Bruzzese, and Weinberg (2002) provide advice to journalists involved in business reporting. Their advice is equally relevant to competitive intelligence professionals.

How far can one go in getting information about the competition? While eavesdropping, bribery, and acts of deception are clear violations of law and ethics, there are many gray areas in which actions are questionable but not obviously illegal or unethical. Competitive intelligence professionals must be careful, avoiding any implication of wrongdoing.

Competitive intelligence builds on the free and open exchange of information among business professionals. Trade shows and professional conferences are good places to find people "in the know." But we can find these people online as well. When a company puts information on the web, it posts that information to the world. To protect information of competitive value—proprietary information and trade secrets—firms should be wary of the web. Of course, this is more easily said than done in today's world.

Much of what we know about competitive intelligence in general applies to competitive intelligence online. Some books about online search focus on competitive intelligence sources (Miller 2000a; Vine 2000; Campbell and Swigart 2014). A few researchers discuss competitive intelligence spiders (Chau and Chen 2003; Hemenway and Calishain 2004), and most of what we have learned about focused crawlers or spiders applies to their use in competitive intelligence.

We demonstrate the competitive intelligence process, focusing on Spirit Airlines. Here is the scenario. We are one of the airlines competing with Spirit on the route between Los Angeles and Chicago. We are concerned about Spirit's growth and suspect that its low fares may be attracting many airline passengers. We want to learn as much as we can about Spirit, and we want to learn quickly. We are working online, as usual.[1] A summary of our interactive work is provided in table 12.1.

[1] Data we review here were collected November 4, 2014.

*Table 12.1. Competitive Intelligence Sources for Spirit Airlines*

| *Source* | *Objective (Method)* | *Web Address (Query String)* |
|---|---|---|
| Wikipedia | Overview | `http://en.wikipedia.org/wiki/Spirit_Airlines` |
| LinkedIn | Organization | `https://www.linkedin.com/company/spirit-airlines` |
| indeed.com | Job Postings | `http://www.indeed.com/` |
| | (Advanced Job Search) | `FindJobs:what=company: (SpiritAirlineswhere=blank)` |
| | (Advanced Job Search) | `FindJobs:what=company: (SpiritAirlineswhere=Miami,FL)` |
| | (Advanced Job Search) | `FindJobs:what=company: (SpiritAirlineswhere=Detriot,MI)` |
| | (Advanced Job Search) | `FindJobs:what=company: (SpiritAirlineswhere=Dallas,TX)` |
| Glassdoor | Job Postings | `http://www.glassdoor.com/index.htm` |
| | (Search) | `Jobs.Company:SpiritAirlines` |
| | (Search) | `Salaries` |
| Compete.com | Website Activity | `https://www.compete.com/` |
| | (Search) | `Website:http://www.spirit.com/` |
| Yahoo! Finance | Financial Data | `http://finance.yahoo.com/` |
| | (Search) | `SearchFinance:SAVE` |
| Google Finance | Financial Data | `https://www.google.com/finance` |
| | (Search) | `Search:NASDAQ:SAVE` |
| Bloomberg | Financial News | `http://www.bloomberg.com/` |
| | (Search) | `http://www.bloomberg.com/quote/SAVE:US` |
| Quandl | Financial Data | `https://www.quandl.com/` |
| | (Financial Statements) | `https://www.quandl.com/c/stocks/save` |
| Spirit Airlines | Company Website | `http://www.spirit.com/` |
| | (Investor Relations) | `http://ir.spirit.com/` |

We want to make the process of learning about Spirit Airlines as efficient as possible. We start with an overview from Wikipedia: date of founding, top executives, headquarters, and the like. We also see the top ten airports in its network and the number of daily flights from each.

Next, we move to the organization. We have the name of the President and CEO, Ben Baldanza, so we could start with him and trace his links, mapping an ego-centric network. But suppose we go to a source of such professional links, LinkedIn, and find out how many of Spirit's employees are listed with the site. We can click on "See all" to see them all along with a listing of where they are located: Miami/Fort Lauderdale (656), greater Detroit area (123), Dallas/Fort Worth area (81), and so on. These data may be used to estimate the proportion of employees at each site, providing an idea of the airline's presence in the areas it serves.

Job postings may be a good place to learn about growth plans of the airline. We start with an advanced job search for all jobs at Spirit Airlines. Then we add the locations of the major cities served by the airline. So we go to indeed.com and conduct searches across major cities served by the airline. We see 50 jobs listed, including 27 in Miami, 3 in Detroit, and 3 in Dallas.

Glassdoor is another jobs site. It lists 272 jobs at Spirit Airlines on the date of our search. Clicking on the "Salaries" link, we see the salary breakdown for both salaried and hourly workers. A web developer at Spirit might make $86 thousand a year, but a passenger service agent is making just under $10 an hour. We could continue on to compare these with compensation at other airlines.

Website statistics reflect customer interest in the services provided by a firm. This may be less true for Spirit Airlines because many of its sales will be through travel and ticketing service providers. We go to Compete.com for website data. We note that there were more than 1.8 million unique visitors to the Spirit Airlines site in September 2014. More importantly, we see a level pattern of activity (no growth) over the last year. It is not the website itself we are interested in here. It is the level of activity, which is an indicator of customer interest in the services that the airline provides.

We could also go to social media sites such as Twitter, Facebook, and Google+ to see what people are saying about the airline. This would be best accomplished by using an Application Programming Interface (API) for each of the social media sources (Russell 2014).

Financial data for Spirit Airlines (NASDAQ stock symbol: SAVE) are available through a variety of sources, including Yahoo! Finance Google Finance, and Bloomberg. Even better, we can gather extensive data about the firm from Quandl. Using R, we can download these data in an efficient manner, display time series, and create financial forecasts.

Of course, our competitive work would not be finished without looking at the Spirit Airlines website itself. There we will find extensive information about products and services. We can review the firm's annual report to shareholders to get general information about total revenues and numbers of employees. We can read press releases, such as this one from October 28, 2014: "Spirit Airlines Announces Record Third Quarter 2014 Results: Third Quarter 2014 Adjusted Net Income Increases 27.6 Percent to $73.9 Million."

Perhaps other airlines should have reason to be concerned about Spirit and its low-fare/amenity pricing approach to air travel. Note the trend in stock price over the last four years in figure 12.1.

Sean Campbell and Scott Swigart (2014) of Cascade Insights provide an excellent review of online methods for competitive intelligence, such as those we have demonstrated with the Spirit Airlines example. Research about Spirit Airlines is more fun than flying. The pointing and clicking is easy, and anything we do interactively can be programmed for automated data acquisition from the web. Competitive intelligence is a low-fare trip with no screaming infants.

Exhibit 12.1 shows the R program for gathering competitive intelligence for Spirit Airlines from the web and storing those data for future analysis. Airlines are particularly responsive to economic cycles. The program draws on numerous R packages for financial data acquisition, manipulation, analysis, and display (Grolemund and Wickham 2014; Lang 2014a; Lang 2014b; McTaggart and Daroczi 2014; Ryan 2014; Ryan and Ulrich 2014; Wickham and Chang 2014; Zeileis, Grothendieck, and Ryan 2014).

*Figure 12.1.* Competitive Intelligence: Spirit Airlines Flying High

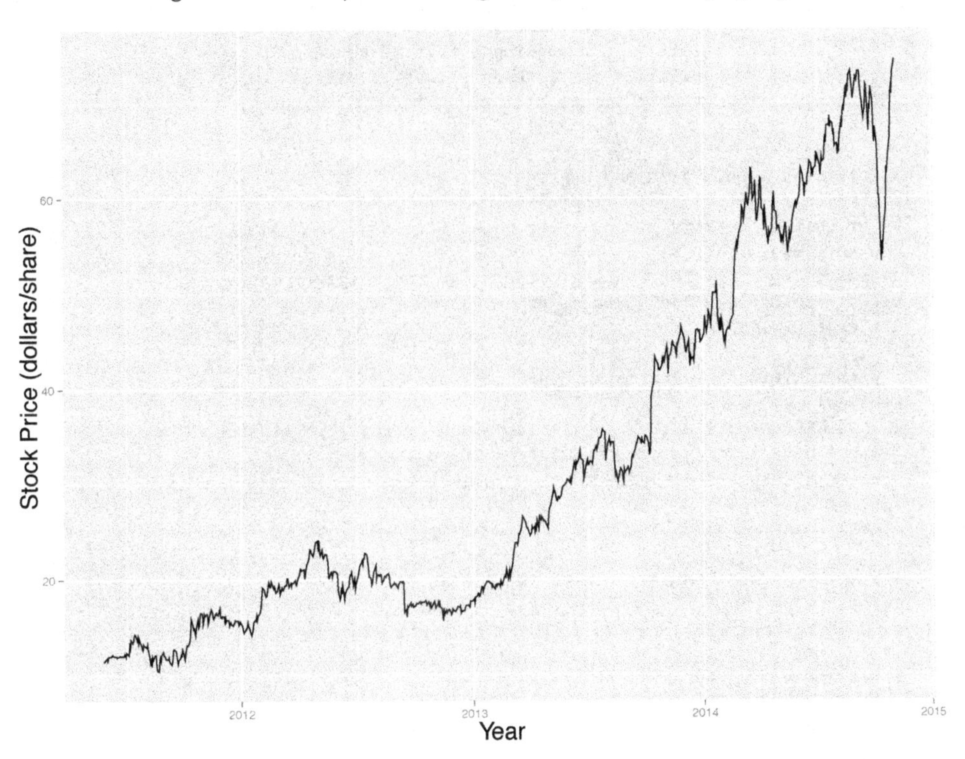

***Exhibit 12.1.*** *Competitive Intelligence: Spirit Airlines Financial Dossier (R)*

```
# Competitive Intelligence: Spirit Airlines Financial Dossier (R)

# install required packages

# bring packages into the workspace
library(RCurl)  # functions for gathering data from the web
library(XML)  # XML and HTML parsing
library(quantmod) # use for gathering and charting economic data
# online documentation for quantmod at <http://www.quantmod.com/>
library(Quandl)  # extensive financial data online
# online documentation for Quandl at <https://www.quandl.com/>
library(lubridate) # date functions
library(zoo)  # utilities for working with time series
library(xts)   # utilities for working with time series
library(ggplot2)  # data visualization

# ----------------------------------
# Text data acquisition and storage
# ----------------------------------
# get current working directory (commands for Mac OS or Linux)
cwd <- getwd()
# create directory for storing competitive intelligence data
ciwd <- paste(cwd, "/ci_data/", sep ="")
dir.create(ciwd)

# gather Wikipedia data using RCurl package
wikipedia_web_page <-
    getURLContent('http://en.wikipedia.org/wiki/Spirit_Airlines')
# store data in directory ciwd for future processing
sink(paste(ciwd, "/", "wikipedia_web_page", sep = ""))
wikipedia_web_page
sink()

# similar procedures may be used for all acquired data
# for the Spirit Airlines competitive intelligence study
# use distinct file names to identify the data sources

# -----------------------------------------------------------------
# Yahoo! Finance for Spirit Airlines (NASDAQ stock symbol: SAVE)
# -----------------------------------------------------------------
# stock symbols for companies can be obtained from Yahoo! Finance
# <http://finance.yahoo.com/lookup>

# get Spirit Airlines stock price data
getSymbols("SAVE", return.class = "xts", src = "yahoo")
print(str(SAVE)) # show the structure of this xts time series object
# plot the series stock price
chartSeries(SAVE,theme="white")
# examine the structure of the R data object
print(str(SAVE))
print(SAVE)
```

```
# convert character string row names to decimal date for the year
Year <- decimal_date(ymd(row.names(as.data.frame(SAVE))))
# Obtain the closing price of Spirit Airlines stock
Price <- as.numeric(SAVE$SAVE.Close)
# create data frame for Spirit Airlines Year and Price for future plots
SPIRIT_data_frame <- data.frame(Year, Price)

# similar procedures may be used for all airline competitors

# -------------------------------------------------------------
# Google! Finance for Spirit Airlines (NASDAQ stock symbol: SAVE)
#   (basically the same data as from Yahoo! Finance)
# -------------------------------------------------------------

# get Spirit Airlines stock price data
getSymbols("SAVE", return.class = "xts", src = "google")
print(str(SAVE)) # show the structure of this xts time series object
# plot the series stock price
chartSeries(SAVE,theme="white")
# examine the structure of the R data object
print(str(SAVE))
print(SAVE)

# ----------------------------------------
# ggplot2 time series plotting, closing
# price of Spirit Airlines common stock
# ----------------------------------------
# with a data frame object in hand... we can go on to use ggplot2
# and methods described in Chang (2013) R Graphics Cookbook

# use data frame defined from Yahoo! Finance
# the Spirit Airlines closing price per share SPIRIT_data_frame
# now we use that data structure to prepare a time series plot
# again, let's highlight the Great Recession on our plot for this time series
plotting_object <- ggplot(SPIRIT_data_frame, aes(x = Year, y = Price)) +
    geom_line() +
    ylab("Stock Price (dollars/share)") +
    ggtitle("Spirit Airlines Stock Price")
print(plotting_object)
# send the plot to an external file (sans title, larger axis labels)
pdf("fig_competitive_intelligence_spirit.pdf", width = 11, height = 8.5)
plotting_object <- ggplot(SPIRIT_data_frame, aes(x = Year, y = Price)) +
    geom_line() + ylab("Stock Price (dollars/share)") +
    theme(axis.title.y = element_text(size = 20, colour = "black")) +
    theme(axis.title.x = element_text(size = 20, colour = "black"))
print(plotting_object)
dev.off()

# ----------------------------------------
# FRED for acquiring general financial data
# ----------------------------------------
# general financial data may be useful in understanding what is
# happening with a company over time... here is how to get those data
```

```
# demonstration of R access to and display of financial data from FRED
# requires a connection to the Internet
# ecomonic research data from the Federal Reserve Bank of St. Louis
# see documentation of tags at http://research.stlouisfed.org/fred2/
# choose a particular series and click on it, a graph will be displayed
# in parentheses in the title of the graph will be the symbol
# for the financial series some time series are quarterly, some monthly
# ... others weekly... so make sure the time series match up in time
# see the documentation for quantmod at
# <http://cran.r-project.org/web/packages/quantmod/quantmod.pdf>

# here we show how to download the Consumer Price Index
# for All Urban Consumers: All Items, Not Seasonally Adjusted, Monthly
getSymbols("CPIAUCNS",src="FRED",return.class = "xts")
print(str(CPIAUCNS)) # show the structure of this xts time series object
# plot the series
chartSeries(CPIAUCNS,theme="white")

# Real Gross National Product in 2005 dollars
getSymbols("GNPC96",src="FRED",return.class = "xts")
print(str(GNPC96)) # show the structure of this xts time series object
# plot the series
chartSeries(GNPC96,theme="white")

# National Civilian Unemployment Rate,
#    not seasonally adjusted (monthly, percentage)
getSymbols("UNRATENSA",src="FRED",return.class = "xts")
print(str(UNRATENSA)) # show the structure of this xts time series object
# plot the series
chartSeries(UNRATENSA,theme="white")

# University of Michigan: Consumer Sentiment,
#    not seasonally adjusted (monthly, 1966 = 100)
getSymbols("UMCSENT",src="FRED",return.class = "xts")
print(str(UMCSENT)) # show the structure of this xts time series object
# plot the series
chartSeries(UMCSENT,theme="white")

# New Homes Sold in the US, not seasonally adjusted (monthly, thousands)
getSymbols("HSN1FNSA",src="FRED",return.class = "xts")
print(str(HSN1FNSA)) # show the structure of this xts time series object
# plot the series
chartSeries(HSN1FNSA,theme="white")

# ---------------------------------------
# Multiple time series plots
# ---------------------------------------
# let's try putting consumer sentiment and new home sales on the same plot

# University of Michigan Index of Consumer Sentiment (1Q 1966 = 100)
getSymbols("UMCSENT", src="FRED", return.class = "xts")
ICS <- UMCSENT # use simple name for xts object
dimnames(ICS)[2] <- "ICS" # use simple name for index
```

```
chartSeries(ICS, theme="white")
ICS_data_frame <- as.data.frame(ICS)
ICS_data_frame$date <- ymd(rownames(ICS_data_frame))
ICS_time_series <- ts(ICS_data_frame$ICS,
  start = c(year(min(ICS_data_frame$date)), month(min(ICS_data_frame$date))),
  end = c(year(max(ICS_data_frame$date)),month(max(ICS_data_frame$date))),
  frequency=12)

# New Homes Sold in the US, not seasonally adjusted (monthly, millions)
getSymbols("HSN1FNSA",src="FRED",return.class = "xts")
NHS <- HSN1FNSA
dimnames(NHS)[2] <- "NHS" # use simple name for index
chartSeries(NHS, theme="white")
NHS_data_frame <- as.data.frame(NHS)
NHS_data_frame$date <- ymd(rownames(NHS_data_frame))
NHS_time_series <- ts(NHS_data_frame$NHS,
  start = c(year(min(NHS_data_frame$date)),month(min(NHS_data_frame$date))),
  end = c(year(max(NHS_data_frame$date)),month(max(NHS_data_frame$date))),
  frequency=12)

# define multiple time series object
economic_mts <- cbind(ICS_time_series,
  NHS_time_series)
  dimnames(economic_mts)[[2]] <- c("ICS","NHS") # keep simple names
modeling_mts <- na.omit(economic_mts) # keep overlapping time intervals only

# examine the structure of the multiple time series object
# note that this is not a data frame object
print(str(modeling_mts))

# -----------------------------------------
# Prepare data frame for ggplot2 work
# -----------------------------------------
# for zoo examples see vignette at
# <http://cran.r-project.org/web/packages/zoo/vignettes/zoo-quickref.pdf>
modeling_data_frame <- as.data.frame(modeling_mts)
modeling_data_frame$Year <- as.numeric(time(modeling_mts))

# examine the structure of the data frame object
# notice an intentional shift to underline in the data frame name
# this is just to make sure we keep our object names distinct
# also you will note that programming practice for database work
# and for work with Python is to utilize underlines in variable names
# so it is a good idea to use underlines generally
print(str(modeling_data_frame))
print(head(modeling_data_frame))

# -------------------------------------------
# ggplot2 time series plotting of economic data
# -------------------------------------------
# according to the National Bureau of Economic Research the
# Great Recession extended from December 2007 to June 2009
# using our Year variable this would be from 2007.917 to 2009.417
```

```
# let's highlight the Great Recession on our plot
plotting_object <- ggplot(modeling_data_frame, aes(x = Year, y = ICS)) +
    geom_line() +
    annotate("rect", xmin = 2007.917, xmax = 2009.417,
        ymin = min(modeling_data_frame$ICS),
        ymax = max(modeling_data_frame$ICS),
        alpha = 0.3, fill = "red") +
    ylab("Index of Consumer Sentiment") +
    ggtitle("Great Recession and Consumer Sentiment")
print(plotting_object)

# ----------------------------------------------------------------
# Quandl for Spirit Airlines (NASDAQ stock symbol: SAVE)
# obtain more extensive financial data for Spirit Airlines
# more documentation at http://blog.quandl.com/blog/using-quandl-in-r/
# ----------------------------------------------------------------
Spirit_Price <- Quandl("GOOG/NASDAQ_SAVE", collapse="monthly", type="ts")
plot(stl(Spirit_Price[,4],s.window="periodic"),
    main = "Time Series Decomposition for Spirit Airlines Stock Price")

# Suggestions for the student: Employ search and crawling code to access
# all of the competitive intelligence reports cited in the chapter.
# Save these data in the directory, building a text corpus for further study.
# Scrape and parse the text documents using XPath and regular expressions.
# Obtain stock price series for all the major airlines and compare those
# series with Spirit's. See if there are identifiable patterns in these
# data and if those patterns in any way correspond to economic conditions.
# (Note that putting the economic/monthly and stock price/daily data
#   into the same analysis will require a periodicity change using the
#   xts/zoo data for Spirit Airlines. Refer to documentation at
#   <http://www.quantmod.com/Rmetrics2008/quantmod2008.pdf>
#   or from Quandl at <http://blog.quandl.com/blog/using-quandl-in-r/>.)
# Conduct a study of ticket prices on a round trips between two cities
# (non-stop flights between Los Angeles and Chicago, perhaps).
# Crawl and scrape the pricing data to create a pricing database across
# alternative dates into the future, taking note of the day and time
# of each flight and the airline supplying the service. Develop a
# competitive pricing model for airline travel between these cities.
# Select one of the competitive airlines as your client, and report
# the results of your competitive intelligence and competitive pricing
# research. Make strategic recommendations about what to do about
# the low-fare/amenities pricing approach of Spirit Airlines.
```

# 13 Predicting Sales

"Gentlemen, you can't fight in here! This is the War Room!"

—PETER SELLERS AS PRESIDENT MERKIN MUFFLEY IN *Dr. Strangelove or: How I Learned to Stop Worrying and Love the Bomb* (1964)

We are nearing the end of our journey through marketing data science. Soon it will be time for the beach.

I do not fit into the muscle-beach scene at Venice or Malibu, and I find Santa Monica too pricey. Redondo would be fine, Manhattan as well. But between those two is my favorite beach in the world, Hermosa Beach.

I first visited Hermosa Beach in my late twenties. It was during a month-long winter-term break from teaching. I rented an apartment above a bar on Pier Avenue and listened to rock music late into every night. I played volleyball with the locals and learned how to jump on the sand. I bought a hot bicycle for $25 and rode along the strand ten to fifteen miles a day.

When I returned to Hermosa many years later, I was surprised to see how little it had changed. The ocean has a way of promoting continuity. I am thankful for that.

And location? There is much to be said about the value of location.

Making predictions about sales is the job of marketing data science. For some problems we are asked to predict future sales for a company or make sales forecasts for product lines—this is time series sales forecasting, which is covered in appendix A (page 286). Here we focus on predicting sales in retail site selection.

Site selection problems involve predicting sales by location. A retailer with fifty stores in one geographic area wants to open stores in a new area. The retailer wants the new stores to be profitable, with sales revenue much higher than costs. Consumer demographics and business data can be used to guide the selection of new sites, and it is common to organize data by location and perform a cross-sectional analysis.

Site selection problems usually involve data sets with many more explanatory variables than there are stores. The challenge is to find the right combination of explanatory variables to predict sales at existing stores and then to use that combination of variables to get accurate sales forecasts for new stores.

Each store or potential store site, geocoded by longitude and latitude, represents a point on a map and may be associated with thousands of variables about population, housing, and economic conditions. We use geographical information systems to estimate population within a certain distance of the store. We get drive-time-based measures such as median income for households within five minutes of the store. We describe store sites by size and layout, location relative to nearby highways, signage, and parking. We collect data about the business environment, nearby retailers, and potential competitors. And when consumer data are available, we compute trade-area-based measures such as the percentage of families who shop at the store among the ten thousand families residing closest to the store. All of these can be used as explanatory variables in models for predicting store sales.

In site selection, we often employ cross-sectional models, ignoring the spatial data aspects of the problem. We use a store's census identifier or zip code to link sales and census data for each store or site. An example of the cross-sectional approach is shown for the thirty-three sites known as Studenmund's Restaurants, as discussed in appendix C (page 373).

*Table 13.1.* *Fitted Regression Model for Restaurant Sales*

| | Response: Sales |
|---|---|
| Competition | -9.075e+034*** |
| Population | 3.547e-01*** |
| Income | 1.288e+00*** |
| Constant | 1.022e+05*** |
| Observations | 33 |
| $R^2$ | 0.6182 |
| Adjusted $R^2$ | 0.5787 |
| Residual Std. Error | $14540 (df = 29)$ |
| F statistic | $15.65^{***} (df = 3; 29)$ |
| *Notes:* | ***Significant at the 0.001 level. |

*Table 13.2.* *Predicting Sales for New Restaurant Sites*

| *competition* | *population* | *income* | *predicted sales* |
|---|---|---|---|
| 2 | 50,000 | 25,000 | 133,975 |
| 3 | 200,000 | 22,000 | 174,236 |
| 5 | 220,000 | 19,000 | 159,317 |

We begin as we often do, with exploratory data analysis, plotting univariate histograms, densities, and empirical cumulative distribution functions. Then we move to looking at pairwise relationships, as shown in figures 13.1 and 13.2.

The linear regression of sales on competition, population, and income is summarized in table 13.1. Variance inflation factors computed for this problem show that multicollinearity is not an issue. Diagnostic plots in figure 13.3 indicate no special issues regarding the fitted model.

Applying the fitted model to prospective restaurant sites involves finding the associated explanatory variable values for each site and computing the predicted sales. Table 13.2 shows the results for three prospective store sites. The middle site has the highest predicted sales, so this would be the recommended site for the next new restaurant.

*Figure 13.1. Scatter Plot Matrix for Restaurant Sales and Explanatory Variables*

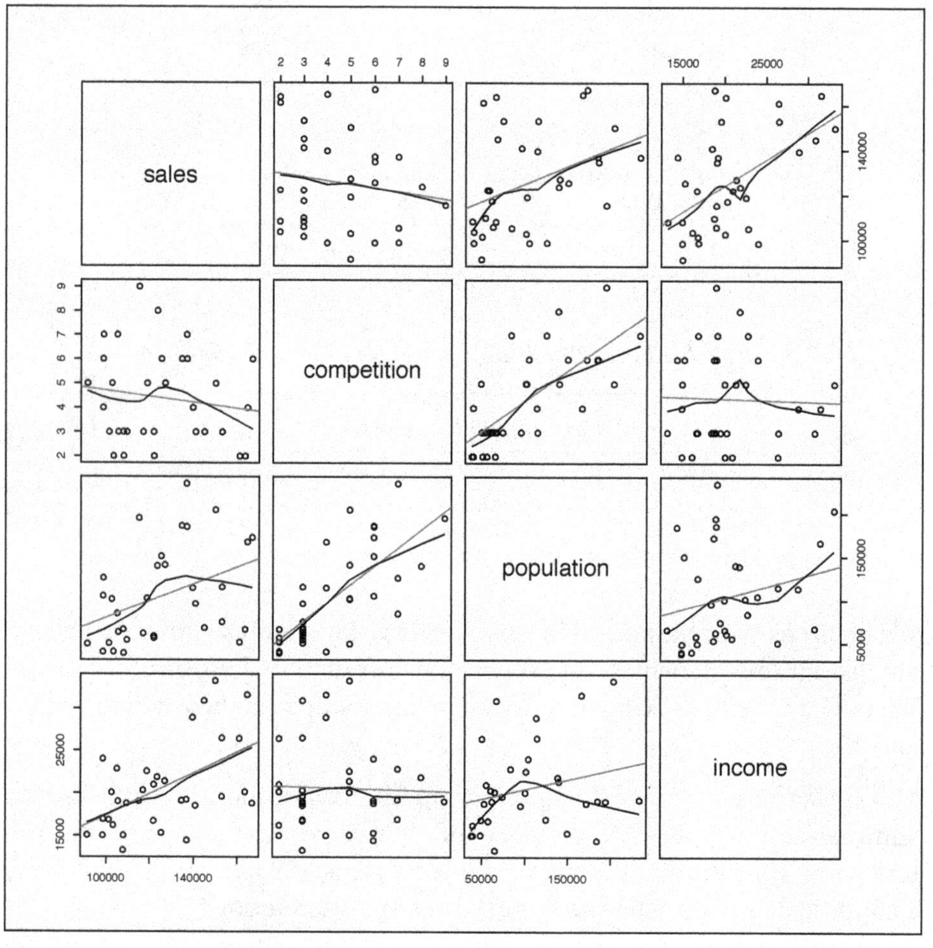

*Figure 13.2.* *Correlation Heat Map for Restaurant Sales and Explanatory Variables*

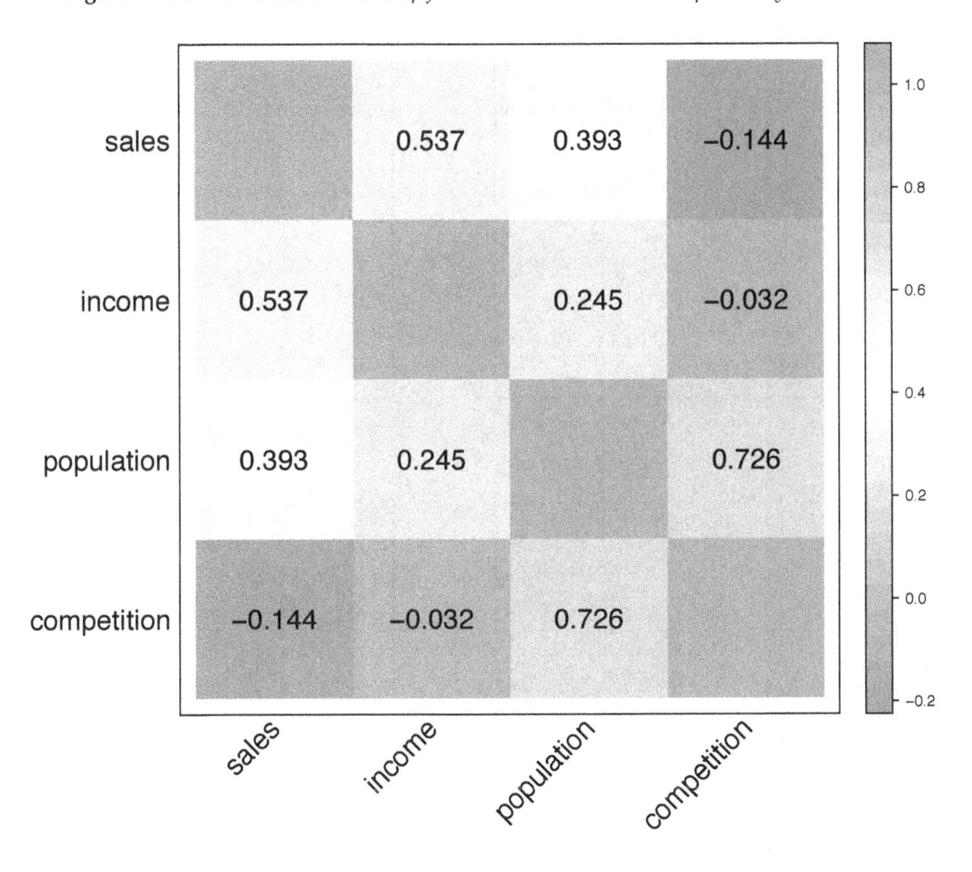

***Figure 13.3.*** *Diagnostics from Fitted Regression Model*

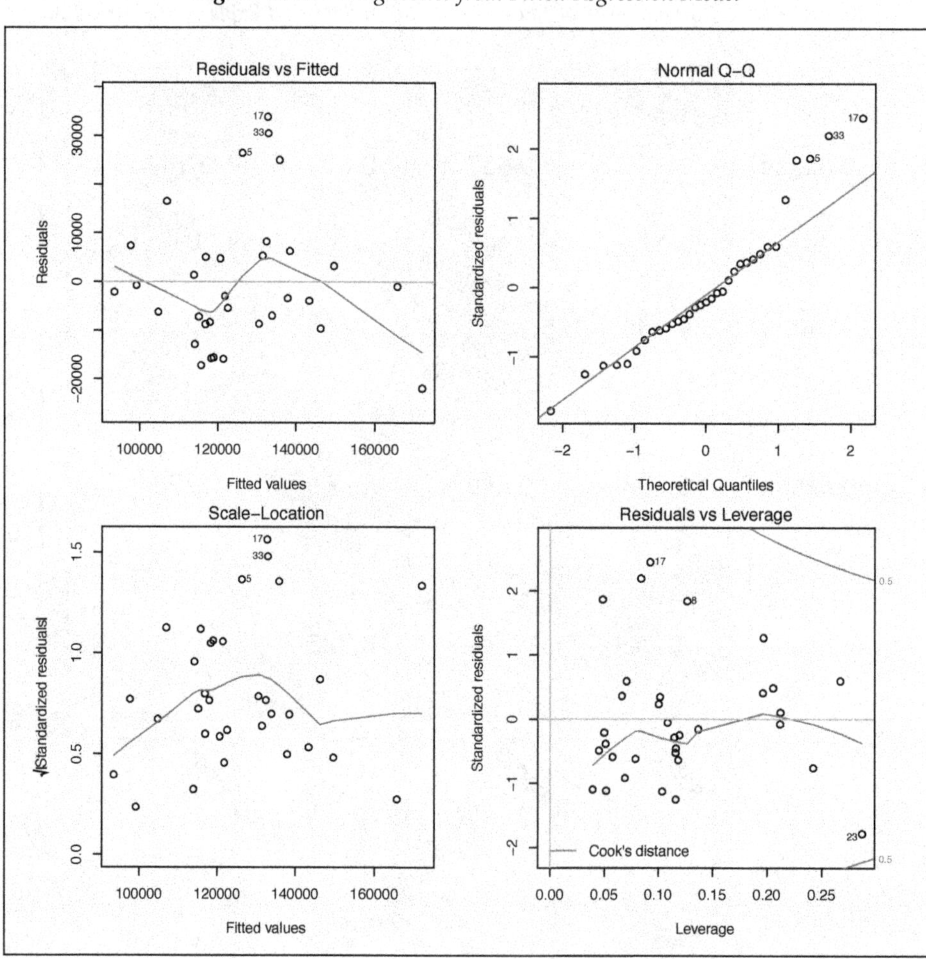

Studenmund's Restaurants is a small site selection problem that shows what is possible with ordinary least squares regression for predicting sales response. Moving to larger problems, we can take what is possible with one small subset of explanatory variables and extend it to many small subsets of explanatory variables—we can develop regression ensembles.

Site selection problems often involve thousands of potential explanatory variables and small numbers of current sites. While there are formal statistical methods for dealing with high-dimensional data (Bühlmann and van de Geer 2011), these are of little use in dealing with problems on the scale we encounter with site selection. What are needed instead are trustworthy model-building heuristics, techniques for partitioning the explanatory variable space into smaller subspaces, and then working with each subspace in turn. Let us review this divide-and-conquer approach.

Partitioning of explanatory variables into meaningful groups can be a first step in dealing with large numbers of explanatory variables. Taken together, these variable groups cover the full range of the explanatory variable space. Some variables relate to demographics, others to business characteristics. Some measures are set at a one-mile radius from the site, others at two-miles, five-miles, and so on. Working with road network overlays in a geographic information system, we can also derive drive-time measures, which may be partitioned into sets at one minute, five minutes, and ten minutes from the site, for example.

Next, working with each group of explanatory variables, we can use techniques such as tree-structured regression and random forests to provide lists of the most important explanatory variables—the best predictors of sales response. With a set of thirty or so best explanatory variables, we go on to employ all possible regressions, identifying the very best subset model within each group of explanatory variables.

Repeating the subset selection and best-possible-regression search for each group of explanatory variables, we arrive at an ensemble of predictive models. The ensemble predictor can be a simple average of the predictors or a composite with weights proportional to the anticipated out-of-sample predictive accuracy of the individual models. We can use linear regression or any number of machine learning methods in hybrid component models.

Site selection problems challenge us to practice the art as well as the science of modeling. The real power of the heuristic approach we call divide-and-conquer flows from the fact that the final predictive model is an ensemble, covering the full range of the explanatory variable space. Furthermore, the entire process described here can be embedded within an internal cross-validation scheme to protect against over-fitting.

The edited volumes by Davies and Rogers (1984) and Wrigley (1988) review the historical development of cross-sectional methods of inquiry for site selection. Peterson (2004) reviews commercial supplier work in this area. Alternative approaches include nearest-neighbor methods and spatial data models based on point processes, grid, or lattice structures (Cressie 1993; Bivand, Pebesma, and Gómez-Rubio 2008).

Gravity models for retail site selection posit that shoppers are influenced by the size and location of competing stores. Shoppers are thought to be attracted to larger stores due to greater variety or lower prices. Shoppers are thought to be attracted to nearby stores because travel times to nearby stores are shorter than travel times to distant stores. Tayman and Pol (1995) and Lilien and Rangaswamy (2003) describe gravity models for retail site selection.

We can perform location-based site selection and time series sales forecasting simultaneously with longitudinal or panel data models. Frees and Miller (2004) illustrate the process using data from the Wisconsin Lottery Sales case in appendix C (page 389).

Site selection research and location-based marketing are supported by geo-demographic data aggregators and providers of geographic information systems. Alteryx provides commercial solutions integrated with the R software environment.

Exhibit 13.1 shows an R program that analyzes data from Studenmund's Restaurants, drawing on regression tools from Fox (2014). The corresponding Python program is in exhibit 13.2.

***Exhibit 13.1.*** *Restaurant Site Selection (R)*

```
# Restaurant Site Selection (R)

# brind packages into workspace
library(car)  # regression tools
library(lattice)  # needed for correlation _heat_map function

load("correlation_heat_map.RData")  # from R utility programs

# read data for Studenmund's Restaurants
# creating data frame restdata
restdata <- read.csv("studenmunds_restaurants.csv", header = TRUE)

# examine the data frame
print(str(restdata))
print(restdata)

# compute summary statistics
print(summary(restdata))

# exploratory data analysis... graphics for discovery
# cumulative distribution function of the sales response
pdf(file = "fig_selecting_sites_hist.pdf",
    width = 8.5, height = 8.5)
with(restdata, hist(sales/1000,
    xlab="Sales (thousands)",
    ylab="Frequency",
    main = "", las = 1))
dev.off()

pdf(file = "fig_selecting_sites_cdf.pdf",
    width = 8.5, height = 8.5)
with(restdata, plot(sort(sales/1000),
    (1:length(sales))/length(sales),
    type="s", ylim=c(0,1), las = 1,
    xlab="Restaurants Ordered by Sales (thousands)",
    ylab="Proportion of Restaurants with Lower Sales"))
dev.off()

# scatter plot matrix with simple linear regression
# models and lowess smooth fits for variable pairs
pdf(file = "fig_selecting_sites_scatter_plot_matrix.pdf",
    width = 8.5, height = 8.5)
pairs(restdata,
    panel = function(x, y) {
        points(x, y)
        abline(lm(y ~ x), lty = "solid", col = "red")
        lines(lowess(x, y))
        }
    )
dev.off()
```

```
# correlation heat map
pdf(file = "fig_selecting_sites_correlation_heat_map.pdf",
    width = 8.5, height = 8.5)
restdata_cormat <-
    cor(restdata[,c("sales","competition","population","income")])
correlation_heat_map(cormat = restdata_cormat)
dev.off()

# specify regression model
restdata_model <- {sales ~ competition + population + income}

# fit linear regression model
restdata_fit <- lm(restdata_model, data = restdata)

# report fitted linear model
print(summary(restdata_fit))

# examine multicollinearity across explanatory variables
# ensure that all values are low (say, less than 4)
print(vif(restdata_fit))

# default residuals plots. . . diagnostic graphics
pdf(file = "fig_selecting_sites_residuals.pdf",
    width = 8.5, height = 8.5)
par(mfrow=c(2,2),mar=c(4, 4, 2, 2))
plot(restdata_fit)
dev.off()

# define data frame of sites for new restaurants
sites <- data.frame(sales = c(NA, NA, NA),
    competition = c(2, 3, 5),
    population = c(50000, 200000, 220000),
    income = c(25000, 22000, 19000))

# obtain predicted sales for the new restaurants
# rounding to the nearest dollar
sites$predicted_sales <- round(predict(restdata_fit, newdata = sites),0)
print(sites)

# Suggestions for the student: Employ alternative methods of regression
# to predict sales response. Compare results with those obtained from
# ordinary least squares regression. Examine the out-of-sample predictive
# power of models within a cross-validation framework.
# Having predicted sales for a cross-sectional/site selection problem,
# try a time series forecasting problem, working with one of the
# cases provided for this purpose: Lydia Pinkham's Medicine Company or
# Wisconsin Lottery Sales.
```

***Exhibit* 13.2.** *Restaurant Site Selection (Python)*

```
# Restaurant Site Selection (Python)

# prepare for Python version 3x features and functions
from __future__ import division, print_function

# import packages for analysis and modeling
import pandas as pd  # data frame operations
import numpy as np  # arrays and math functions
import statsmodels.api as sm  # statistical models (including regression)
import statsmodels.formula.api as smf  # statistical models (including regression)

# read data for Studenmund's Restaurants
# creating data frame restdata
restdata = pd.read_csv('studenmunds_restaurants.csv')

# print the first five rows of the data frame
print(pd.DataFrame.head(restdata))

# specify regression model
my_model = str('sales ~ competition + population + income')

# fit the model to the data
my_model_fit = smf.ols(my_model, data = restdata).fit()
# summary of model fit to the training set
print(my_model_fit.summary())
# predictions from the model fit to the data for current stores
restdata['predict_sales'] = my_model_fit.fittedvalues

# compute the proportion of response variance accounted for
print('\nProportion of Test Set Variance Accounted for: ',\
    round(np.power(restdata['sales'].corr(restdata['predict_sales']),2),3))

# define DataFrame of sites for new restaurants
sites_data = {'sales': [0,0,0],
             'competition': [2, 3, 5],
             'population': [50000, 200000, 220000],
             'income': [25000, 22000, 19000]}

sites = pd.DataFrame(sites_data)

# obtain predicted sales for the new restaurants
# rounding to the nearest dollar
sites['sales_pred'] = my_model_fit.predict(sites)
print('\nNew sites with predicted sales', sites, '\n')
```

# 14 Redefining Marketing Research

"The whole purpose of places like Starbucks is for people with no decision-making ability whatsoever to make six decisions just to buy one cup of coffee. So people who don't know what the hell they're doing or who on earth they are can, for only $2.95, get not just a cup of coffee but an absolutely defining sense of self: Tall. Decaf. Cappuccino."

—TOM HANKS AS JOE FOX IN *You've Got Mail* (1998)

Data science has emerged as a new discipline. A combination of business, information technology, and statistics, it speaks to the needs of organizations in a data-rich, data-intensive world. Analytics help us turn data into information and information into intelligence.

The world has been transformed by information technology and instant communications. Some say that data science is the new statistics. Perhaps it is the new IT and business as well. And it is fast becoming the new look of marketing research.

The World Wide Web is both an information store and the medium through which firms gather new information. Research providers and clients alike employ methods for automated data acquisition on the web, gathering data from surfing, crawling, scraping, online surveys, blogs, wikis, and social media. There is continued growth in online research, which has proven to be an efficient modality for data collection and reporting.

With techniques for flexible, data-adaptive modeling and sophisticated software for statistical graphics and data reduction, we are more productive in providing information to management. We see expanding demand for information services, with corporations collecting more customer data than ever before. Firms pursue new markets that require exploration and understanding. There are opportunities in the areas of multicultural and multinational research. There is a growing need for people who can design research studies, make sense out of data, and communicate with management. These are exciting times for data science in general and for the specialization we call marketing data science.

These are also challenging times for researchers. We have seen reductions in response rates across a variety of media. Do-not-call regulations affect research as well as their intended targets of telemarketing. There are public concerns about privacy. Management makes demands for accountability and return on research investments. There is pressure to execute studies as quickly and cheaply as possible while maintaining scientific rigor. And, as always, we must take care in designing marketing measures, fitting models, describing research findings, and recommending actions to management.

There are providers of marketing information—data scientists and research firms. Providers can be internal or external to the organization. There are clients or users of information—business leaders and marketing managers.

Consider the plight of an external provider, a traditional marketing research firm, as summarized in figure 14.1. The analysis draws on the economics of industrial organization and Michael Porter's (1980) five-forces model.

The custom research provider converts physical and human resources into information products. Physical inputs to the information production function include computer and communications hardware and software, offices, and office equipment. Human resources, especially important to information production, include clerical personnel, technical writers, telephone and face-to-face interviewers, focus group moderators and participants, survey respondents, database administrators, business analysts, management consultants, field researchers, and, of course, data scientists. Because there are many suppliers of labor, individual suppliers have little bargaining power.

*Figure 14.1.* *Competitive Analysis for the Custom Research Provider*

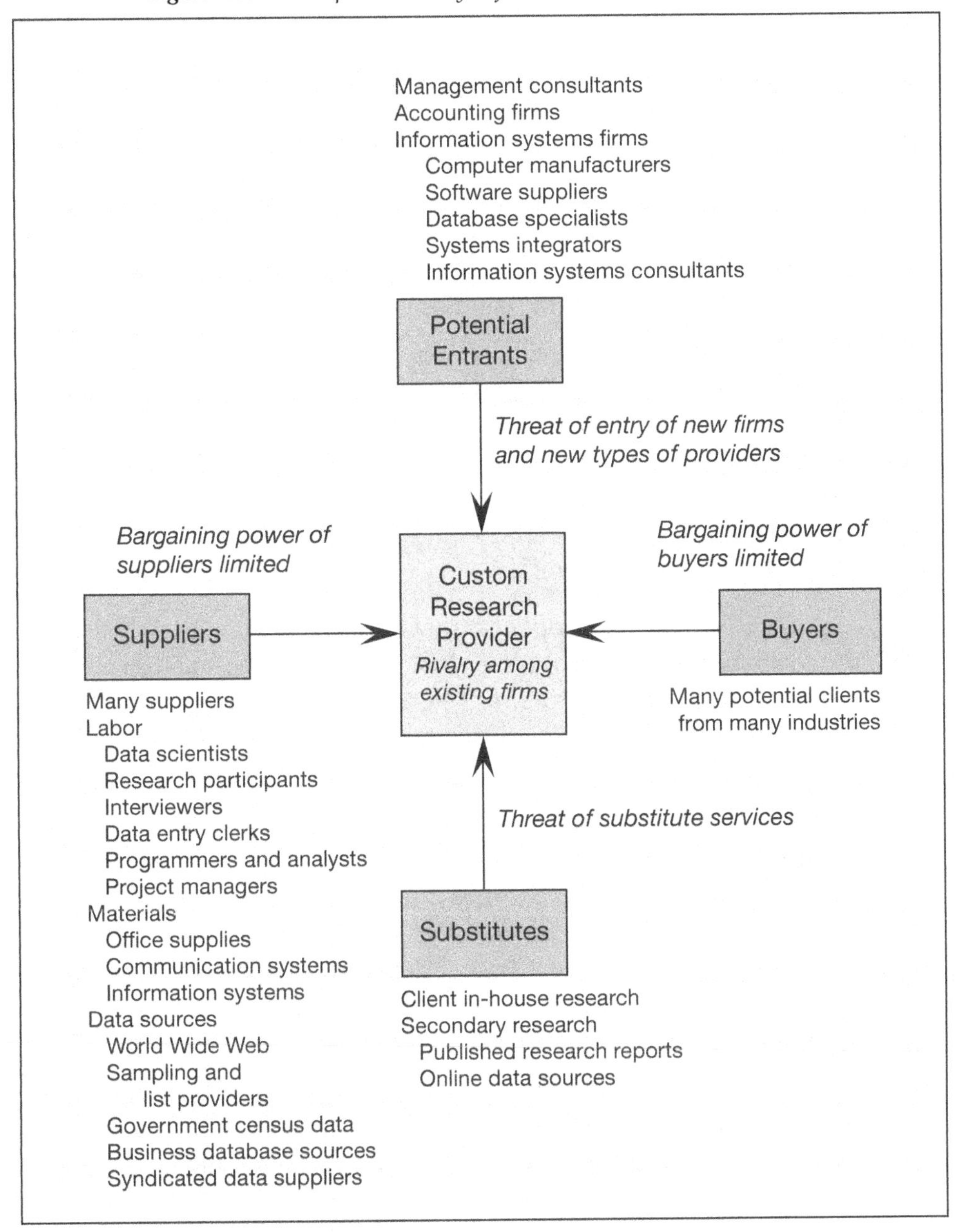


Adapted from Miller (2008).

Likewise, there are many suppliers of computer and communications equipment. Prices of data storage and processing continue to drop. And the availability of cloud-based services allows firms to convert fixed information systems costs into variable costs.

The bargaining power of labor varies with the type of labor. The supply of clerical workers, telephone interviewers, and data entry clerks is plentiful, and many unskilled laborers can be trained to fill these positions. Research respondents constitute a source of labor. These are great in number and hold little bargaining power. But variable costs of recruitment have increased in recent years as a result of reduced response rates. The supply of data scientists and research professionals is less plentiful.

To supplement the supply of data scientists and research professionals graduating from business and professional schools, many firms turn to graduates from the social sciences and humanities—students with strong communication skills, which are critical to success in marketing data science. Research skills are often learned on the job or through courses provided through professional organizations, commercial training firms, and massive open online courses (MOOCs).

There are many sources of supply, so that no individual source has power over the custom research provider. Just the same, the custom research provider has little power over its suppliers. In the language of economics, the custom research provider is a price-taker in the labor and capital markets of its supply.

What about the buyers of custom research? Custom research addresses the needs of organizational clients. These are the buyers in Porter's model, and there are many buyers. Sometimes information buyers and sellers lock themselves into multiyear contracts, but more often, they set up short-term contracts on a project-by-project basis. Because there are many potential buyers of business and economic information, individual buyers have limited bargaining power.

In the market for research services, there are few barriers to entry and many potential entrants. Management consulting and information technology firms can offer custom research. Political, social, and public opinion research firms can expand their research practices to include business and consumer research. Advertising, media, and publishing firms can supply research services. Firms specializing in governmental and nonprofit research can, with relatively little effort, add private sector service practices.

Less vulnerable to threats of entry are large syndicated data providers. There can be significant barriers to entry in the scanner data arena and in certain highly specialized research practices. Scanner data firms often maintain long-term contracts with their clients and have a lock on scanner data acquisition across nationwide retail outlets. There is an infrastructure in place for collecting, aggregating, and analyzing scanner data, and there are considerable economies of scale with operations of this type.

What are the alternatives to custom research from an external provider? Perhaps the most serious substitution threat comes from the clients themselves, who can decide to conduct research using internal resources rather than contracting with external providers. Do-it-yourself research is on the rise, given the availability of facilitating software for online surveys, open-source software for analytics, and technologies for automated data collection from the web.

In many, but not all contexts, secondary research represents a potential substitute for custom primary research. It is less expensive to acquire a copy of past research than to do new research. Web-based information and library searches substitute for custom research. Research reports and syndicated data sources substitute for custom research. Secondary research may be conducted with a client's internal as well as external data sources.

There is rivalry among existing research firms. For most custom research projects, hundreds, if not thousands of firms would be capable of providing the services clients require. For large multinational projects there are fewer bidders, but still a sufficient number to ensure a fiercely competitive environment. This is a fact of life of which provider firms are well aware.

In recent years, we have seen many research firms die or stop growing, not because the need for research and information services has declined, but because those firms became ensnared in the trappings of primary research. With all their interviewers, moderators, and tabulators, custom research providers were defined by what they did to collect data, rather than by the value they added to data.

Microeconomic theory teaches us that firms in competitive industries are price takers. When there are many firms and few barriers to entry, profit margins are low. To move away from competition toward a position of market power, a firm must distinguish itself from other firms. Building barriers to entry is one way to do this. Some firms succeed by having control over essential technologies, resources, or trade secrets. Some succeed through economies of scale. Some succeed by being in a market with economic barriers to entry, some by legal barriers to entry, such as trademarks, copyrights, and patents. And many firms succeed through product differentiation.

Research, information, and consulting services are intangible, intellectual products, defined by the people who deliver them. The data scientist, analyst, or researcher adds value to information. Each analysis reflects the capabilities and biases of the analyst, each report the insight of its author, each recommendation the integrity of the consultant. For many projects this is precisely what the client wants—the skills of a data scientist, the diligence of a careful researcher, the knowledge of a well-read scholar, the wisdom of a seasoned professional.

For specialized analytics, clients seek academic experts with knowledge about particular modeling techniques. For problems in selected application areas, such as new product testing, advertising research, brand development, or customer satisfaction, clients seek firms with demonstrated expertise in those areas of application.

Some firms succeed by focusing on particular client industries or geographical areas. Proprietary research product offerings, methods, measures, and norms flow from years of experience in selected product and service areas. Product differentiation fully nurtured and developed becomes a trade secret and a barrier to entry.

Industry analysis à la Porter is a useful first step in competitive analysis for provider research firms. What we learn about rivalry, threats of entry, substitute products, suppliers, and buyers helps us to understand the industry. And an understanding of the industry helps us to define business strategy.

To succeed as research providers in a data-driven, data-intensive world, managers need to think about a firm's current competitive position relative to other research providers. Managers need to think about prospects

*Figure 14.2.* *A Model for Strategic Planning*

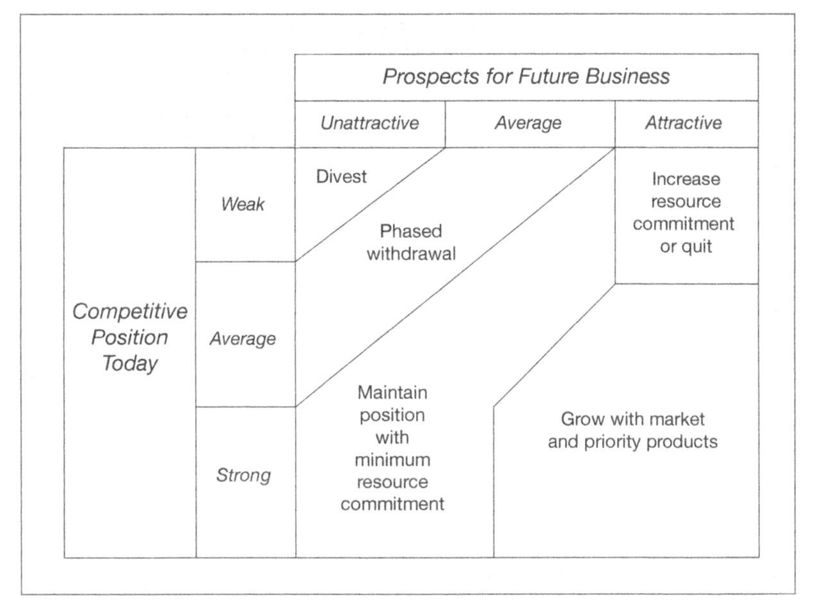


Adapted from Sutton (1980).

for future business. Figure 14.2 provides guidance in deciding on research projects and opportunities to pursue. The product portfolio strategies depicted in this figure have been utilized by major corporations dating back to the 1970s (Day 1986).

Just as we can describe businesses in terms of what they do with dollars and cents, euros or yen, we can describe businesses in terms of what they do with information. Information exchanges accompany business transactions as buyers and sellers interact.

Many of today's businesses are information-centric or information-driven. Customers find suppliers while searching the web. Business relationships grow through information exchange. Firms involved with electronic commerce grow with the information they gather, store, and analyze. Firms add value to information and generate additional information using traditional statistical models, machine learning methods, mathematical programming solutions, and simulation studies.

*Figure 14.3. Data Sources in the Information Supply Chain*

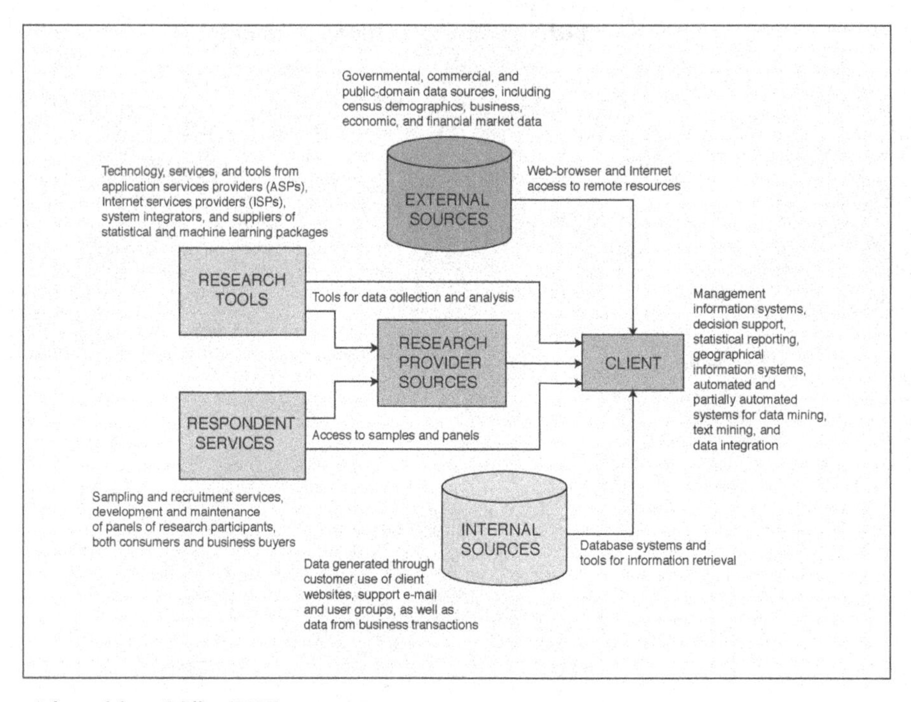


Adapted from Miller (2005).

Search engine and e-mail service providers Google, Yahoo!, and Microsoft are key players in the information landscape. Market-makers eBay and Amazon.com use electronic media to display information needed for efficient transactions between buyers and sellers. Having many eyes looking at computer screens translates into revenues for these information-centric firms. Information begets information.

Marketing data science relies on an information supply chain, as illustrated in figure 14.3. What has changed over the last twenty years is the extent to which client firms utilize external databases. Custom research providers have also modified their practices, becoming integrators of secondary data as well as collectors of primary data. Much of today's marketing research is conducted from the web and through the web. Think of the World Wide Web is a huge data repository, a path to the world's knowledge, and the research medium through which we develop new knowledge (Miller 2015c).

***Figure 14.4.*** *Client Information Sources and the World Wide Web*

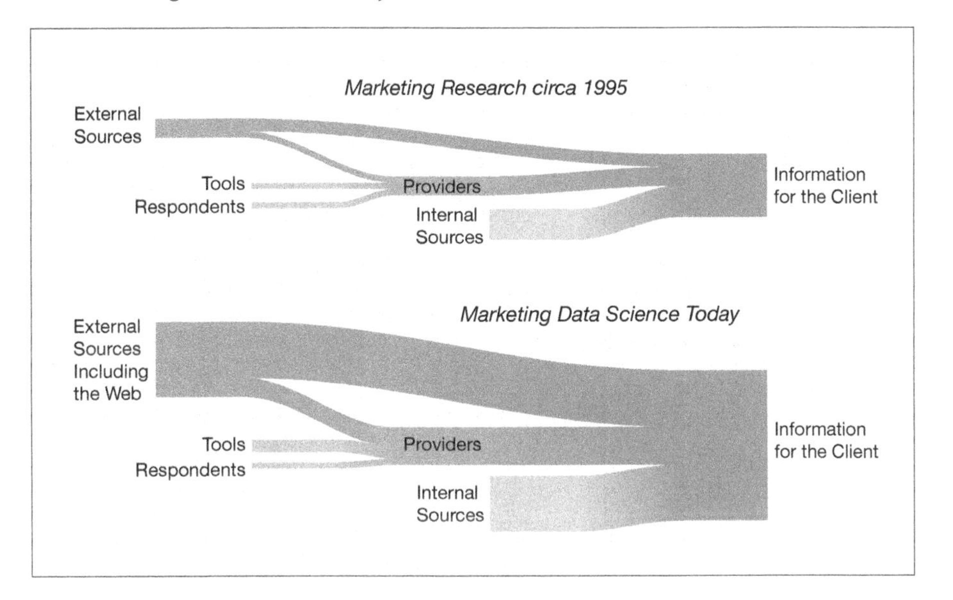


The story of the World Wide Web, recounted by its inventor, Tim Berners-Lee (1999), is the story of open, standards-based communications and information sharing. The web has indeed changed the world of business and, with those changes, the way marketing research is conducted. Figure 14.4 illustrates the direction of change in data sources across the information supply chain.

Extensive data are available to clients from internal and external sources. Clients and research providers alike are less reliant on primary research and more reliant on external, secondary data sources, including the web. New provider businesses emerge as aggregators of secondary data sources. Social media, mediated by the web, are added to the mix. Meanwhile, client internal databases grow as companies move significant portions of their businesses online, another consequence of the web.

The world of marketing data science holds great promise. With the proliferation of secondary data sources and the growing need for marketing information and competitive intelligence, we can expect new types of firms to emerge.

*Figure 14.5. Networks of Research Providers, Clients, and Intermediaries*

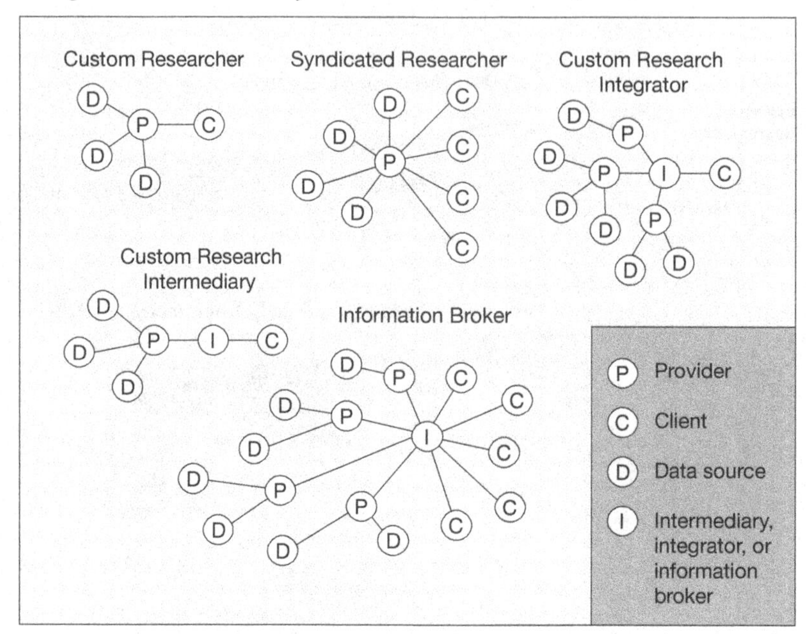


Adapted from Miller (2008).

Some firms will take on information brokering roles, offering subscription services focused on research reports, legal or business periodical sources. Other firms will fill the gap between subscription and public domain sources. There will be thousands of independent information brokers offering customized, individualized services.

Looking to the future, we can imagine a variety of business-to-business models of information exchange, as suggested by figure 14.5. The information supply chain or network includes intermediaries, integrators, or brokers working as agents between providers and clients. The path between the initial primary or secondary data source and the client can take many forms. The challenge for each firm or link along the chain will be to establish its unique contribution or information value added.

# A Data Science Methods

As marketing data scientists, we must speak the language of business—accounting, finance, marketing, and management. We need to know about information technology, including data structures, algorithms, and object-oriented programming. We must understand statistical modeling, machine learning, mathematical programming, and simulation methods. These are the things that we do:

- **Information search.** We begin by learning what others have done before, learning from the literature. We draw on the work of academics and practitioners in many fields of study, contributors to predictive analytics and data science.
- **Preparing text and data.** Text is unstructured or partially structured data that must be prepared for analysis. We extract features from text. We define measures. Quantitative data are often messy or missing. They may require transformation prior to analysis. Data preparation consumes much of a data scientist's time.
- **Looking at data.** We do exploratory data analysis, data visualization for the purpose of discovery. We look for groups in data. We find outliers. We identify common dimensions, patterns, and trends.
- **Predicting how much.** We are often asked to predict how many units or dollars of product will be sold, the price of financial securities or real estate. Regression techniques are useful for making these predictions.

- **Predicting yes or no.** Many business problems are classification problems. We use classification methods to predict whether or not a person will buy a product, default on a loan, or access a web page.
- **Testing it out.** We examine models with diagnostic graphics. We see how well a model developed on one data set works on other data sets. We employ a training-and-test regimen with data partitioning, cross-validation, or bootstrap methods.
- **Playing what-if.** We manipulate key variables to see what happens to our predictions. We play what-if games in simulated marketplaces. We employ sensitivity or stress testing of mathematical programming models. We see how values of input variables affect outcomes, payoffs, and predictions. We assess uncertainty about forecasts.
- **Explaining it all.** Data and models help us understand the world. We turn what we have learned into an explanation that others can understand. We present project results in a clear and concise manner.

Prediction is distinct from explanation. We may not know why models work, but we need to know when they work and when to show others how they work. We identify the most critical components of models and focus on the things that make a difference.[1]

Data scientists are methodological eclectics, drawing from many scientific disciplines and translating the results of empirical research into words and pictures that management can understand. These presentations benefit from well-constructed data visualizations. In communicating with management, data scientists need to go beyond formulas, numbers, definitions of terms, and the magic of algorithms. Data scientists convert the results of predictive models into simple, straightforward language that others can understand.

Data scientists are knowledge workers par excellence. They are communicators playing a critical role in today's data-intensive world. Data scientists turn data into models and models into plans for action.

The approach we have taken in this and other books in the *Modeling Techniques* series has been to employ both classical and Bayesian methods. And

[1] Statisticians distinguish between explanatory and predictive models. Explanatory models are designed to test causal theories. Predictive models are designed to predict new or future observations. See Geisser (1993), Breiman (2001), and Shmueli (2010).

sometimes we dispense with traditional statistics entirely and rely on machine learning algorithms.

Within the statistical literature, Seymour Geisser introduced an approach best described as *Bayesian predictive inference* (Geisser 1993). In emphasizing the success of predictions in marketing data science, we are in agreement with Geisser. But our approach is purely empirical and in no way dependent on classical or Bayesian thinking. We do what works, following a simple premise:

**The value of a model lies in the quality of its predictions.**

We learn from statistics that we should quantify our uncertainty. On the one hand, we have confidence intervals, point estimates with associated standard errors, significance tests, and *p*-values—that is the classical way. On the other hand, we have posterior probability distributions, probability intervals, prediction intervals, Bayes factors, and subjective (perhaps diffuse) priors—the path of Bayesian statistics.

The role of data science in business has been discussed by many (Davenport and Harris 2007; Laursen and Thorlund 2010; Davenport, Harris, and Morison 2010; Franks 2012; Siegel 2013; Maisel and Cokins 2014; Provost and Fawcett 2014). In-depth reviews of methods include those of Izenman (2008), Hastie, Tibshirani, and Friedman (2009), and Murphy (2012).

Doing data science means implementing flexible, scalable, extensible systems for data preparation, analysis, visualization, and modeling. We are empowered by the growth of open source. Whatever the modeling technique or application, there is likely a relevant package, module, or library that someone has written or is thinking of writing. Doing data science with open-source tools is discussed in Conway and White (2012), Putler and Krider (2012), James et al. (2013), Kuhn and Johnson (2013), Lantz (2013), and Ledoiter (2013). Additional discussion of data science, modeling techniques in predictive analytics, and open-source tools is provided in other books in the *Modeling Techniques* series (Miller 2015a, 2015b, and 2015c).

This appendix identifies classes of methods and reviews selected methods in databases and data preparation, statistics, machine learning, data visualization, and text analytics. We provide an overview of these methods and cite relevant sources for further reading.

## A.1 Database Systems and Data Preparation

There have always been more data than we have time to analyze. What is new today is the ease of collecting data and the low cost of storing data. Data come from many sources. There are unstructured text data from online systems. There are pixels from sensors and cameras. There are data from mobile phones, tablets, and computers worldwide, located in space and time. Flexible, scalable, distributed systems are needed to accommodate these data.

Relational databases have a row-and-column table structure, similar to a spreadsheet. We access and manipulate these data using structured query language (SQL). Because they are transaction-oriented with enforced data integrity, relational databases provide the foundation for sales order processing and financial accounting systems.

It is easy to understand why non-relational (NoSQL) databases have received so much attention. Non-relational databases focus on availability and scalability. They may employ key-value, column-oriented, document-oriented, or graph structures. Some are designed for online or real-time applications, where fast response times are key. Others are well suited for massive storage and off-line analysis, with map-reduce providing a key data aggregation tool.

Many firms are moving away from internally owned, centralized computing systems and toward distributed cloud-based services. Distributed hardware and software systems, including database systems, can be expanded more easily as the data management needs of organizations grow.

Doing data science means being able to gather data from the full range of database systems, relational and non-relational, commercial and open source. We employ database query and analysis tools, gathering information across distributed systems, collating information, creating contingency tables, and computing indices of relationship across variables of interest. We use information technology and database systems as far as they can take us, and then we do more, applying what we know about statistical inference and the modeling techniques of predictive analytics.

Regarding analytics, we acknowledge an unwritten code in data science. We do not select only the data we prefer. We do not change data to conform to what we would like to see or expect to see. A two of clubs that destroys the meld is part of the natural variability in the game and must be played with the other cards. We play the hand that is dealt. The hallmarks of science are an appreciation of variability, an understanding of sources of error, and a respect for data. Data science is science.

Raw data are unstructured, messy, and sometimes missing. But to use data in models, they must be organized, clean, and complete. We are often asked to make a model out of a mess. Management needs answers, and the data are replete with miscoded and missing observations, outliers and values of dubious origin. We use our best judgement in preparing data for analysis, recognizing that many decisions we make are subjective and difficult to justify.

Missing data present problems in applied research because many modeling algorithms require complete data sets. With large numbers of explanatory variables, most cases have missing data on at least one of the variables. Listwise deletion of cases with missing data is not an option. Filling in missing data fields with a single value, such as the mean, median, or mode, would distort the distribution of a variable, as well as its relationship with other variables. Filling in missing data fields with values randomly selected from the data adds noise, making it more difficult to discover relationships with other variables. Multiple imputation is preferred by statisticians.

Garcia-Molina, Ullman, and Widom (2009) and Connolly and Begg (2015) review database systems with a focus on the relational model. Worsley and Drake (2002) and Obe and Hsu (2012) review PostgreSQL. White (2011), Chodorow (2013), and Robinson, Webber, and Eifrem (2013) review selected non-relational systems. For MongoDB document database examples, see Copeland (2013) and Hoberman (2014). For map-reduce operations, see Dean and Ghemawat (2004) and Rajaraman and Ullman (2012).

Osborne (2013) provides an overview of data preparation issues, and the edited volume by McCallum (2013) provides much needed advice about what to do with messy data. Missing data methods are discussed in various sources (Rubin 1987; Little and Rubin 1987; Schafer 1997; Lumley 2010; Snijders and Bosker 2012), with methods implemented in R packages from Gelman et al. (2014), Honaker, King, and Blackwell (2014), and Lumley (2014).

## A.2 Classical and Bayesian Statistics

How shall we draw inferences from data? Formal scientific method suggests that we construct theories and test those theories with sample data. The process involves drawing statistical inferences as point estimates, interval estimates, or tests of hypotheses about the population. Whatever the form of inference, we need sample data relating to questions of interest. For valid use of statistical methods we desire a random sample from the population.

Which statistics do we trust? Statistics are functions of sample data, and we have more faith in statistics when samples are representative of the population. Large random samples, small standard errors, and narrow confidence intervals are preferred.

Classical and Bayesian statistics represent alternative approaches to inference, alternative ways of measuring uncertainty about the world. Classical hypothesis testing involves making null hypotheses about population parameters and then rejecting or not rejecting those hypotheses based on sample data. Typical null hypotheses (as the word *null* would imply) state that there is no difference between proportions or group means, or no relationship between variables. Null hypotheses may also refer to parameters in models involving many variables.

To test a null hypothesis, we compute a special statistic called a test statistic along with its associated $p$-value. Assuming that the null hypothesis is true, we can derive the theoretical distribution of the test statistic. We obtain a $p$-value by referring the sample test statistic to this theoretical distribution. The $p$-value, itself a sample statistic, gives the probability of rejecting the null hypothesis under the assumption that it is true.

Let us assume that the conditions for valid inference have been satisfied. Then, when we observe a very low $p$-value (0.05, 0.01, or 0.001, for instance), we know that one of two things must be true: either (1) an event of very low probability has occurred under the assumption that the null hypothesis is true or (2) the null hypothesis is false. A low $p$-value leads us to reject the null hypothesis, and we say the research results are statistically significant. Some results are statistically significant and meaningful. Others are statistically significant and picayune.

For applied research in the classical tradition, we look for statistics with low $p$-values. We define null hypotheses as straw men with the intention of rejecting them. When looking for differences between groups, we set up a null hypothesis that there are no differences between groups. In studying relationships between variables, we create null hypotheses of independence between variables and then collect data to reject those hypotheses. When we collect sufficient data, testing procedures have statistical power.

Variability is both our enemy and our friend. It is our enemy when it arises from unexplained sources or from sampling variability—the values of statistics vary from one sample to the next. But variability is also our friend because, without variability, we would be unable to see relationships between variables.[2]

While the classical approach treats parameters as fixed, unknown quantities to be estimated, the Bayesian approach treats parameters as random variables. In other words, we can think of parameters as having probability distributions representing of our uncertainty about the world.

The Bayesian approach takes its name from Bayes' theorem, a famous theorem in statistics. In addition to making assumptions about population distributions, random samples, and sampling distributions, we can make assumptions about population parameters. In taking a Bayesian approach, our job is first to express our degree of uncertainty about the world in the form of a probability distribution and then to reduce that uncertainty by collecting relevant sample data.

How do we express our uncertainty about parameters? We specify prior probability distributions for those parameters. Then we use sample data and Bayes' theorem to derive posterior probability distributions for those same parameters. The Bayesian obtains conditional probability estimates from posterior distributions.

Many argue that Bayesian statistics provides a logically consistent approach to empirical research. Forget the null hypothesis and focus on the research question of interest—the scientific hypothesis. There is no need to talk

[2] To see the importance of variability in the discovery of relationships, we can begin with a scatter plot of two variables with a high correlation. Then we restrict the range of one of the variables. More often than not, the resulting scatter plot within the window of the restricted range will exhibit a lower correlation.

about confidence intervals when we can describe uncertainty with a probability interval. There is no need to make decisions about null hypotheses when we can view all scientific and business problems from a decision-theoretic point of view (Robert 2007). A Bayesian probabilistic perspective can be applied to machine learning as well as traditional statistical models (Murphy 2012).

It may be a challenge to derive mathematical formulas for posterior probability distributions. Indeed, for many research problems, it is impossible to derive formulas for posterior distributions. This does not stop us from using Bayesian methods, however, because computer programs can generate or estimate posterior distributions. Markov chain Monte Carlo simulation is at the heart of Bayesian practice (Tanner 1996; Albert 2009; Robert and Casella 2009; Suess and Trumbo 2010).

Bayesian statistics is alive and well today because it helps us solve real-world problems (McGrayne 2011; Flam 2014). In the popular press, Silver (2012) makes a strong argument for taking a Bayesian approach to predictive models. As Efron (1986) points out, however, there are good reasons why everyone is not a Bayesian.

There are many works from which to learn about classical inference (Fisher 1970; Fisher 1971; Snedecor and Cochran 1989; Hinkley, Reid, and Snell 1991; Stuart, Ord, and Arnold 2010; O'Hagan 2010; Wasserman 2010). There are also many good sources for learning about Bayesian methods (Geisser 1993; Gelman, Carlin, Stern, and Rubin 1995; Carlin and Louis 1996; Robert 2007).

When asked if the difference between two groups could have arisen by chance, we might prefer a classical approach. We estimate a $p$-value as a conditional probability, given a null hypothesis of no difference between the groups. But when asked to estimate the probability that the share price of Apple stock will be above $100 at the beginning of the next calendar year, we may prefer a Bayesian approach. Which is better, classical or Bayesian? It does not matter. We need both. Which is better, Python or R? It does not matter. We need both.

## A.3 Regression and Classification

Much of the work of data science involves a search for meaningful relationships between variables. We look for relationships between pairs of continuous variables using scatter plots and correlation coefficients. We look for relationships between categorical variables using contingency tables and the methods of categorical data analysis. We use multivariate methods and multi-way contingency tables to examine relationships among many variables. And we build predictive models.

There are two main types of predictive models: *regression* and *classification.* Regression is prediction of a response of meaningful magnitude. Classification involves prediction of a class or category. In the language of machine learning, these are methods of supervised learning.

The most common form of regression is *least-squares regression,* also called ordinary least-squares regression, linear regression, or multiple regression. When we use ordinary least-squares regression, we estimate regression coefficients so that they minimize the sum of the squared residuals, where residuals are differences between the observed and predicted response values. For regression problems, we think of the response as taking any value along the real number line, although in practice the response may take a limited number of distinct values. The important thing for regression is that the response values have meaningful magnitude.

*Poisson regression* is useful for counts. The response has meaningful magnitude but takes discrete (whole number) values with a minimum value of zero. Log-linear models for frequencies, grouped frequencies, and contingency tables for cross-classified observations fall within this domain.

For models of events, duration, and survival, as in *survival analysis,* we must often accommodate censoring, in which some observations are measured precisely and others are not. With left censoring, all we know about imprecisely measured observations is that they are less than some value. With right censoring, all we know about imprecisely measured observations is that they are greater than some value.

A good example of a duration or survival model in marketing is customer lifetime estimation. We know the lifetime or tenure of a customer only after that person stops being our customer. For current customers, lifetime is imprecisely measured—it is right censored.

Most traditional modeling techniques involve *linear models* or linear equations. The response or transformed response is on the left-hand side of the linear model. The *linear predictor* is on the right-hand side. The linear predictor involves explanatory variables and is linear in its parameters. That is, it involves the addition of coefficients or the multiplication of coefficients by the explanatory variables. The coefficients we fit to linear models represent estimates of population parameters.

Generalized linear models, as their name would imply, are generalizations of the classical linear regression model. They include models for choices and counts, including logistic regression, multinomial logit models, log-linear models, ordinal logistic models, Poisson regression, and survival data models. To introduce the theory behind these important models, we begin by reviewing the classical linear regression model.

We can write the classical linear regression model in matrix notation as

$$\mathbf{y} = \mathbf{X}\mathbf{b} + \mathbf{e}$$

$\mathbf{y}$ is an $n \times 1$ vector of responses. $\mathbf{X}$ is an $n \times p$ matrix, with $p$ being the number of parameters being estimated. Often the first column of $\mathbf{X}$ is a column of ones for the constant or intercept term in the model; additional columns are for parameters associated with explanatory variables. $\mathbf{b}$ is a $p \times 1$ vector of parameter estimates. That is to say that $\mathbf{Xb}$ is linear predictor in matrix notation. The error vector $\mathbf{e}$ represents independent and identically distributed errors; it is common to assume a Gaussian or normal distribution with mean zero.

The assumptions of the classical linear regression model give rise to classical methods of statistical inference, including tests of regression parameters and analyses of variance. These methods apply equally well to observational and experimental studies. Parameters in the classical linear regression model are estimated by ordinary least squares. There are many variations on the theme, including generalized least squares and a variety of econometric models for time-series regression, panel (longitudinal) data models, and hierarchical models in general. There are also Bayesian alternatives for most classical models. For now, we focus on classical inference and the simplest error structure—independent, identically distributed (*iid*) errors.

Let $y$ be one element from the response vector, corresponding to one observation from the sample, and let $\mathbf{x}$ be its corresponding row from the matrix $\mathbf{X}$. Because the mean or expected value of the errors is zero, we observe that

$$E[y] = \mu = \mathbf{xb}$$

That is, $\mu$, the mean of the response, is equal to the linear predictor. Generalized linear models build from this fact. If we were to write in functional notation $g(\mu) = \mathbf{xb}$, then, for the Gaussian distribution of classical linear regression, $g$ is the identity function: $g(\mu) = \mu$.

Suppose $g(\mu)$ were the logit transformation. We would have the logistic regression model:

$$g(\mu) = log\left(\frac{\mu}{(1-\mu)}\right) = \mathbf{xb}$$

Knowing that the exponential function is the inverse of the natural logarithm, we solve for $\mu$ as

$$\mu = \frac{e^{\mathbf{xb}}}{1 + e^{\mathbf{xb}}}$$

For every observation in the sample, the expected value of the binary response is a proportion. It represents the probability that one of two events will occur. It has a value between zero and one. In generalized linear model parlance, $g$ is called the link function. It links the mean of the response to the linear predictor.

In most of the choice studies in this book, each observation is a binary response. Customers choose to stay with their current telephone service or switch to another service. Commuters choose to drive their cars or take the train. Probability theorists think of binary responses as Bernoulli trials with the proportion or probability $\mu$ representing the mean of the response. For $n$ observations in a sample, we have mean $n\mu$ and variance $\frac{\mu(1-\mu)}{n}$. The distribution is binomial.[3]

[3] In many statistical treatments of this subject, $\pi$, rather than $\mu$ is used to represent the mean of the response. We use $\mu$ here to provide a consistent symbol across the class of generalized linear models.

*Table A.1. Three Generalized Linear Models*

| Distribution | Link Function | Variance Function |
|---|---|---|
| Gaussian (normal) | $\mu$ | 1 |
| Poisson | $\log(\mu)$ | $\mu$ |
| Binomial | $log\left(\frac{\mu}{(1-\mu)}\right)$ | $\frac{\mu(1-\mu)}{n}$ |

Table A.1 provides an overview of the most important generalized linear models for work in business and economics. Classical linear regression has an identity link. Logistic regression uses the logit link. Poisson regression and log-linear models use a log link. We work Gaussian (normal), binomial, and Poisson distributions, which are in the exponential family of distributions. Generalized linear models have linear predictors of the form **Xb**. This is what makes them linear models; they involve functions of explanatory variables that are linear in their parameters.

Generalized linear models help us model what are obvious nonlinear relationships between explanatory variables and responses. Except for the special case of the Gaussian or normal model, which has an identity link, the link function is nonlinear. Also, unlike the normal model, there is often a relationship between the mean and variance of the underlying distribution.

The binomial distribution builds on individual binary responses. Customers order or do not order, respond to a direct marketing mailing or not. Customers choose to stay with their current telephone service or switch to another service. This type of problem lends itself to logistic regression and the use of the logit link. Note that the multinomial logit model is a natural extension of logistic regression. Multinomial logit models are useful in the analysis of multinomial response variables. A customer chooses Coke, Pepsi, or RC Cola. A commuter drives, takes the train or bus, or walks to work.

When we record choices over a period of time or across a group of individuals, we get counts or frequencies. Counts arranged in multi-way contingency tables comprise the raw data of categorical data analysis. We also use the Poisson distribution and the log link for categorical data analysis and log-linear modeling. As we have seen from our discussion of the logit transformation, the log function converts a variable defined on the domain of positive reals into a variable defined on the range of all real numbers. This is why it works for counts.

The Poisson distribution is discrete on the domain of non-negative integers. It is used for modeling counts, as in Poisson regression. The insurance company counts the number of claims over the past year. A retailer counts the number of customers responding to a sales promotion or the number of stock units sold. A nurse counts the number of days a patient stays in the hospital. An auto dealer counts the number of days a car stays on the lot before it sells.

Linear regression is a special generalized linear model. It has normally distributed responses and an identity link relating the expected value of responses to the linear predictor. Linear regression coefficients may be estimated by ordinary least squares. For other members of the family of generalized linear models we use maximum likelihood estimation. With the classical linear model we have analysis of variance and F-tests. With generalized linear models we have analysis of deviance and likelihood ratio tests, which are asymptotic chi-square tests.

There are close connections among generalized linear models for the analysis of choices and counts. Alternative formulations often yield comparable results. The multinomial model looks at the distribution of counts across response categories with a fixed sum of counts (the sample size). For the Poisson model, counts, being associated with individual cases, are random variables, and the sum of counts is not known until observed in the sample. But we can use the Poisson distribution for the analysis of multinomial data. Log-linear models make no explicit distinction between response and explanatory variables. Instead, frequency counts act as responses in the model. But, if we focus on appropriate subsets of linear predictors, treating response variables as distinct from explanatory variables, log-linear models yield results comparable to logistic regression. The Poisson distribution and the log link are used for log-linear models.

When communicating with managers, we often use R-squared or the coefficient of determination as an index of goodness of fit. This is a quantity that is easy to explain to management as the proportion of response variance accounted for by the model. An alternative index that many statisticians prefer is the *root-mean-square error* (RMSE), which is an index of badness or lack of fit. Other indices of badness of fit, such as the percentage error in prediction, are sometimes preferred by managers.

The method of *logistic regression*, although called "regression," is actually a classification method. It involves the prediction of a binary response. Ordinal and multinomial logit models extend logistic regression to problems involving more than two classes. Linear discriminant analysis is another classification method from the domain of traditional statistics. The benchmark study of text classification in the chapter on sentiment analysis employed logistic regression and a number of machine learning algorithms for classification.

Evaluating classifier performance presents a challenge because many problems are low base rate problems. Fewer than five percent of customers may respond to a direct mail campaign. Disease rates, loan default, and fraud are often low base rate events. When evaluating classifiers in the context of low base rates, we must look beyond the percentage of events correctly predicted. Based on the four-fold table known as the *confusion matrix*, figure A.1 provides an overview of various indices available for evaluating binary classifiers.

Summary statistics such as Kappa (Cohen 1960) and the area under the receiver operating characteristic (ROC) curve are sometimes used to evaluate classifiers. Kappa depends on the probability cut-off used in classification. The area under the ROC curve does not.

The area under the ROC curve is a preferred index of classification performance for low-base-rate problems. The ROC curve is a plot of the true positive rate against the false positive rate. It shows the tradeoff between sensitivity and specificity and measures how well the model separates positive from negative cases.

***Figure A.1.*** *Evaluating the Predictive Accuracy of a Binary Classifier*

Confusion Matrix

| | | Actual Binary Response: YES | Actual Binary Response: NO | |
|---|---|---|---|---|
| Predicted Binary Response | YES | **True Positive** **a** | **False Positive** **b** | a + b |
| | NO | **False Negative** **c** | **True Negative** **d** | c + d |
| | | a + c | b + d | n = a + b + c + d |

Expected accuracy by chance based upon marginal totals of the confusion matrix

$$E = \left[\frac{(a + b)(a + c) + (b + d)(c + d)}{n^2}\right]$$

$$\text{True Positive Rate} = \frac{a}{a + c}$$

The true positive rate, also called sensitivity, shows the proportion of positives (YES responses) that are correctly identified as positive. We want this to be high.

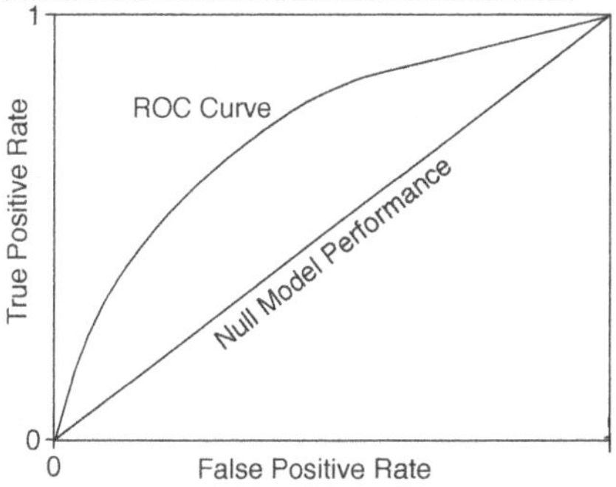

ROC Curve

Null Model Performance

True Positive Rate

False Positive Rate

1

0

0

$$\text{False Positive Rate} = \frac{b}{b + d}$$

The false positive rate shows the proportion of negatives (NO responses) incorrectly identified as positive. We want this to be low.

Observed predictive accuracy = proportion correctly classified
O = (a + d) / (a + b + c + d)

Precision = proportion of positive predictions that are true positives
= a / (a + b)

Specificity = true negative rate
= proportion of negatives that are correctly classified as negatives
= d / (b + d)
= 1 - (false positive rate)

Kappa = accuracy relative to accuracy expected by chance
= (O - E) / (1 - E)

The receiver operating characteristic (ROC) curve shows the performance of a binary classifier across the full range of decision criteria or cutoffs. Perfect performance would correspond to the point (0,1), that is, a false positive rate of 0 and a true positive rate of 1. The area under the curve provides a general index of classification performance.

The area under the ROC curve provides an index of predictive accuracy independent of the probability cut-off that is being used to classify cases. Perfect prediction corresponds to an area of 1.0 (curve that touches the top-left corner). An area of 0.5 depicts random (null-model) predictive accuracy.

Useful references for linear regression include Draper and Smith (1998), Harrell (2001), Chatterjee and Hadi (2012), and Fox and Weisberg (2011). Data-adaptive regression methods and machine learning algorithms are reviewed in Berk (2008), Izenman (2008), and Hastie, Tibshirani, and Friedman (2009). For traditional nonlinear models, see Bates and Watts (2007).

Of special concern to data scientists is the structure of the regression model. Under what conditions should we transform the response or selected explanatory variables? Should interaction effects be included in the model? Regression diagnostics are data visualizations and indices we use to check on the adequacy of regression models and to suggest variable transformations. Discussion may be found in Belsley, Kuh, and Welsch (1980) and Cook (1998). The base R system provides many diagnostics, and Fox and Weisberg (2011) provide additional diagnostics. Diagnostics may suggest that transformations of the response or explanatory variables are needed in order to meet model assumptions or improve predictive performance. A theory of power transformations is provided in Box and Cox (1964) and reviewed by Fox and Weisberg (2011).

When defining parametric models, we would like to include the right set of explanatory variables in the right form. Having too few variables or omitting key explanatory variables can result in biased predictions. Having too many variables, on the other hand, may lead to over-fitting and high out-of-sample prediction error. This bias-variance tradeoff, as it is sometimes called, is a statistical fact of life.

Shrinkage and regularized regression methods provide mechanisms for tuning, smoothing, or adjusting model complexity (Tibshirani 1996; Hoerl and Kennard 2000). Alternatively, we can select subsets of explanatory variables to go into predictive models. Special methods are called into play when the number of parameters being estimated is large, perhaps exceeding the number of observations (Bühlmann and van de Geer 2011). For additional discussion of the bias-variance tradeoff, regularized regression, and subset selection, see Izenman (2008) and Hastie, Tibshirani, and Friedman (2009).

Graybill (1961, 2000) and Rencher and Schaalje (2008) review linear models. Generalized linear models are discussed in McCullagh and Nelder (1989) and Firth (1991). Kutner, Nachtsheim, Neter, and Li (2004) provide a comprehensive review of linear and generalized linear models, including discussion of their application in experimental design. R methods for the estimation of linear and generalized linear models are reviewed in Chambers and Hastie (1992) and Venables and Ripley (2002).

The standard reference for generalized linear models is McCullagh and Nelder (1989). Firth (1991) provides additional review of the underlying theory. Hastie (1992) and Venables and Ripley (2002) give modeling examples with S/SPlus, most which are easily duplicated in R. Lindsey (1997) discusses a wide range of application examples. See Christensen (1997), Le (1998), and Hosmer, Lemeshow, and Sturdivant (2013) for discussion of logistic regression. Lloyd (1999) provides an overview of categorical data analysis. See Fawcett (2003) and Sing et al. (2005) for further discussion of the ROC curve. Discussion of alternative methods for evaluating classifiers is provided in Hand (1997) and Kuhn and Johnson (2013).

For Poisson regression and the analysis of multi-way contingency tables, useful references include Bishop, Fienberg, and Holland (1975), Cameron and Trivedi (1998), Fienberg (2007), Tang, He, and Tu (2012), and Agresti (2013). Reviews of survival data analysis have been provided by Andersen, Borgan, Gill, and Keiding (1993), Le (1997), Therneau and Grambsch (2000), Harrell (2001), Nelson (2003), Hosmer, Lemeshow, and May (2013), and Allison (2010), with programming solutions provided by Therneau (2014) and Therneau and Crowson (2014). Wassertheil-Smoller (1990) provides an elementary introduction to classification procedures and the evaluation of binary classifiers. For a more advanced treatment, see Hand (1997). Burnham and Anderson (2002) review model selection methods, particularly those using the Akaike information criterion or AIC (Akaike 1973).

We sometimes consider robust regression methods when there are influential outliers or extreme observations. Robust methods represent an active area of research using statistical simulation tools (Fox 2002; Koller and Stahel 2011; Maronna, Martin, and Yohai 2006; Maechler 2014b; Koller 2014). Huet et al. (2004) and Bates and Watts (2007) review nonlinear regression, and Harrell (2001) discusses spline functions for regression problems.

## A.4 Data Mining and Machine Learning

Recommender systems, collaborative filtering, association rules, optimization methods based on heuristics, as well as a myriad of methods for regression, classification, and clustering fall under the rubric of machine learning.

We use the term "machine learning" to refer to the methods or algorithms that we use as an alternative to traditional statistical methods. When we apply these methods in the analysis of data, we use the term "data mining." Rajaraman and Ullman (2012) describe data mining as the "discovery of models for data." Regarding the data themselves, we are often referring to massive data sets.

With traditional statistics, we define the model specification prior to working with the data. With traditional statistics, we often make assumptions about the population distributions from which the data have been drawn. Machine learning, on the other hand, is data-adaptive. The model specification is defined by applying algorithms to the data. With machine learning, few assumptions are made about the underlying distributions of the data.

Machine learning methods often perform better than traditional linear or logistic regression methods, but explaining why they work is not easy. Machine learning models are sometimes called *black box models* for a reason. The underlying algorithms can yield thousands of formulas or nodal splits fit to the training data.

Extensive discussion of machine learning algorithms may be found in Duda, Hart, and Stork (2001), Izenman (2008), Hastie, Tibshirani, and Friedman (2009), Kuhn and Johnson (2013), Tan, Steinbach, and Kumar (2006), and Murphy (2012). Bacon (2002) describes their application in marketing.

Hothorn et al. (2005) review principles of benchmark study design, and Schauerhuber et al. (2008) show a benchmark study of classification methods. Alfons (2014a) provides cross-validation tools for benchmark studies. Benchmark studies, also known as statistical simulations or statistical experiments, may be conducted with programming packages designed for this type of research (Alfons 2014b; Alfons, Templ, and Filzmoser 2014).

Duda, Hart, and Stork (2001), Tan, Steinbach, and Kumar (2006), Hastie, Tibshirani, and Friedman (2009), and Rajaraman and Ullman (2012) introduce clustering from a machine learning perspective. Everitt, Landau,

Leese, and Stahl (2011), Kaufman and Rousseeuw (1990) review traditional clustering methods. Izenman (2008) provides a review of traditional clustering, self-organizing maps, fuzzy clustering, model-based clustering, and biclustering (block clustering).

Within the machine learning literature, cluster analysis is referred to as *unsupervised learning* to distinguish it from classification, which is *supervised learning*, guided by known, coded values of a response variable or class. Association rules modeling, frequent itemsets, social network analysis, link analysis, recommender systems, and many multivariate methods as we employ them in data science represent unsupervised learning methods.

An important multivariate method, principal component analysis, draws on linear algebra and gives us a way to reduce the number of measures or quantitative features we use to describe domains of interest. Long a staple of measurement experts and a prerequisite of factor analysis, principal component analysis has seen recent applications in latent semantic analysis, a technology for identifying important topics across a document corpus (Blei, Ng, and Jordan 2003; Murphy 2012; Ingersoll, Morton, and Farris 2013).

When some observations in the training set have coded responses and others do not, we employ a *semi-supervised learning* approach. The set of coded observations for the supervised component can be small relative to the set of uncoded observations for the unsupervised component (Liu 2011).

Leisch and Gruen (2014) describe programming packages for various clustering algorithms. Methods developed by Kaufman and Rousseeuw (1990) have been implemented in R programs by Maechler (2014a), including silhouette modeling and visualization techniques for determining the number of clusters. Silhouettes were introduced by Rousseeuw (1987), with additional documentation and examples provided in Kaufman and Rousseeuw (1990) and Izenman (2008).

Thinking more broadly about machine learning, we see it as a subfield of artificial intelligence (Luger 2008; Russell and Norvig 2009). Machine learning encompasses biologically-inspired methods, genetic algorithms, and heuristics, which may be used to address complex optimization, scheduling, and systems design problems. (Mitchell 1996; Engelbrecht 2007; Michalawicz and Fogel 2004; Brownlee 2011).

## A.5 Data Visualization

Data visualization is critical to the work of data science. Examples in this book demonstrate the importance of data visualization in discovery, diagnostics, and design. We employ tools of exploratory data analysis (discovery) and statistical modeling (diagnostics). In communicating results to management, we use presentation graphics (design).

Statistical summaries fail to tell the story of data. To understand data, we must look beyond data tables, regression coefficients, and the results of statistical tests. Visualization tools help us learn from data. We explore data, discover patterns in data, identify groups of observations that go together and unusual observations or outliers. We note relationships among variables, sometimes detecting underlying dimensions in the data.

Graphics for exploratory data analysis are reviewed in classic references by Tukey (1977) and Tukey and Mosteller (1977). Regression graphics are covered by Cook (1998), Cook and Weisberg (1999), and Fox and Weisberg (2011). Statistical graphics and data visualization are illustrated in the works of Tufte (1990, 1997, 2004, 2006), Few (2009), and Yau (2011, 2013). Wilkinson (2005) presents a review of human perception and graphics, as well as a conceptual structure for understanding statistical graphics. Cairo (2013) provides a general review of information graphics. Heer, Bostock, and Ogievetsky (2010) demonstrate contemporary visualization techniques for web distribution. When working with very large data sets, special methods may be needed, such as partial transparency and hexbin plots (Unwin, Theus, and Hofmann 2006; Carr, Lewin-Koh, and Maechler 2014; Lewin-Koh 2014).

R is particularly strong in data visualization. An R graphics overview is provided by Murrell (2011). R lattice graphics, discussed by Sarkar (2008, 2014), build on the conceptual structure of an earlier system called S-Plus Trellis™ (Cleveland 1993; Becker and Cleveland 1996). Wilkinson's (2005) "grammar of graphics" approach has been implemented in the Python ggplot package (Lamp 2014) and in the R ggplot2 package (Wickham and Chang 2014), with R programming examples provided by Chang (2013). Cairo (2013) and Zeileis, Hornik, and Murrell (2009, 2014) provide advice about colors for statistical graphics. Ihaka et al. (2014) show how to specify colors in R by hue, chroma, and luminance.

## A.6 Text and Sentiment Analysis

Text analytics draws from a variety of disciplines, including linguistics, communication and language arts, experimental psychology, political discourse analysis, journalism, computer science, and statistics. And, given the amount of text being gathered and stored by organizations, text analytics is an important and growing area of predictive analytics.

We have discussed web crawling, scraping, and parsing. The output from these processes is a document collection or text corpus. This document collection or corpus is in the natural language. The two primary ways of analyzing a text corpus are the *bag of words* approach and *natural language processing*. We parse the corpus further, creating commonly formatted expressions, indices, keys, and matrices that are more easily analyzed by computer. This additional parsing is sometimes referred to as text annotation. We extract features from the text and then use those features in subsequent analyses.

Natural language is what we speak and write every day. Natural language processing is more than a matter of collecting individual words. Natural language conveys meaning. Natural language documents contain paragraphs, paragraphs contain sentences, and sentences contain words. There are grammatical rules, with many ways to convey the same idea, along with exceptions to rules and rules about exceptions. Words used in combination and the rules of grammar comprise the linguistic foundations of text analytics as shown in figure A.2.

Linguists study natural language, the words and the rules that we use to form meaningful utterances. "Generative grammar" is a general term for the rules; "morphology," "syntax," and "semantics" are more specific terms. Computer programs for natural language processing use linguistic rules to mimic human communication and convert natural language into structured text for further analysis.

Natural language processing is a broad area of academic study itself, and an important area of computational linguistics. The location of words in sentences is a key to understanding text. Words follow a sequence, with earlier words often more important than later words, and with early sentences and paragraphs often more important than later sentences and paragraphs.

*Figure A.2.* *Linguistic Foundations of Text Analytics*

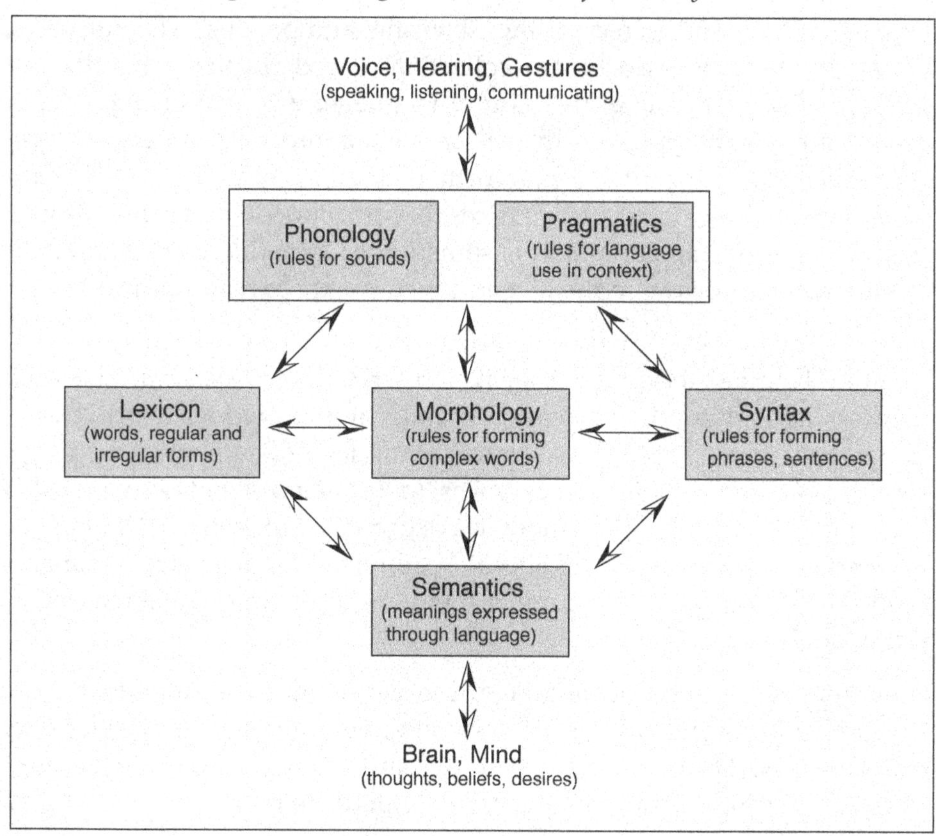


*Source:* Adapted from Pinker (1999).

Words in the title of a document are especially important to understanding the meaning of a document. Some words occur with high frequency and help to define the meaning of a document. Other words, such as the definite article "the" and the indefinite articles "a" and "an," as well as many prepositions and pronouns, occur with high frequency but have little to do with the meaning of a document. These *stop words* are dropped from the analysis.

The features or attributes of text are often associated with terms—collections of words that mean something special. There are collections of words relating to the same concept or word stem. The words "marketer," "marketeer," and "marketing" build on the common word stem "market." There are syntactic structures to consider, such as adjectives followed by nouns and nouns followed by nouns. Most important to text analytics are sequences of words that form terms. The words "New" and "York" have special meaning when combined to form the term "New York." The words "financial" and "analysis" have special meaning when combined to form the term "financial analysis." We often employ *stemming*, which is the identification of word stems, dropping suffixes (and sometimes prefixes) from words. More generally, we are parsing natural language text to arrive at structured text.

In English, it is customary to place the subject before the verb and the object after the verb. In English, verb tense is important. The sentence "Daniel carries the Apple computer," can have the same meaning as the sentence "The Apple computer is carried by Daniel." "Apple computer," the object of the active verb "carry" is the subject of the passive verb "is carried." Understanding that the two sentences mean the same thing is an important part of building intelligent text applications.

A key step in text analysis is the creation of a terms-by-documents matrix (sometimes called a lexical table). The rows of this data matrix correspond to words or word stems from the document collection, and the columns correspond to documents in the collection. The entry in each cell of a terms-by-documents matrix could be a binary indicator for the presence or absence of a term in a document, a frequency count of the number of times a term is used in a document, or a weighted frequency indicating the importance of a term in a document.

Figure A.3 illustrates the process of creating a terms-by-documents matrix. The first document comes from Steven Pinker's *Words and Rules* (1999, p. 4), the second from Richard K. Belew's *Finding Out About* (2000, p. 73). Terms correspond to words or word stems that appear in the documents. In this example, each matrix entry represents the number of times a term appears in a document. We treat nouns, verbs, and adjectives similarly in the definition of stems. The stem "combine" represents both the verb "combine" and the noun "combination." Likewise, "function" represents the verb, noun, and adjective form "functional." An alternative system might distinguish among parts of speech, permitting more sophisticated syntactic searches across documents. After being created, the terms-by-documents matrix is like an index, a mapping of document identifiers to terms (keywords or stems) and vice versa. For information retrieval systems or search engines we might also retain information regarding the specific location of terms within documents.

Typical text analytics applications have many more terms than documents, resulting in sparse rectangular terms-by-documents matrices. To obtain meaningful results for text analytics applications, analysts examine the distribution of terms across the document collection. Very low frequency terms, those used in few documents, are dropped from the terms-by-documents matrix, reducing the number of rows in the matrix.

Unsupervised text analytics problems are those for which there is no response or class to be predicted. Rather, as we showed with the movie taglines, the task is to identify common patterns or trends in the data. As part of the task, we may define text measures describing the documents in the corpus.

For supervised text analytics problems there is a response or class of documents to be predicted. We build a model on a training set and test it on a test set. Text classification problems are common. Spam filtering has long been a subject of interest as a classification problem, and many e-mail users have benefitted from the efficient algorithms that have evolved in this area. In the context of information retrieval, search engines classify documents as being relevant to the search or not. Useful modeling techniques for text classification include logistic regression, linear discriminant function anal-

***Figure A.3.*** *Creating a Terms-by-Documents Matrix*

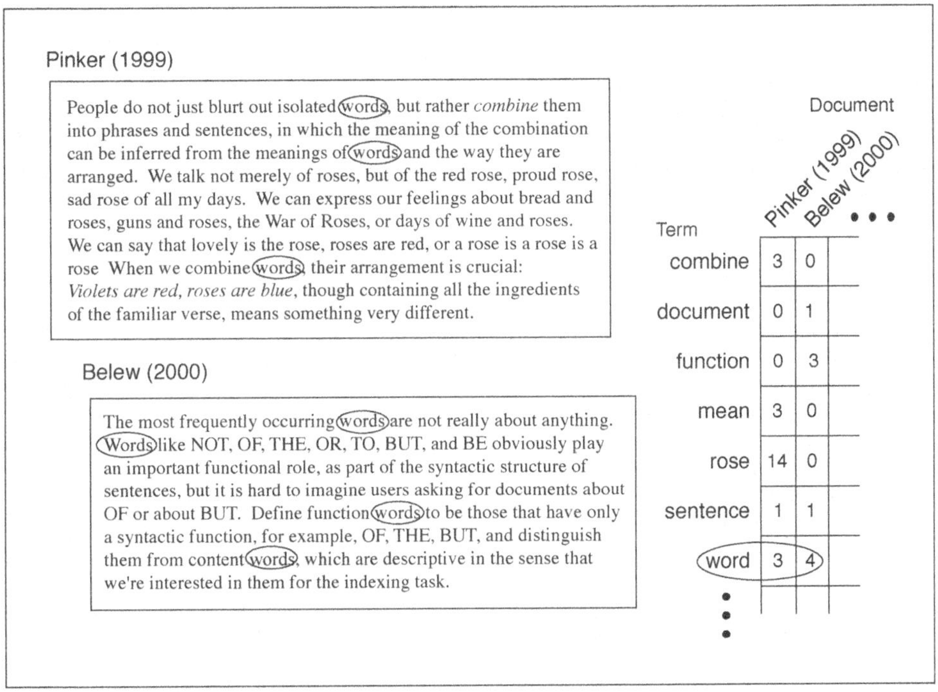

Pinker (1999)

People do not just blurt out isolated words, but rather *combine* them into phrases and sentences, in which the meaning of the combination can be inferred from the meanings of words and the way they are arranged. We talk not merely of roses, but of the red rose, proud rose, sad rose of all my days. We can express our feelings about bread and roses, guns and roses, the War of Roses, or days of wine and roses. We can say that lovely is the rose, roses are red, or a rose is a rose is a rose When we combine words, their arrangement is crucial: *Violets are red, roses are blue*, though containing all the ingredients of the familiar verse, means something very different.

Belew (2000)

The most frequently occurring words are not really about anything. Words like NOT, OF, THE, OR, TO, BUT, and BE obviously play an important functional role, as part of the syntactic structure of sentences, but it is hard to imagine users asking for documents about OF or about BUT. Define function words to be those that have only a syntactic function, for example, OF, THE, BUT, and distinguish them from content words, which are descriptive in the sense that we're interested in them for the indexing task.

Document

| Term | Pinker (1999) | Belew (2000) | • • • |
|---|---|---|---|
| combine | 3 | 0 | |
| document | 0 | 1 | |
| function | 0 | 3 | |
| mean | 3 | 0 | |
| rose | 14 | 0 | |
| sentence | 1 | 1 | |
| word | 3 | 4 | |
| ⋮ | | | |

*Source:* Adapted from Miller (2005).

ysis, classification trees, and support vector machines. Various ensemble or committee methods may be employed.

Automatic text summarization is an area of research and development that can help with information management. Imagine a text processing program with the ability to read each document in a collection and summarize it in a sentence or two, perhaps quoting from the document itself. Today's search engines are providing partial analysis of documents prior to their being displayed. They create automated summaries for fast information retrieval. They recognize common text strings associated with user requests. These applications of text analysis comprise tools of information search that we take for granted as part of our daily lives.

Programs with syntactic processing capabilities, such as IBM's Watson, provide a glimpse of what intelligent agents for text analytics are becoming. These programs perform grammatical parsing with an understanding of the roles of subject, verb, object, and modifier. They know parts of speech (nouns, verbs, adjective, adverbs). And, using identified entities representing people, places, things, and organizations, they perform relationship searches.

Sentiment analysis is measurement-focused text analysis. Sometimes called opinion mining, one approach to sentiment analysis is to draw on positive and negative word sets (lexicons, dictionaries) that convey human emotion or feeling. These word sets are specific to the language being spoken and the context of application. Another approach to sentiment analysis is to work directly with text samples and human ratings of those samples, developing text scoring methods specific to the task at hand.

A semi-supervised machine learning regimen can be especially useful in sentiment analysis. We work two sets of text samples. One sample, often a small sample (because it is expensive and time-consuming to obtain human ratings of text), associates a rating or score with each text document. Another much larger sample is unrated but comes from the same content domain. We learn the direction of scoring from the first sample, and we learn about the text domain (including term frequencies in context) from both samples.

The objective of sentiment analysis is to score text for affect, feelings, attitudes, or opinions. Sentiment analysis and text measurement in general hold promise as technologies for understanding consumer opinion and markets. Just as political researchers can learn from the words of the public, press, and politicians, business researchers can learn from the words of customers and competitors. There are customer service logs, telephone transcripts, and sales call reports, along with user group, listserv, and blog postings. And we have ubiquitous social media from which to build document collections for text and sentiment analysis.

Precursors to sentiment analysis may be found in content analysis, thematic, semantic, and network text analysis (Roberts 1997; Popping 2000; West 2001; Leetaru 2011; Krippendorff 2012). These methods have seen a wide range of applications within the social sciences, including analysis of political discourse. An early computer implementation of content analysis is found in the General Inquirer program (Stone et al. 1966; Stone 1997). Buvač and Stone (2001) describe a version of the program that provides text measures based upon word counts across numerous semantic categories.

Text measures flow from a measurement model (algorithms for scoring) and a dictionary, both defined by the researcher or analyst. A dictionary in this context is not a traditional dictionary; it is not an alphabetized list of words and their definitions. Rather, the dictionary used to construct text measures is a repository of word lists, such as synonyms and antonyms, positive and negative words, strong and weak sounding words, bipolar adjectives, parts of speech, and so on. The lists come from expert judgments about the meaning of words. A text measure assigns numbers to documents according to rules, with the rules being defined by the word lists, scoring algorithms, and modeling techniques in predictive analytics.

Sentiment analysis and text measurement in general hold promise as technologies for understanding consumer opinion and markets. Just as political researchers can learn from the words of the public, press, and politicians, marketing data scientists can learn from the words of customers and competitors.

For the marketing data scientist interested in understanding consumer opinions about brands and products, there are substantial sources from which to draw samples. We have customer service logs, telephone transcripts, and sales call reports, along with user group, listserv, and blog postings. And we have ubiquitous social media from which to build document collections for text and sentiment analysis.

The measurement story behind opinion and sentiment analysis is an important story that needs to be told. Sentiment analysis, like all measurement, is the assignment of numbers to attributes according to rules. But what do the numbers mean? To what extent are text measures reliable or valid? To demonstrate content or face validity, we show that the content of the text measure relates to the attribute being measured. We examine word sets, and we try to gain agreement (among subject matter experts, perhaps) that they measure a particular attribute or trait. Sentiment research often involves the testing of word sets within specific contexts and, when possible, testing against external criteria. To demonstrate predictive validity, we show that a text measure can be used for prediction.

Regarding Twitter-based text measures, there have been various attempts to predict the success of movies prior to their being distributed to theaters nationwide (Sharda and Delen 2006; Delen, Sharda, and Kumar 2007). Most telling is work completed at HP Labs that utilized chat on Twitter as a predictor of movie revenues (Asur and Huberman 2010). Bollen, Mao, and Zeng (2011) utilize Twitter sentiment analysis in predicting stock market movements. Taddy's (2013b, 2014) sentiment analysis work builds on the inverse regression methods of Cook (1998, 2007). Taddy (2013a) uses Twitter data to examine political sentiment.

Some have voiced concerns about unidimensional measures of sentiment. There have been attempts to develop more extensive sentiment word sets, as well as multidimensional measures (Turney 2002; Asur and Huberman 2010). Recent developments in machine learning and quantitative linguistics point to sentiment measurement methods that employ natural language processing rather than relying on positive and negative word sets (Socher et al. 2011).

Among the more popular measurement schemes from the psychometric literature is Charles Osgood's semantic differential (Osgood, Suci, and Tannenbaum 1957; Osgood 1962). Exemplary bipolar dimensions include the positive–negative, strong–weak, and active–passive dimensions. Schemes like Osgood's set the stage for multidimensional measures of sentiment.

We expect sentiment analysis to be an active area of research for many years. Reviews of sentiment analysis methods have been provided by Liu (2010, 2011, 2012) and Feldman (2013). Other books in the *Modeling Techniques* series provide examples of sentiment analysis using document collections of movie reviews Miller (2015b, 2015a, 2015c).

Ingersoll, Morton, and Farris (2013) provide an introduction to the domain of text analytics for the working data scientist. Those interested in reading further can refer to Feldman and Sanger (2007), Jurafsky and Martin (2009), Weiss, Indurkhya, and Zhang (2010), and the edited volume by Srivastava and Sahami (2009). Reviews may be found in Trybula (1999), Witten, Moffat, and Bell (1999), Meadow, Boyce, and Kraft (2000), Sullivan (2001), Feldman (2002b), and Sebastiani (2002). Hausser (2001) gives an account of generative grammar and computational linguistics. Statistical language learning and natural language processing are discussed by Charniak (1993), Manning and Schütze (1999), and Indurkhya and Damerau (2010).

The writings of Steven Pinker (1994, 1997, 1999) provide insight into grammar and psycholinguistics. Maybury (1997) reviews data preparation for text analytics and the related tasks of source detection, translation and conversion, information extraction, and information exploitation. Detection relates to identifying relevant sources of information; conversion and translation involve converting from one medium or coding form to another.

Belew (2000), Meadow, Boyce, and Kraft (2000) and the edited volume by Baeza-Yates and Ribeiro-Neto (1999) provide reviews of technologies for information retrieval, which depend on text classification, among other technologies and algorithms.

Authorship identification, a problem addressed a number of years ago in the statistical literature by Mosteller and Wallace (1984), continues to be an active area of research (Joula 2008). Merkl (2002) provides discussion of clustering techniques, which explore similarities between documents and the grouping of documents into classes. Dumais (2004) reviews latent semantic analysis and statistical approaches to extracting relationships among terms in a document collection.

## A.7 Time Series and Market Response Models

Sales forecasts are a critical component of business planning and a first step in the budgeting process. Models and methods that provide accurate forecasts can be of great benefit to management. They help managers to understand the determinants of sales, including promotions, pricing, advertising, and distribution. They reveal competitive position and market share.

There are many approaches to forecasting. Some are judgmental, relying on expert opinion or consensus. There are top-down and bottom-up forecasts, and various techniques for combining the views of experts. Other approaches depend on the analysis of past sales data.

Certain problems in business have a special structure, and if we pay attention to that structure, we can find our way to a solution. Sales forecasting is one of those problems. Sales forecasts can build on the special structure of sales data as they are found in business. These are data organized by time and location, where location might refer to geographical regions or sales territories, stores, departments within stores, or product lines.

A sales forecasting model can and should be organized by time periods useful to management. These may be days, weeks, months, or whatever intervals make sense for the problem at hand. Time dependencies can be noted in the same manner as in traditional time-series models. Autoregressive terms are useful in many contexts. Time-construed covariates, such as day of the week or month of the year, can be added to provide additional predictive power. And we can include promotion, pricing, and advertising variables organized in time.

An analyst can work with time series data, using past sales to predict future sales, noting overall trends and cyclical patterns in the data. Exponential smoothing, moving averages, and various regression and econometric methods may be used with time-series data.

Forecasting by location provides detail needed for management action. And organizing data by location contributes to a model's predictive power. Location may itself be used as a factor in models. In addition, we can search for explanatory variables tied to location. With geographic regions, for example, we might include consumer and business demographic variables known to relate to sales.

Sales dollars per time period is the typical response variable of interest in sales forecasting studies. Alternative response variables include sales volume and time-to-sale. Related studies of market share require information about the sales of other firms in the same product category.

Forecasting is a large area of application deserving of its own professional conferences and journals. An overview of business forecasting methods is provided by Armstrong (2001). Time-series, panel (longitudinal) data, financial, and econometric modeling methods are especially relevant to this area of application (Judge et al. 1985; Hamilton 1994; Zivot and Wang 2003). Frees and Miller (2004) describe a sales forecasting method that utilizes a mixed modeling approach, reflecting the special structure of sales data.

Having gathered the data and looked at the plots, we turn to forecasting the future. For these economic time series, we use autoregressive integrated moving average (ARIMA) models or what are often called Box-Jenkins models (Box, Jenkins, and Reinsel 2008).

Drawing on software from Hyndman et al. (2014) and working with one measure at a time, we use programs to search across large sets of candidate models, including autoregressive, moving-average, and seasonal components. We try to select the very best model in terms of the Akaike Information Criterion (AIC) or some other measure that combines goodness-of-fit and parsimony. Then, having found the model that the algorithm determines as the best for each measure, we use that model to forecast future values of the economic measure. In particular, we ask for forecasts for a time horizon that management requests. We obtain the forecast mean as well as a prediction interval around that mean for each time period of the forecasting horizon.

Forecasting uncertainty is estimated around the forecasted values. There is often much uncertainty about the future, and the further we look into the future, the greater our uncertainty. The value of a model lies in the quality of its predictions, and sales forecasting presents challenging problems for the marketing data scientist.

When working with multiple time series, we might fit a multivariate time series or vector autoregressive (VAR) model to the four time series. Alternatively, we could explore dynamic linear models, regressing one time series on another. We could utilize ARIMA transfer function models or state space models with regression components. The possibilities are as many as the modeling issues to be addressed.

There is a subtle but important distinction to be made here. The term *time series regression* refers to regression analysis in which the organizing unit of analysis is time. We look at relationships among economic measures organized in time. Much economic analysis concerns time series regression. Special care must be taken to avoid what might be called spurious relationships, as many economic time series are correlated with one another because they depend upon underlying factors, such as population growth or seasonality.

In time series regression, we use standard linear regression methods. We check the residuals from our regression to ensure that they are not correlated in time. If they are correlated in time (autocorrelated), then we use a method such as generalized least squares as an alternative to ordinary least squares. That is, we incorporate an error data model as part of our modeling process. Longitudinal data analysis or panel data analysis is an example of a mixed data method with a focus on data organized by cross-sectional units and time.

When we use the term *time series analysis*, however, we are not talking about time series regression. We are talking about methods that start by focusing on one economic measure at a time and its pattern across time. We look for trends, seasonality, and cycles in that individual time series. Then, after working with that single time series, we look at possible relationships with other time series. If we are concerned with forecasting or predicting the future, as we often are in predictive analytics, then we use methods of time series analysis. Recently, there has been considerable interest in state space models for time series, which provide a convenient mechanism for incorporating regression components into dynamic time series models (Commandeur and Koopman 2007; Hyndman, Koehler, Ord, and Snyder 2008; Durbin and Koopman 2012).

There are myriad applications of time series analysis in marketing, including marketing mix models and advertising research models. Along with sales forecasting, these fall under the general class of market response models, as reviewed by Hanssens, Parsons, and Schultz (2001). Marketing mix models look at the effects of price, promotion, and product placement in retail establishments. These are multiple time series problems.

Advertising research looks for cumulative effectiveness of advertising on brand and product awareness, as well as sales. Exemplary reviews of advertising research methods and findings have been provided by Berndt (1991) and Lodish et al. (1995). Much of this research employs defined measures such as "advertising stock," which attempt to convert advertising impressions or rating points to a single measure in time. The thinking is that messages are most influential immediately after being received, decline in influence with time, but do not decline completely until many units in time later. Viewers or listeners remember advertisements long after initial exposure to those advertisements. Another way of saying this is to note that there is a carry-over effect from one time period to the next. Needless to say, measurement and modeling on the subject of advertising effectiveness presents many challenges for the marketing data scientist.

Similar to other data with which we work, sales and marketing data are organized by observational unit, time, and space. The observational unit is typically an economic agent (individual or firm) or a group of such agents as in an aggregate analysis. It is common to use geographical areas as a basis for aggregation. Alternatively, space (longitude and latitude) can be used directly in spatial data analyses. Time considerations are especially important in macroeconomic analysis, which focuses upon nationwide economic measures.

Baumohl (2008) provides a review of economic measures that are commonly thought of as leading indicators. Kennedy (2008) provides an introduction to the terminology of econometrics. Key references in the area of econometrics include Judge et al.(1985), Berndt (1991), Enders (2010), and Greene (2012). Reviews of time series modeling and forecasting methods are provided by Holden, Peel, and Thompson (1990) and in the edited volume by Armstrong (2001).

More detailed discussion of time series methods is provided by Hamilton (1994), Makridakis, Wheelwright, and Hyndman (2005), Box, Jenkins, and Reinsel (2008), Hyndman et al. (2008), Durbin and Koopman (2012), and Hyndman and Athanasopoulos (2014). Time-series, panel (longitudinal) data, financial, and econometric modeling methods are especially relevant in demand and sales forecasting. Frees and Miller (2004) present a longitudinal sales forecasting method, reflecting the special structure of sales data in space and time. Hierarchical and grouped time series methods are discussed by Athanasopoulos, Ahmed, and Hyndman (2009) and Hyndman et al. (2011).

For gathering economic data with R, we can build on foundation code provided by Ryan (2014). Useful for programming with dates are R functions provided by Grolemund and Wickham (2011, 2014). Associated sources for econometric and time series programming are Kleiber and Zeileis (2008), Hothorn et al. (2014), Cowpertwait and Metcalfe (2009), Petris, Petrone, and Campagnoli (2009), and Tsay (2013). Most useful for time series forecasting is code from Hyndman et al.(2014), Petris (2010), Petris and Gilks (2014), and Szymanski (2014).

The Granger test of causality, a test of temporal ordering, was introduced in the classic reference by Granger (1969). The interested reader should also check out a delightful article that answers the perennial question *Which came first, the chicken or the egg?* (Thurman and Fisher 1988).

Applications of traditional methods and models in economics, business, and market research are discussed by Leeflang et al. (2000), Franses and Paap (2001), Hanssens, Parsons, and Schultz (2001), and Frees and Miller (2004). Lilien, Kotler, and Moorthy (1992) and Lilien and Rangaswamy (2003) focus upon marketing models. For a review of applications from the research practitioner's point of view, see Chakrapani (2000).

# B Marketing Data Sources

Doing marketing data science means working with many data sources. There is the World Wide Web, a path to extensive secondary research data. We work with corporate databases. And we are often asked to engage in primary research, running experiments and conducting surveys. Which is to say that marketing data science is preceded by data collection.

It is a mistake to think that managers demand information per se. What they demand is the right information. They want information reduced and synthesized, information that is directly relevant to decisions at hand.

How do we find our way to the right information? Through a variety of methods, including online searches, library and secondary research, and primary research. We collect survey data, run experiments, conduct focus groups, and observe consumer behavior.

What people buy depends on where they come from, who they are, their friends and relationships, what they need, and what they like. We measure attitudes and behavior. We observe demographics and lifestyles. We see consumers' social networks. What people buy also depends on the availability of products in the marketplace. We note the business environment—competitive products, substitute products, and the prices of those products.

Measurements come in many forms, and they can involve observation of consumer characteristics, observation of the business characteristics, or both, as shown in figure B.1. This appendix reviews measurement theory, sampling, and methods of data collection in marketing data science.

*Figure B.1.* *A Framework for Marketing Measurement*

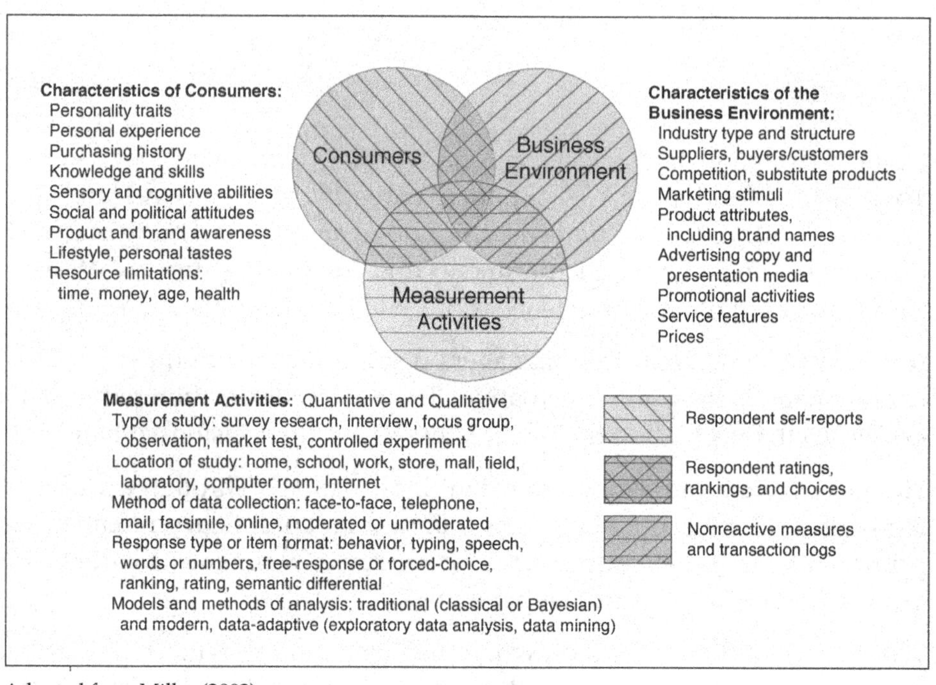


Adapted from Miller (2002).

## B.1 Measurement Theory

The classic article by Campbell and Fiske (1959) provides a clear definition of reliability and validity: *Reliability is the agreement between two efforts to measure the same trait through maximally similar methods. Validity is represented in the agreement between two attempts to measure the same trait through maximally different methods.* (83)

The prototypical validation study involves a multitrait-multimethod matrix, as described by Campbell and Fiske (1959). This correlation matrix provides a structure for demonstrating construct validity. The matrix is partitioned by measurement methods. Within each method there are rows and columns associated with traits (attributes). Each element of the matrix represents a trait-method unit. The components of the matrix are the reliability diagonal, validity diagonals, heterotrait-monomethod triangles, and heterotrait-heteromethod triangles.

Suppose we want to demonstrate that extroverts rate comedies more highly than other types of movies. To carry out research in this area, we could develop measures of the personality trait *introversion-extroversion* and the consumer trait *preference for comedies*. Suppose further that we develop two measures for each trait, one based on a rating scale and the other on a text measure. With higher scores on measures of *introversion-extroversion* implying greater extroversion and higher scores on measures of *preference for comedies* indicating higher preference for comedies, we might expect to observe results such as those in figure B.2.

What do we want to observe in a multitrait-multimethod matrix? We would like to see high indices of internal consistency on the reliability diagonals. We want different measures of the same trait to correlate highly with one another, yielding high correlations on the validity diagonals. We want measures of the same trait to correlate more highly with one another than with measures of different traits. Accordingly, we should see higher correlations on the validity diagonals than in either the heterotrait-monomethod or the heterotrait-heteromethod triangles. How high is "high"? What precise pattern of correlations should we expect? To answer these questions, we refer to theory and to prior empirical research.

*Figure B.2. Hypothetical Multitrait-Multimethod Matrix*

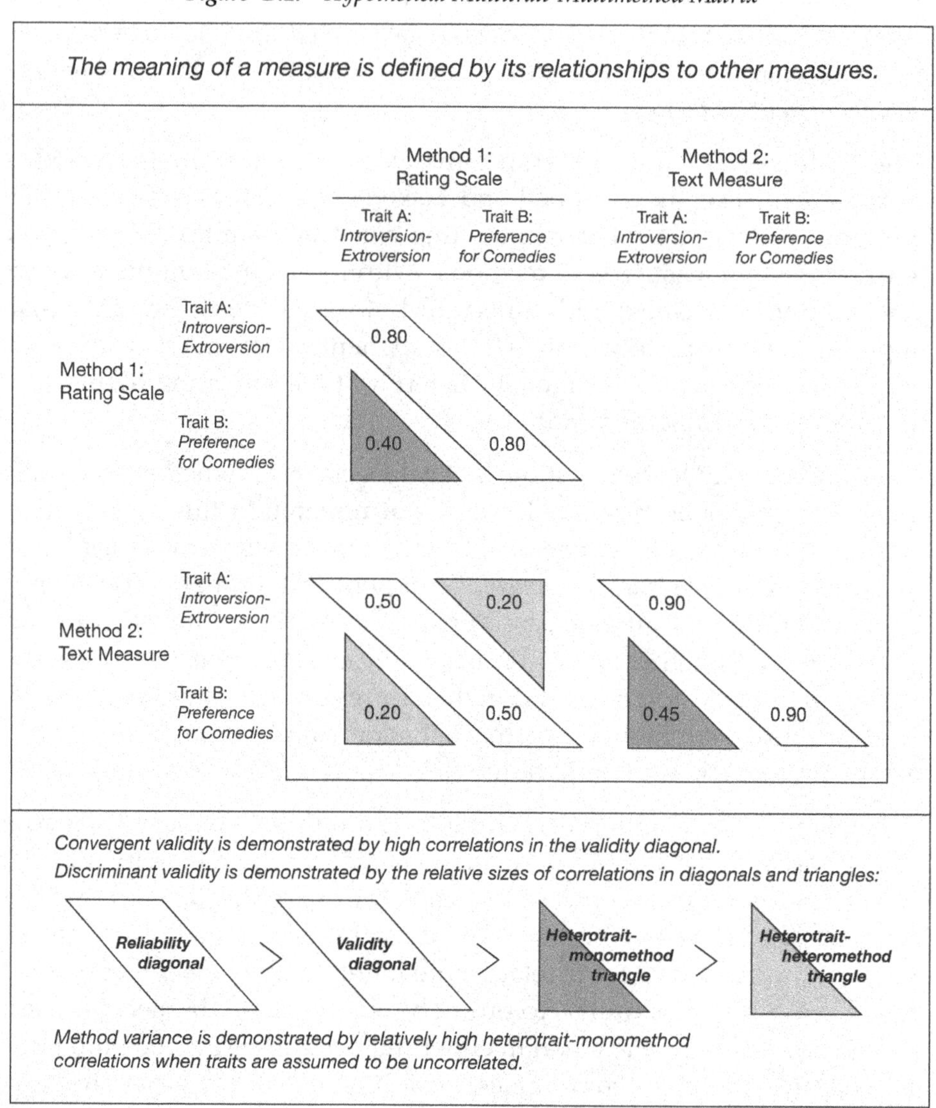


Source: Adapted from Miller (2015a) and Miller (2015b).

Campbell and Fiske (1959) talk about convergent validity and discriminant validity. Convergent validity refers to the notion that different measures of the same trait should converge. That is, different measures of the same trait or attribute should have relatively high correlations. Discriminant validity refers to the notion that measures of different traits should diverge. In other words, measures of different traits should have lower correlations than measures of the same trait. Convergent and discriminant validation are part of what we mean by construct validation. The meaning of a measure is defined in terms of its relationship to other measures.

Multitrait-multimethod matrices have been applied widely in the social sciences and consumer research (Bagozzi and Yi 1991). Cronbach (1995) talks about method variance and the fact that measures of different traits that utilize the same measurement technique can be correlated as a result of the way subjects respond to the technique itself. In other words, there can be individual differences in response style.

Validation is rarely easy. Low correlations between response and explanatory variables are common in social research. There are many variables to consider, many possible explanatory variables for every response. And there is considerable individual variability from one administration of a survey to the next and from one situation to the next.

Discussions of validity touch on fundamental issues in the philosophy of science, issues of theory construction, measurement, and testability. There are no easy answers here. If the theory is correct and the measures valid, then the pattern of relationships among the measures should be similar to the pattern predicted by theory. To the extent that this is true for observed data, we have partial confirmation of the theory and, at the same time, demonstration of construct validity. But what if the predictions do not pan out? Then we are faced with a dilemma: the theory could be wrong, one or more of the measures could be invalid, or we could have observed an event of low probability with correct theory and valid measures.

## B.2 Levels of Measurement

S. S. Stevens (1946) wrote "On the Theory of Scales of Measurement," an influential article that identified four general types of measures: nominal, ordinal, interval, and ratio. It was the strength of Stevens' convictions, perhaps more than the strength of his argument, that influenced generations of researchers. The words he chose to describe levels of measurement seemed to carry the force of law. He talked about the formal properties of scales and "permissible statistics," arguing that "the statistical manipulations that can legitimately be applied to empirical data depend on the type of scale" (Stevens 1946, 677). Stevens argued, we could use more powerful statistical methods with these measures. Ordinal measures, on the other hand, were to be analyzed with less powerful methods.

Table B.1 summarizes scale types or levels of measurement from Stevens (1946). The formal definition of a scale type follows from its mathematical properties, or what Stevens called its "mathematical group structure." This refers to the set of data transformations that, when used on the original measures, will create new measures with the same scale properties as the original measures.

For nominal scales, any one-to-one transformation will preserve the number of categories and, hence, the scale's essential property. For ordinal scales, any one-to-one monotonic transformation will preserve the property of order. For interval scales, any one-to-one linear transformation, a function of the form $y = ax + b$, will preserve the properties of the scale. Ratio scales are similar to linear scales, except that the zero-point must be preserved. Accordingly, for ratio scales, a data transformation that preserves its properties must have the form $y = ax$.

Researchers following Stevens' dictums constitute the weak measurement school. They argue that most measures in the social sciences are ordinal rather than interval and that statistics relying on sums or differences, including means and variances, would be inappropriate for ordinal measures.

Researchers following the strong statistics school, on the other hand, argue that statistical methods make no explicit assumptions about the meaning of measurements or their relationship to underlying dimensions. Strong statistics can be used with weak measurements. Reviews of the weak measurements versus strong statistics controversy and its relevance (or lack of relevance) to science and statistical inference have been presented by Baker, Hardyck, and Petrinovich (1966) and Velleman and Wilkinson (1993).

For practical purposes, we ask whether or not a variable has meaningful magnitude. If a variable is categorical, it lacks meaningful magnitude. One further observation is appropriate for categorical data: we note whether the variable is binary (taking only two possible values) or multinomial (taking more than two possible values). If we can make these simple distinctions across measures, we can do much useful research.

As data scientists, we need to understand levels of measurement, but we should not be bound by the prescriptions of S. S. Stevens regarding appropriate statistics. For example, there are many situations in which computing a mean of makes good sense. Models of statistical inference draw on assumptions about the distribution of measurements but have little to say about the way measurements are made.

For references on measurement reliability and validity, we refer to the literature of psychometrics (Gulliksen 1950; Cronbach 1951; Ghiselli 1964; Nunnally 1967; Nunnally and Bernstein 1994; Lord and Novick 1968; Fiske 1971; Brown 1976). Betz and Weiss (2001) and Allen and Yen (2002) introduce concepts of measurement theory. Item response theory is discussed by Rogers, Swaminathan, and Hambleton (1991). Articles in the volume edited by Shrout and Fiske (1995) provide many examples of multitrait-multimethod matrices and review quantitative methods available for the analysis of such matrices. Lumley (2010) discusses sample survey design and analysis in the R programming environment. For R functions in psychometrics, see Revelle (2014).

Measurement theory applies equally well to text measures and designed surveys. We work with term frequencies within documents and term frequencies adjusted by overall corpus term frequencies. We assign scores to text messages and posts. We utilize methods of natural language processing to detect features in text and to annotate documents (Bird, Klein, and Loper 2009; Pustejovsky and Stubbs 2013). These are measurements.

*Table B.1.* *Levels of measurement*

| Level of Measurement (Scale Type) | Basic Empirical Operations | Mathematical Group Structure | Examples of Permissible Statistics |
|---|---|---|---|
| Nominal | equality, numbers like names | one-to-one correspondence | number of cases in class, frequency table, modal class |
| Ordinal | greater than, less than | one-to-one monotonic | median, percentiles, rank-order correlation |
| Interval | equality of intervals | one-to-one linear | mean, standard deviation, product-moment correlation |
| Ratio | equality of ratios | one-to-one linear, preserving the zero point | same statistics as interval level |

*Source:* Adapted from Stevens (1946).

## B.3 Sampling

Sampling methods are important to all methods of data collection. The target population is the buyer group of interest. The sample—the group we actually observe—is a subset of the population. Statistics are characteristics of samples, whereas parameters are characteristics of populations. We use sample statistics to make inferences about the characteristics of the population (population parameters). To trust our inferences, we want the sample to be representative of the population. Practical applications of sampling in research and information services involve sampling without replacement from finite populations. Whatever the target population or method of sampling, we want the sample to be representative of the target.

A census gathers data from the entire population. The federal government conducts a census of population every ten years, as mandated by the *United States Constitution*. Many firms keep records of all business transactions. These are censuses. As methods of inquiry, however, censuses are rare. More likely, when we do business and consumer research, we will use a sample.

There are times when a researcher has access to a list of members of a target population. For example, a researcher in a mail-order firm may have a list of names, telephone numbers, and addresses of the firm's customers. Many of the firm's customers may opt in, indicating a willingness to be contacted for surveys. In the jargon of statistics, the list of names for the target population is called the sampling frame.

There are two general types of samples: probability samples and nonprobability samples. For a probability sample each member of the sampling frame has a known and nonzero probability of being included in the sample. For a nonprobability sample, sometimes called a judgment sample, the chances of a member being selected are unknown.

When drawing an inference from a sample to a population, we think of the data themselves as being a random sample from the population. To draw a random sample, we employ a systematic procedure that ensures that each member of the population has an equal probability of being included in the sample.

Taking a simple random sample from the population is easy if we have a sampling frame or list of prospective survey respondents. To take a simple random sample without replacement, for example, we can use a computer program to randomly order the names of the sampling frame and then select the first $n$ names from the randomly ordered list, where $n$ is the sample size.

Convenience samples are nonprobability samples, careless samples that result in biased estimates of population characteristics. When little thought is given to the sample and its relationship to the population, we cannot trust inferences about the population.

Many surveys are conducted without a sampling frame. They are convenience samples, administered to people who just happen to be available. Online pop-up surveys, for example, are presented to those who just happen to be clicking on corporate home pages. Online polls are administered to those who just happen to respond to banner advertisements broadcasted to a subset of individuals who just happen to be browsing a website. Furthermore, little care is taken to see that respondents respond once and only once to many online surveys.

Probability samples employ the random selection of sample units. There are many techniques for obtaining probability samples. We can choose a particular technique based on its efficiency or its ease of implementation. Alternatives to simple random sampling include stratified, cluster, and adaptive sampling. Like simple random sampling, these techniques require an initial sampling frame or listing of population members.

In taking a stratified sample, we divide the target population into parts on the basis of some variable of interest and then take random samples from each part. The parts are called strata. We often use stratification when our objective is to make comparisons among the strata. Sometimes strata of interest have a low frequency of occurrence or base rate in the population. We want members of low-base-rate strata to have a higher probability of being included in samples than members of other strata. Stratification can be a cost-effective and practical way of getting representative samples for all strata.

Sampling efficiency relates to the reduction in standard error associated with an increase in the sample size. If stratified sampling yields a lower standard error than a simple random sampling for the same total sample size (and cost of sampling), then a stratified sampling will be more efficient than simple random sampling.

Cluster sampling, another common alternative to simple random sampling, usually involves two random selections. The members of the population are organized into groups or clusters. We randomly select the cluster and then randomly select members within the cluster.

Adaptive sampling, another alternative to simple random sampling, involves selecting later sampling units by using information from earlier sampling units. Rarely used in social or business research, adaptive sampling procedures may prove to be useful in online research. With online surveys we should be able to use computers to dynamically adjust the sampling method as a survey progresses.

Suppose a researcher groups potential respondents into high-, middle-, and low-income strata. After two days of initial data collection with equal-probability sampling across the three strata, the researcher observes lower response rates in the low-income strata than in the other two strata. An adaptive sampling plan could adjust to differential response rates by recruiting low-income individuals with higher probability. The result of such an adaptive sampling plan would be a sample with a more equal distribution of respondents across income strata.

Budgetary and timing considerations force researchers to make hard choices regarding the size and composition of samples used in survey research. For statistical inference, larger samples are better than smaller samples, but larger samples are more expensive. Probability samples are better than non-probability samples, but probability samples may be difficult to obtain.

Nonsampling errors present additional problems for survey researchers. Sampling error refers to errors associated with sampling variability—how much sample statistics vary from one sample to the next. Nonsampling error refers to all other types of errors, including lack of measurement control and standardization, limited access to online media, respondent refusals, nonresponse, inaccurate response, incomplete survey response, and coverage problems. Nonsampling errors can lead to biased samples.

Are we covering the entire target population? This is an important question for researchers given changes in communication media in recent years. Coverage errors can arise from problems of access and control.

Undercoverage is a particular problem for consumer surveys because many consumers in the target population have little or no chance of being included in samples. Online surveys require special care in their development. Results from convenience or nonrepresentative samples can provide misleading information about target populations. Online surveys are subject to coverage problems.

The online population is large and similar to the general adult population. Nevertheless, there are differences in demographics between the online population and the general population. If we want to use online samples to make inferences about the general population, then we must recognize that fewer households have access to the Internet than to mail or telephone. Senior citizens, less-educated people and lower-income individuals may be excluded from online sample surveys because they lack access to the Internet.

Telephone surveys are also subject to coverage problems. The land-line telephone used to be a part of almost every household, with most households having one phone. Random digit dialing and telephone interviewing was the standard practice for sample surveys. Today, with wireless, cable, and Internet-based alternatives to land-lines and with the number of phones varying greatly from one household to the next, coverage presents a major problem for telephone surveys.

Improper coverage can also result when members outside the target population or outside the sampling frame are mistakenly included in the sample. Because online surveys are self-administered, researchers need to define rigorous controls regarding access to surveys. Targeted respondents may try to access an online survey more than once or give access to family members, friends, or colleagues.

Many people with access to a research medium choose not to respond to a survey. The opt-out and opt-in rates can be expected to differ across media and target populations, and these rates are highly influenced by incentives provided to respondents.

Obtaining logically correct and complete responses is also likely to vary across survey media. Telephone interviewers can prompt for responses and give explanations of survey questions where appropriate. Self-administered surveys offer little assistance to survey takers.

Defining the sampling frame presents a problem in today's media-diverse world. Even if we were willing to settle for a target population of people with household access to the Internet and a willingness to participate in online research, we would have a difficult time reaching them through the Internet. There is no central listing of e-mail addresses and no online analogue to the telephone book. Nor is there an online analogue to random-digit dialing. Furthermore, sending unsolicited e-mail to potential survey participants (spamming) is an inappropriate research practice.

Undercoverage is a fact of life in survey research. Bickart and Schmittlein (1999) describe research showing that fewer than twenty-five percent of adults accounted for all survey responses (traditional and online) among the general adult population in the United States. Most people are inaccessible or unwilling to respond to surveys by any medium. Practical survey design calls for a recognition of media access and nonresponse problems, which are evident in traditional and online surveys.

For many research problems and target populations, it may be difficult to employ random sampling. When a research organization has no list of phone numbers, and no physical or e-mail addresses for the target population, then probability sampling becomes a difficult and expensive, if not impossible, proposition. With coverage and nonresponse problems adding to the survey user's list of woes, many research organizations have turned to the use of respondent panels. Increasingly, researchers ask whether recruited panels of respondents can yield trustworthy results.

When using panels, we want to ensure that panel members are who they claim to be and that each member responds only once to each survey. By assigning personal identification numbers and passwords to potential respondents, we safeguard against respondent misidentification and duplication. In addition, respondents may be asked preliminary questions designed to verify their identities. To increase response rates, we can use mixed-mode surveys. Telephone recruitment followed by online survey administration, for example, can be a useful, although expensive, mixed-mode approach. Panel respondents usually receive financial incentives for their participation, and they are encouraged to participate in various surveys.

Providers of telephone or online panels strive for demographic balance with the general population. They try to show that their panels are representative of the general population or of a particular target population, often by using quotas for groups of individuals defined on the basis of demographic characteristics such as age, sex, income, education, and geographic location.

There are commercial research organizations that specialize in the creation of preselected panels, groups of respondents who are supposed to be representative of certain target populations. Panels are used extensively in business and consumer research.

Some research professionals have expressed concerns about the growing use of panels in survey research. Panel respondents who self-select, volunteering to participate in numerous surveys, could be thought of as professional respondents. Financial incentives may be their primary reason for volunteering. Furthermore, the responses of panel members may be affected by the fact that participants complete numerous surveys. There could be panel conditioning, with survey responses affected by carry-over, memory, or training effects from previous surveys.

Good research design attempts to reduce bias and sampling variability. It is the researcher's responsibility to be aware of problems with sampling bias and variability and to correct problems. To reduce bias, we seek representative samples, using random sampling whenever possible. To reduce variance, we use larger samples.

How large a sample is needed to obtain useful information? The answer to this question depends on the nature of the research question, the nature of the population, the statistic we are computing, the degree of precision, and degree of confidence required in making statistical inferences. All other things being equal, larger samples are better than smaller samples.

Although costs associated with sample recruitment and panel maintenance are often similar for online and traditional research, costs of data collection are lower for online research. Larger samples are less expensive with online methods, and larger samples translate into lower sampling variability. Unfortunately, what we gain with a reduction of sampling variability we may lose with an increase in nonsampling error (bias due to undercoverage). Lacking appropriate external criteria for the evaluation of research, it is often difficult to choose among alternative methods of inquiry.

Some researchers argue that the method of inquiry makes no difference, that managers will make the same business decisions whether guided by data from online or traditional telephone or mail surveys. When the objective of research is to create a preference ranking of products or services, as we might seek from concept tests, then online and traditional methods may yield similar results. But when the objective is to obtain an accurate forecast of sales or market share, then the method of research can make a big difference (Miller 2001).

## B.4 Marketing Databases

Most companies gather data on sales transactions as part of their day-to-day business. Successful firms gather relevant information about consumers, competitors, and markets, and use that information to guide business decisions. They are information-rich and information-wise.

Sales transaction data are the core of marketing databases. We can append data associated with product registrations, loyalty programs, and surveys. Commercial firms offer more complete data for individuals based on name and address and/or social security number. These data may include aggregated purchasing data from credit cards and credit scoring firms.

Knowing a customer's address, we geocode by longitude and latitude and link to ZIP code or block group information from the U.S. Census. Thousands of geo-demographic variables are available for every U.S. address.

These data are the beginning of a more complete effort to provide a full-blown customer relationship management system, a system that tracks marketing, sales, and support interactions between the customer and the firm.

Customer relationship management (CRM) systems go beyond sales information systems by tracking many customer interactions with a firm. Store visits, sales calls, price quotations, proposals, passes through check-out lines, sales transactions, letters, notes, and complaints—all have the potential of providing an electronic record of a firm's relationship with a customer. The times at which and places where firms interact with customers are called *touch-points.*

The customer relationship management story is a familiar one, reminiscent of the marketing concept explained by marketing academics. The job of marketing is not to sell, but to understand consumers. If we understand the consumer and design goods and services that meet consumer needs, then there is less need for marketing communications, advertising, and sales activities.

Successful CRM systems are usually enterprise-wide systems. In fact, the acronym CRM is often associated with the terms "enterprise resource management" (ERM) and "enterprise resource planning" (ERP). These are ways of saying that CRM systems are pervasive information systems. They affect the way firms conduct business.

At the core of the CRM system is an integrated database that provides current customer information to sales and service representatives within the firm. With its focus upon individual customers and promoting positive long-term relationships with customers, CRM must provide accurate and relevant information to the employees who serve customers. Appropriate information should be available at each touch-point.

Some information for CRM is gleaned from normal business operations (price quotations and sales orders keyed by customer number). These are transaction-oriented touch-points. Some information (sales and service call reports) is entered by people working for the firm and interacting with customers. These are agent-mediated touch-points. Customers enter additional information themselves through written correspondence (notes, letters, and requests for proposal or requests for quotation) and electronic interactions (online orders, e-mail, chat room discussion logs, bulletin board postings, and Weblog entries). These are direct touch-points.

Having touch-points presupposes keeping in touch with customers, using membership and loyalty programs, company-issued credit cards, newsletters, user groups, call centers, and members-only promotions. We utilize electronic communication through e-mail, bulletin boards, blogs, and chat rooms.

A personalized focus on the customer or client can extend from initial product design to product support. Many companies have been turning to real-time feedback to mend relationships with dissatisfied customers and build long-term relationships. Success in having satisfied, loyal customers depends upon a company's ability to take fast action in response to customer complaints and concerns.

A key difference between survey-based consumer research and CRM-based research is the way we interact with research participants. In consumer research there is a tradition of protecting research participants from subsequent contact by sales and service personnel. Confidentiality and privacy protection are part of the code of conduct of consumer researchers. Research articipants opt-in when filling out a survey, and they can just as easily opt-out. They are anonymous consumers. The customers of CRM, on the other hand, are anything but anonymous. They are well identified, with recorded sales transactions and customer service operations. With CRM, research and service activities are intertwined.

Fournier, Dobscha, and Mick (1998) remind us that marketing is about relationships. One-to-one or relationship marketing relies upon detailed knowledge of identifiable customers. Information about individual customers may be put to good use in tailoring products and services. Bessen (1993) argues that the key to marketing success is not to build market share but to dominate a market niche. Schoenfeld (1998) touts the benefits of mass customization, in which products are designed to serve the needs of each individual customer. Zabin, Brebach, and Kotler (2004) talk about precision marketing, which involves delivering precise marketing messages to people in very narrow customer segments. Pal and Rangaswamy (2003) talk about the power of one, referring to the benefits of personalization and customization. Indeed, one-to-one marketing may be viewed as the ultimate differentiation strategy with products designed to meet the needs of individual consumers.

Blattberg, Kim, and Neslin (2008) cover the domain of database marketing, reviewing strategic issues, direct marketing methods, and modeling techniques. They provide extensive discussion of customer lifetime value analysis, direct marketing with RFM models, and the application of machine learning techniques in marketing.

## B.5 World Wide Web

Data scientists spend much of their time surfing the web, looking for information about products, consumers, and competitors. But there are many faster and more efficient ways to gather data from the web. Crawling and scraping are automated. We gather text quickly, then we parse the text so it is easy to read.

A crawler or spider does more than gather data from one web page. Like an actual spider in its web, a World Wide Web crawler or spider traverses links with ease. It follows web links from one web page to another, gathering information from many sources.

Firms such as Google, Microsoft, and Yahoo! crawl the entire web to build indexes for their search engines. Our crawler represents a much more restricted effort, or what is known as a *focused crawl*.

A focused crawl has a starting point or points, usually a list of relevant web addresses. We surf a bit before we crawl in order to identify web addresses from which to start. A focused crawl has a stopping point. We restrict the range of the crawl to named domains. We can stop crawling any given domain after a specified number of pages have been downloaded. We can set selection rules for pages being downloaded. And we can limit the total number of pages to be downloaded in a given crawl. A focused crawl has a defined purpose. Therefore, we retain web pages that meet specific criteria.

With the crawl well in hand, the task turns to scraping or extracting the specific information we want from web pages. Each web page includes HTML tags, defining a hierarchy of nodes within a tree structure. The tree structure is defined by the *Document Object Model (DOM)*. And *XPath*, available through Python and R packages, is a specialized syntax for navigating across the nodes and attributes of the DOM, extracting relevant data. We gather data from the nodes, such as text within paragraph tags.

After extracting text data from a web page, we can parse those data using *regular expressions*. Page formatting codes, unnecessary spaces, and punctuation can be removed from each document. A successful crawl should end with a corpus of relevant documents for text analytics.

Figure B.3 shows a framework for automated data acquisition on the web that is consistent with Python web crawling and scraping software developed by Hoffman et al. (2014). First we crawl, then we scrape, then we parse. This is the work of web-based secondary research. Next, we select an appropriate wrapping for text and associated metadata, JSON or XML perhaps. We build a document store or text corpus for subsequent analysis. Scalable, enterprise-ready databases such as PostgreSQL or MongoDB, and data stores built on the Hadoop distributed file system may be employed for this purpose.

Consider a sports example. Suppose we take the role of the commissioner of a major sports league. We are concerned about injuries affecting the game. We want to learn as much as we can about the science of health and exercise physiology relating to injuries, so we search the web for answers. We also want to consult experts in the areas of medicine and sports, so we search the web for contacts. Concerned that news about injuries, some career-ending, some with ramifications long after the end of an athlete's career, affect public opinion about the sport, we review postings to social media.

We want to know if rule changes in recent years have had the beneficial effect of reducing injuries. We consult various sources regarding this problem, including published studies in medical journals and magazines. Our initial path to these sources is the web.

Each management client has his or her own problem to address. But it is likely that the path to data and resources to address problems begins with the web.

Crawling and scraping work well for data collection. Modeling and analysis begin with data, and the web is a massive store of data. Learning how to extract relevant data in an efficient manner is an essential skill of marketing data science.

The World Wide Web is a huge data repository that changes with each moment. To address business problems, we must find our way to the right links, extracting relevant data, and analyzing them in ways that make sense. Campbell and Swigart (2014) implore us to go beyond Google, and rightly so. See Chakrabarti (2003), Liu (2011), and Miller (2015c) for additional discussion of focused web crawling and data mining.

***Figure B.3.*** *Framework for Automated Data Acquisition*

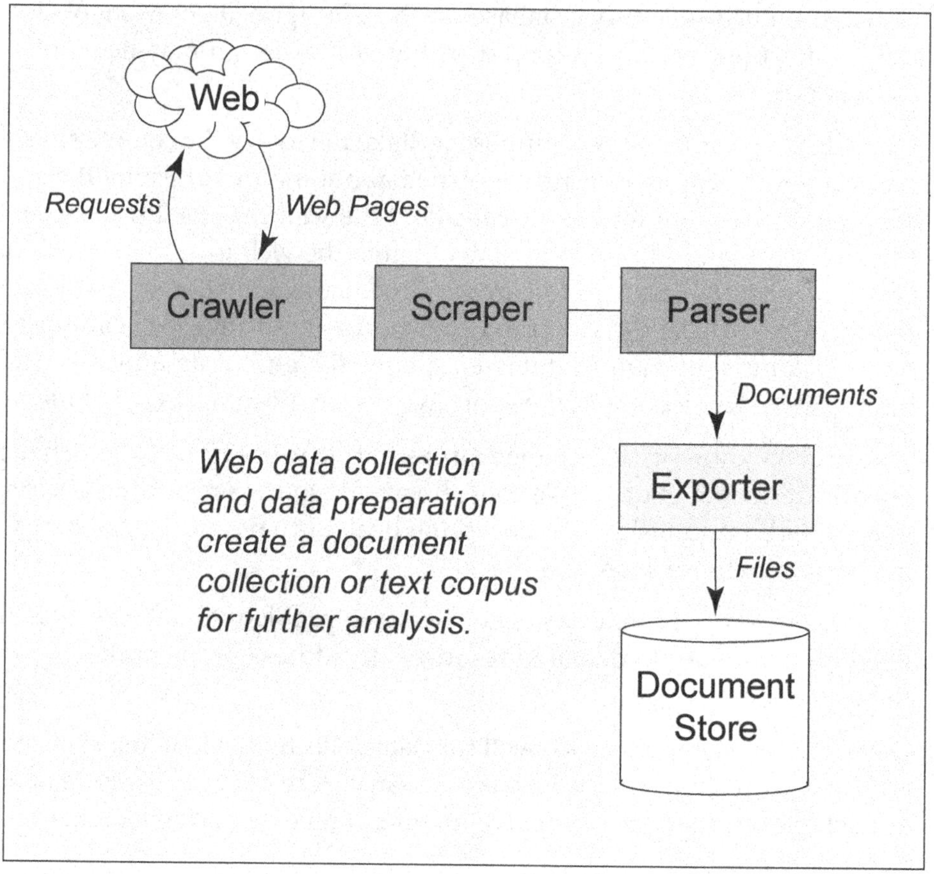

Adapted from Miller (2015c).

## B.6 Social Media

Recognizing growth in the use of electronic social networks and intelligent mobile devices, organizations see opportunities for communicating with and selling to friends of friends. Businesses, nonprofits, and governmental organizations (not to mention political campaigns) are interested in learning from the data of social media. Network data sets are large but accessible. The possibilities are many.

Social media provide two paths to marketing data. There are the connections among social media participants, which lead to social network analysis, and there are the raw data of social media, text and images submitted by participants, which afford opportunities for text measures (sentiment analysis and opinion mining). We begin this section by focusing on network data.

In years past, social network data collection was a painstaking process. Not so today. Social media implemented through the web have rekindled interest in social network analysis. Technologies of the web provide a record of what people do, where they come from, where they go, with whom they communicate, and sometimes transcripts of what they say. There are online friends and followers, tweets and retweets, text messages, e-mail, and blog posts. The data are plentiful. The challenge is to find our way to useful data and make sense of them.

Users of social media generate text data, including short text messages, personal statements, and blog posts. Obtaining these data involves using the various application programming interfaces (APIs) offered by social media providers. Russell (2014) provides Python scripts for using these APIs.

When working with very large social networks and social media sources, processing times and memory requirements can be substantial. We must often sample from large networks to obtain networks of manageable size for modeling and analysis. Sampling presents additional challenges. A good sampling method should yield samples that have values of centrality and connectedness that are similar to the values in the complete network or population. Sampling and missing data issues have been reviewed by numerous researchers (Kossinets 2006; Leskovec and Faloutsos 2006; Handcock and Gile 2007; Huisman 2010; Smith and Moody 2013; Miller 2015c).

## B.7 Surveys

Sample surveys are a common form of business and consumer research. Surveys assess opinion, attitudes, and behavior. They help us understand customers, competitors, and markets. We ask questions. We get answers. And we try to understand those answers through statistical analysis.

Recognizing the importance of surveys in business and consumer research, we provide examples of survey questions from real and hypothetical surveys. The examples below are developed around the common theme of consumer movie-going attitudes and behavior.

We begin with a survey of consumer attitudes and behavior about movie going conducted by Mintel International Group, Ltd. (2001). Advertisers, retailers, and entertainment service providers have many questions about consumer movie-going behavior. What kinds of people go to the movies? How often do they go? What do they expect to spend on a night at the movies? Are there identifiable segments of moviegoers? Does movie-going behavior vary from one part of the country to the next? Mintel's survey was designed to answer such questions as these.

Referring to survey items shown in figure B.4, we see examples of categorical variables, or what Stevens would call nominal measures. Variables such as the presence of children under eighteen and sex of respondent are categorical and binary. Marital status and census region are categorical and multinomial. Income and educational level variables may be thought of as ordinal. Age is the only demographic variable in this example that has interval properties. In fact, age has a meaningful zero-point, so we would call it a ratio measure.

Ordinal measures are common in consumer research. Survey questions are often worded in ways that assure ordinal, rather than interval, responses. This is sometimes intentional. The researcher may think that the respondent can only be trusted to provide answers within certain ranges. Figure B.5 shows the telephone interviewer guide for time-interval and frequency-of-activity questions from the movie-going survey.

Scientific data often arise from counts. We record events as they happen and count the frequency of their occurrence. Behavior logs and business transaction data are rich sources of counts. Sales volume is a count. The

***Figure B.4.*** *Demographic variables from Mintel survey*

| *Variable* | *Description and Coding* |
|---|---|
| Age | Age of respondent in years |
| Sex | Sex of respondent: 1 = male, 2 = female |
| Income | Annual household income level<br>1 = less than $25,000<br>2 = $25,000 to $49,999<br>3 = $50,000 to $74,999<br>4 = $75,000 to $90,999<br>5 = $100,000 or more |
| Marital Status | Marital status<br>1 = married<br>2 = widowed<br>3 = divorced/separated<br>4 = single/never married<br>5 = domestic partnership |
| Children | Presence of children under eighteen in household<br>1 = yes<br>2 = no |
| Region | U.S. census region of household<br>1 = New England<br>2 = Middle Atlantic<br>3 = East North Central<br>4 = West North Central<br>5 = South Atlantic<br>6 = East South Central<br>7 = West South Central<br>8 = Mountain<br>9 = Pacific |

*Source:* Mintel International Group, Ltd. (2001).

*Figure B.5. Sample questions from Mintel movie-going survey*

*Movie Attendance and Refreshments:*
*Guide for Telephone Interviewers*

*ATTENDANCE.* How often do you attend films at an indoor or outdoor theater?

*(DO NOT READ)*

1 *Once a week or more (four or more times per month)*
2 *Two to three times a month*
3 *Once a month*
4 *Once every couple of months*
5 *Less than once every two months*
N *Never* *SKIP REFRESHMENTS QUESTION*
D *Don't know*
R *Refused*

*REFRESHMENTS.* Would you say that you regularly, occasionally, seldom, or never buy refreshments at movie theaters?

4 *Regularly*
3 *Occasionally*
2 *Seldom*
1 *Never*
D *Don't know*
R *Refused*

*Source:* Mintel International Group, Ltd. (2001).

number of times a consumer goes shopping and the number of items in a shopping cart are counts. Grocery scanner data show counts of checkout items by Universal Product Code (UPC). The list goes on and on.

Counts take non-zero integer values and may be thought of as ratio measures. Another group of ratio measures in business and marketing research are monetary variables, such as prices, costs, and sales revenue.

It is common to include open-ended or free-response items as part of attitude or behavior surveys. This is how we learn what people are thinking when they make product choices. And when open-ended questions dominate our research, as they do for in-depth interviews and focus groups, we move from quantitative to qualitative research.

Examples of open-ended questions are shown in figure B.6. Notice that some open-ended questions, such as a person's age in years, may be coded directly as interval or ratio responses. Others, such as the title of a person's favorite movie, represent nominal or categorical variables with many possible values. The other questions from this figure would require the definition of categories before measures or numbers could be assigned to responses, unless perchance the researcher were to use the number of respondent words as the response measure.

Guided open-ended questions are often used in telephone interviews. The interviewer poses the question and guides the respondent to an acceptable response. Figure B.7 shows a guided open-ended question about movie spending.

Check lists, click lists, and binary response tasks are among the easiest of item formats for consumers. *Do you expect to vote in the next election? Are you thinking of buying a computer within the next year? Do you shop at West Towne Mall? Are you the head of the household? Are you over fifty years of age? Do you have children living with you? Will you be gambling on this visit to Las Vegas?*

What could be easier? Ask a consumer if he does something. Ask if something is true. Get a self-report: yes or no. This is a check list. Check lists are used in many research contexts. They are often the first items we encounter in a survey or interview because they are used in screening respondents.

Suppose our job is to study consumer attitudes about movies and consumer movie-going behavior. Before we ask about people's feelings about movies,

***Figure B.6.*** *Open-Ended Questions*

What is your age in years? [ ]

What is your favorite movie of all time?

[ ]

What do you like about going to the movies?

[ ]

What do you think of Quinten Terantino as a director?

[ ]

How did you feel about *Lovely and Amazing?*

[ ]

What were you thinking about when you decided to rent *The Producers* rather than *The Phantom of the Opera?*

[ ]

What was your opinion of Sean Penn in *All the King's Men?*

[ ]

***Figure B.7.*** *Guided Open-Ended Question*

*Movie Spending Question:*
*Guide for Telephone Interviewers*

When you go to the movie theater, how much would you expect to spend in total on a ticket and refreshments per person?

*(INSERT $ AND CENTS. GET BEST ESTIMATE.*
*DO NOT ACCEPT RANGE.*
*IF MORE THAN $20, CLARIFY "Is that per person?"*
*IF RESPONDENT INDICATES "IT VARIES,"*
*ASK FOR THEATER VISITED MOST OFTEN)*

*INSERT AMOUNT $ _ _ . _ _*

*DD (DO NOT READ) Don't know*
*RR (DO NOT READ) Refused*

*Source:* Mintel International Group, Ltd. (2001).

***Figure B.8.*** *Behavior Check List*

Check the movies you have seen within the last five years:

- ☐ *A Clockwork Orange*
- ☐ *Barbarella*
- ☐ *The Blob*
- ☐ *Cast Away*
- ☐ *Chocolat*
- ☐ *Dick*
- ☐ *Dirty Dancing*
- ☐ *Dr. Strangelove*
- ☐ *First Blood*
- ☐ *Frenzy*
- ☐ *Garden State*
- ☐ *High Art*
- ☐ *Hollywood Homicide*
- ☐ *Lone Star*
- ☐ *Matewan*
- ☐ *Monster*
- ☐ *Mystic Pizza*
- ☐ *Network*
- ☐ *North by Northwest*
- ☐ *Pretty Woman*
- ☐ *Riding in Cars with Boys*
- ☐ *Shopgirl*
- ☐ *Sleepless in Seattle*
- ☐ *Southern Comfort*
- ☐ *The Virgin Suicides*
- ☐ *Zelig*

we may want to know which movies they have actually seen. Figure B.8 shows a behavior check list designed for this purpose.

A check list becomes a click list in a computerized or online survey. Figure B.9 shows what a click list for movie experiences might look like as a computerized survey. Respondents often welcome tasks like this because they can be performed quickly and easily. Using on-screen images or icons for the movies, rather than merely the movie titles, can help people remember the movies.

An adjective check list asks the respondent to evaluate a stimulus using a series of yes-no responses to adjectives. In consumer research, the stimulus is usually a product or service to be evaluated. For example, having determined that a movie-goer has seen the movie *One Flew over the Cuckoo's Nest*, we use an adjective check list, as shown in figure B.10, to measure her attitudes about the movie.

Check lists and click lists are examples of the general class of binary response tasks, which call for simple yes-no or agree-disagree responses. Fig-

*Figure B.9.* From Check List to Click List

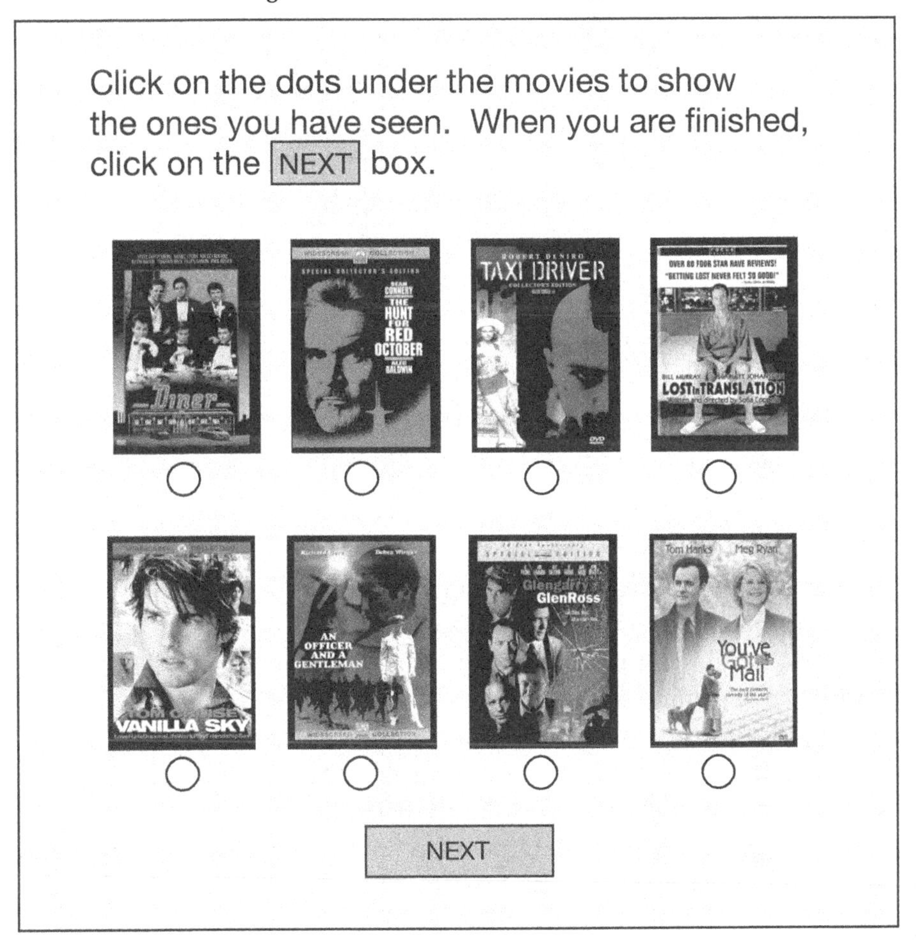

*Figure B.10.* *Adjective Check List*

Consider the movie *One Flew over the Cuckoo's Nest.*
Check all of the adjectives below that describe that movie.

| | |
|---|---|
| ☐ artistic | ☐ nervy |
| ☐ beautiful | ☐ obscure |
| ☐ comic | ☐ pious |
| ☐ dramatic | ☐ quixotic |
| ☐ exciting | ☐ realistic |
| ☐ flamboyant | ☐ sadistic |
| ☐ grating | ☐ tragic |
| ☐ humane | ☐ uplifting |
| ☐ ironic | ☐ violent |
| ☐ jubilant | ☐ warm |
| ☐ kinky | ☐ X-rated |
| ☐ liberal | ☐ youthful |
| ☐ mechanical | ☐ zestful |

ure B.11 shows a variety of questions calling for binary agree-disagree responses.

Rating scales are among the most popular of measurement methods in consumer research. Respondents may be asked to assign words or numbers to stimuli representing products, services, suppliers, or brands. Scale scores are thought to relate to attributes of the object being evaluated. Alternatively, we can make statements about products, services, suppliers, or brands, and ask consumers to respond to these statements along an agree-disagree scale. Figure B.12 shows a rating scale for assessing the importance of selected movie characteristics. Figure B.13 shows a rating scale for agreement or disagreement with statements in the form of movie reviews. Rating scales such these are called Likert scales, after Rensis Likert (1932).

Likelihood-of-purchase scales, common in consumer research, are examples of labeled rating scales. Figure B.14 shows a typical five-level likelihood of purchase scale. Some researchers like to use a even number of scale points. Others prefer a odd number of scale points, so there is a middle point of indifference or no opinion. We can use a number assignment without labels, such as rating on a scale from one to ten. Or we can ask

**Figure B.11.** *Binary Response Questions*

*Agree-Disagree Questions: Instructions for Interviewers*

Please tell me whether or not you agree with
the following statements about refreshments
at movie theaters.

Do you agree or disagree?
*(ASK FOR EACH STATEMENT)*

*1 YES, Agree*
*2 NO, Disagree*
*D Don't know*
*R Refused*

*ROTATE*

1a. If prices on refreshments were raised,
you would stop purchasing them.

1b. You would buy more refreshments
if they were less expensive.

1c. You would like to see healthier food
options at the movie theater.

And now, which of the following statements
do you agree with?

*ROTATE*

2a. You enjoy video/DVD rentals,
but they don't replace movie theaters.

2b. You usually go to a discount theater
or attend movies offering discount tickets.

2c. You feel that movie tickets are reasonably priced.

2d. You would rather attend a multi-screen theater
than a single-screen.

2e. You would be willing to pay higher prices
for better theater facilities
(i.e., stadium seating, higher quality sound system).

*Source:* Mintel International Group, Ltd. (2001).

*Figure B.12. Rating Scale for Importance*

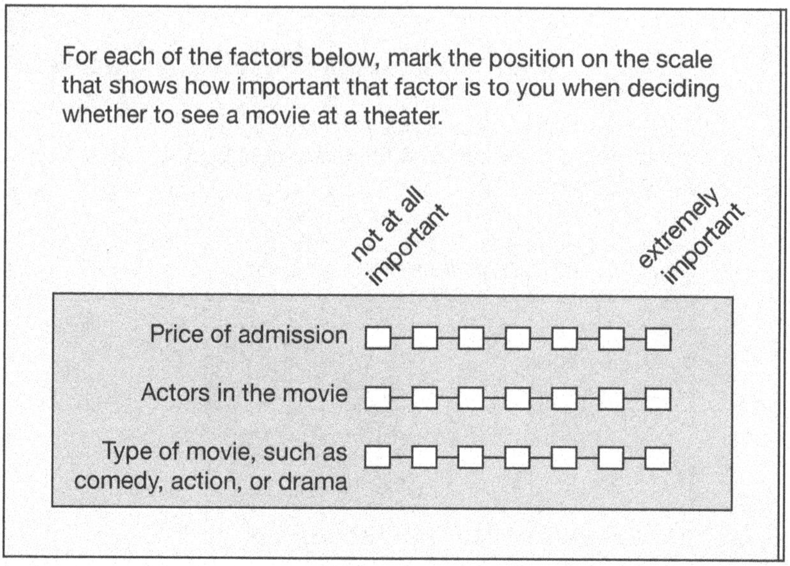

For each of the factors below, mark the position on the scale that shows how important that factor is to you when deciding whether to see a movie at a theater.

| | not at all important | | | | | | extremely important |
|---|---|---|---|---|---|---|---|
| Price of admission | ☐ | ☐ | ☐ | ☐ | ☐ | ☐ | ☐ |
| Actors in the movie | ☐ | ☐ | ☐ | ☐ | ☐ | ☐ | ☐ |
| Type of movie, such as comedy, action, or drama | ☐ | ☐ | ☐ | ☐ | ☐ | ☐ | ☐ |

for a self-report of probability of purchase or an indication of the amount a consumer is willing to pay for a product.

How many scale points should be used? Should verbal labels be used on scales? Should they be used across the entire scale or only at the end points? Are computerized sliding or continuous scales as effective as scales with a discrete number of response options or scale points? There are no clear answers to these questions. Many types of scales—five-point and seven-point, labeled and unlabeled, discrete and continuous—seem to work quite well. Correlations between scales with alternative formats are high. Churchill and Iacobucci (2004) cite research literature addressing these issues.

A semantic differential uses bipolar adjective pairs as the labels for scale endpoints as in the semantic differential developed by Osgood, Suci, and Tannenbaum (1957), as summarized by Osgood (1962). The semantic differential may be thought of as a combination of an adjective check list and a rating scale. It builds on the fact that most adjectives have logical opposites; they are bipolar. The respondent is asked to rate a stimulus or object along a set of scales defined by selected bipolar adjectives. Figure B.15 shows a semantic differential with scales defined by five bipolar adjective pairs.

***Figure* B.13.** *Rating Scale for Agreement/Disagreement*

For each movie review below, indicate the degree to which you agree or disagree with the review. Use a numeric scale as follows:

(1) strongly disagree
(2) disagree
(3) neither agree nor disagree
(4) agree
(5) strongly agree

| | Rating |
|---|---|
| *Romancing the Stone* (1984)<br>"Lots of fun...delightful...pure escapism."<br>- Kevin Thomas, *Los Angeles Times* | |
| *City Slickers* (1991)<br>"Ambitious. Funny. *City Slickers* delivers."<br>- Roger Ebert, *Chicago Sun Times* | |
| *Better than Chocolate* (1999)<br>"Sexy! Delightfully erotic!"<br>- *Curve Magazine* | |

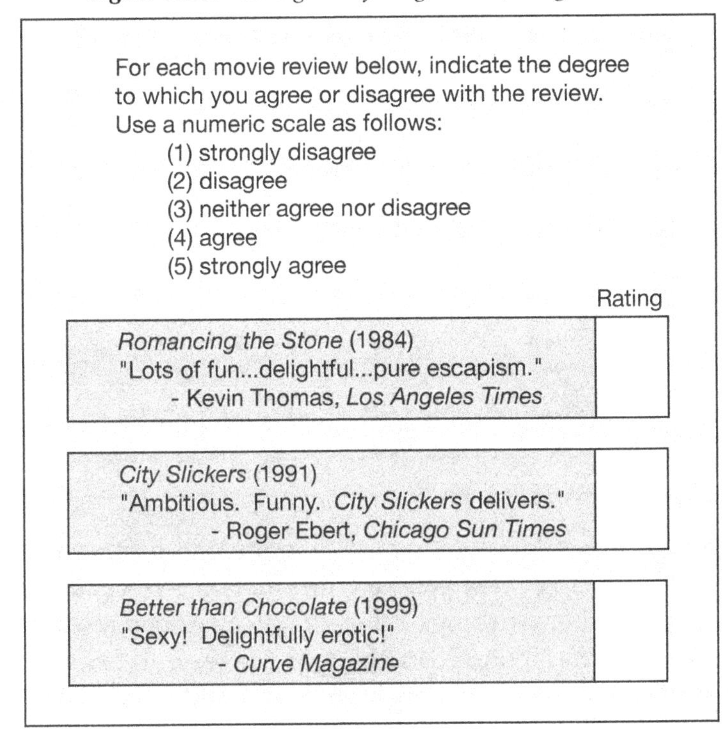

***Figure* B.14.** *Likelihood-of-Purchase Scale*

Click on the box that best describes your likelihood of buying the Blu-ray Disc *Monster's Ball* (2001) with Halle Berry, Billy Bob Thornton, Heath Ledger, and Peter Boyle. Discount price: $19.95

- ☐ Definitely will buy
- ☐ Probably will buy
- ☐ Might or might not buy
- ☐ Probably will not buy
- ☐ Definitely will not buy

*Figure B.15. Semantic Differential*

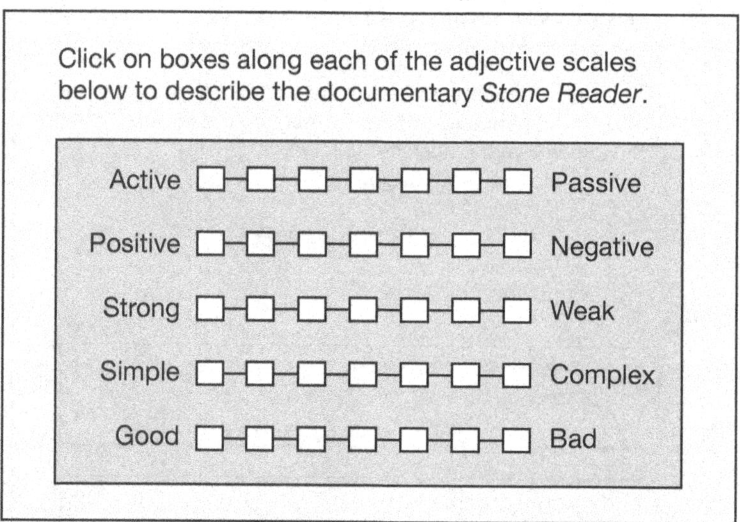


The semantic differential offers a flexible approach to measurement. The object or concept being rated can be a product or service, a firm or brand name, or any number of things relevant to consumer research. Various bipolar adjectives may be used, although it is wise to keep the list short and to use adjectives that have been demonstrated to work in various applications. Nunnally (1967) summarized psychometric studies of semantic differentials, identifying three primary types of adjectives—those associated with evaluation, potency, and activity. Nunnally (1967) suggested that a fourth type of adjective, associated with understandability or familiarity, was observed in some studies. Figure B.16 shows exemplary bipolar adjectives for use in consumer research.

Most past applications of the semantic differential have used paper-and-pencil administration of discrete five- or seven-point scales. But the technique could easily be adapted for administration by computer with continuous or sliding scales, as shown in Figure B.17.

In consumer research there are times when we need to assess a person's thinking or decision-making process. We present detailed information and ask the respondent to rate or rank product profiles, make a choice, perform a calculation, or write an opinion. Through well designed surveys, we learn about price-quality, price-feature, and feature-by-feature trade-offs in con-

*Figure B.16.* Bipolar Adjectives

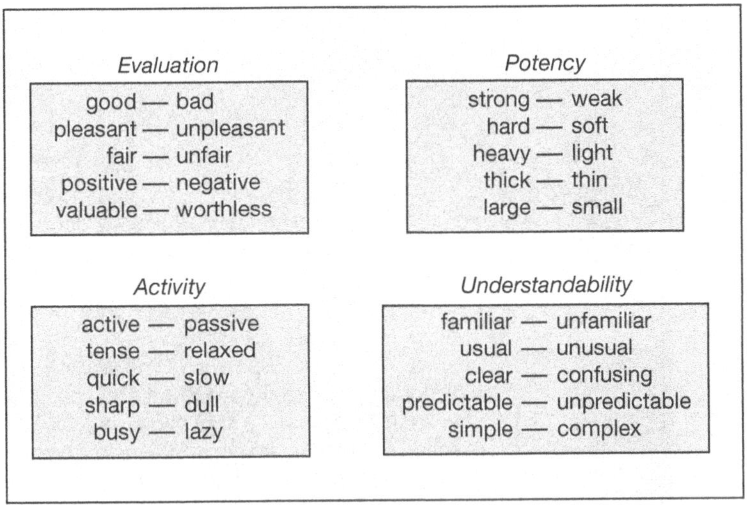

| Evaluation | Potency |
| --- | --- |
| good — bad | strong — weak |
| pleasant — unpleasant | hard — soft |
| fair — unfair | heavy — light |
| positive — negative | thick — thin |
| valuable — worthless | large — small |

| Activity | Understandability |
| --- | --- |
| active — passive | familiar — unfamiliar |
| tense — relaxed | usual — unusual |
| quick — slow | clear — confusing |
| sharp — dull | predictable — unpredictable |
| busy — lazy | simple — complex |

*Source:* Nunnally (1967).

sumer evaluation of new products. We learn how consumers feel about brands.

Suppose we want to assess people's preferences for movies. Figure B.18 shows a conjoint measurement item. The consumer respondent is asked to rate a DVD movie product on a scale from 1 (low) to 7 (high). A brief description of the DVD movie is provided along with its price. This is a straightforward rating task. It would be used in conjunction with many other rating tasks. Respondent ratings will have meaningful magnitude, and the likely analysis is linear regression. This is an example of a traditional conjoint study item. Notice that price is included in this likelihood-to-purchase item.

Figure B.19 shows a more complicated task. A pair of movies is presented side-by-side, and the respondent is asked to compare them in terms of interest or preference. Degree of preference is on a sliding scale. Two measures are obtained from each item of this type, with the measures being conditional on one another. The researcher is not forcing a choice between the movies as in a traditional paired comparison item.

*Figure B.17. Semantic Differential with Sliding Scales*

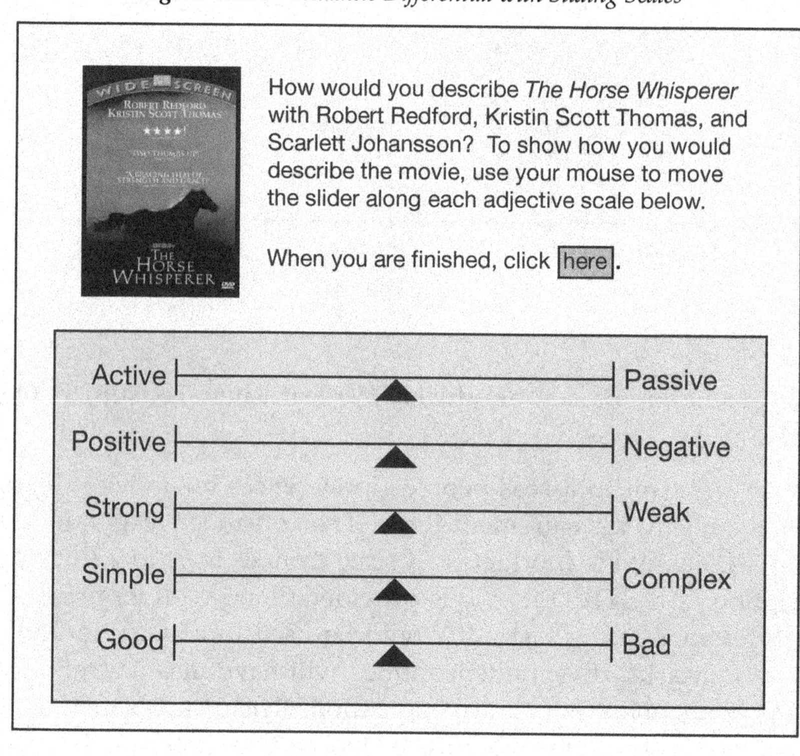

***Figure B.18.*** *Conjoint Degree-of-Interest Rating*

Consider the movie experience described below.
On a scale from 1 (low) to 7 (high), rate your degree of interest in buying this movie DVD.
Use scale extreme values defined as:

(1) not at all interested
(7) extremely interested

Type your rating in the space provided.

*Duck Soup* (1933) with the Marx Brothers
Black-and-white movie
Length: 68 minutes
DVD media
Viewed at home
Purchase price: $5.99

***Figure B.19.*** *Conjoint Sliding Scale for Profile Pairs*

Below you see descriptions of two movies that you could watch. Use your mouse to move the slider along the scale, showing the degree to which you prefer one movie over the other.

| *Gorky Park* (1983) | *The Year of Living Dangerously* (1982) |
|---|---|
| The film opens with the discovery of three mutilated bodies in the snow of Moscow's Gorky Park. The Chief Investigator of the Moscow Militia is assigned to unravel this horrific crime. The trail leads to Irina, a beautiful Soviet dissident who soon becomes his lover, and a rich businessman who will stop at nothing to corner the international sable market. | Guy Hamilton, an Australian reporter on assignment in politically unstable Indonesia, joins forces with a savvy photographer to expose government corruption. But it's Hamilton's connection to the beautiful Jill Bryant, an attaché with the British embassy, that complicates his life. |

***Strongly prefer movie at left*** ***No preference*** ***Strongly prefer movie at right***

Traditional conjoint studies can also utilize rank orders generated from paired comparison choices and multiple-rank-orders items. Exhaustive ranking tasks, especially those with large numbers of objects to be ranked, present special difficulties for respondents. They may seem repetitive or artificial. Respondents complain that such tasks are unnecessarily tedious, even when using Q-sorts to simplify the task of ranking. Nonetheless, such tasks are sometimes used in consumer research.

One variation on the ranking task is an online stacking-and-ranking task, as shown in figure B.20. This mode of eliciting ranks gives the respondent a chance to manipulate icons or images of products in the process of ranking.

A set of simple paired comparison items is shown in figure B.21. The respondent chooses one movie title from each pair, indicating preference. No description of the movies is provided. A paired comparison is simultaneously a ranking and a choice—a choice between two. An exhaustive set of paired comparisons across $k$ movies would require $\frac{k(k-1)}{2}$ paired comparisons, with each movie being paired once with each other movie. Fortunately, individual paired comparisons can be made quickly.

Marketing analysts can also present movies in sets of three or more in what are called multiple-rank-orders, as shown in figure B.22. Presented in sets of three movies, each multiple-rank-order item is equivalent to three paired comparisons. For a complete enumeration, then, multiple-rank-order triads require one-third as many items as paired comparisons. Each item, however, takes a little longer to complete.

Like multiple-rank-order items, best-worst items present objects in sets of three or more. For each item, the respondent is asked to choose a most preferred and least preferred object. A set of three-object best-worst items is equivalent to a set of multiple-rank-order triads, yielding a full set of paired comparisons. When presented in groups of four or more objects, however, best-worst items yield partial paired comparisons.

Figure B.23 shows a best-worst item with four movies. Notice how it is possible to extract five out of the possible six paired comparisons from this item. Best-worst surveys, sometimes called max-diff scaling, also have similarities to choice tasks. Instead of being asked to choose only the most preferred object in a set of objects, the respondent chooses two, the best and worst.

***Figure B.20.*** *A Stacking-and-Ranking Task*

Consider the four movies at the left. To show your interest in watching a movie, click on the movie to highlight it and use the arrow keys in the center to move it from left to right. All the movies you would consider watching should be moved to the right. Then, for the movies you have moved to the right, use the arrow keys at the far right to move them up and down the stack. Rank the movies in order of your interest in watching them. The movie at the top of the stack should be the one you would most like to watch. When finished, click on the NEXT box.

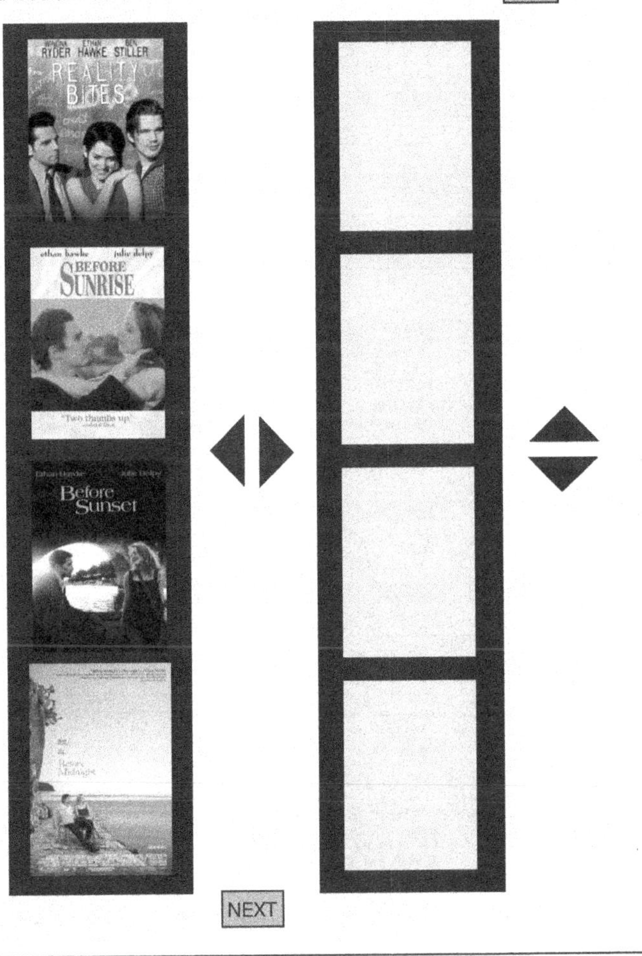

*Figure B.21. Paired Comparisons*

For each pair of movies, click on the box next to the movie you prefer.

| | |
|---|---|
| ☐ *Five Easy Pieces*<br>☐ *Pieces of April* | ☐ *Blow*<br>☐ *Traffic* |
| ☐ *Mona Lisa Smile*<br>☐ *Playing Mona Lisa* | ☐ *Basic Instinct*<br>☐ *Femme Fatale* |
| ☐ *Chicago*<br>☐ *Bugsy Malone* | ☐ *The Talented Mr. Ripley*<br>☐ *Good Will Hunting* |
| ☐ *My Dog Skip*<br>☐ *The Sandlot* | ☐ *Kids*<br>☐ *Thirteen* |
| ☐ *Breaking Away*<br>☐ *The Flamingo Kid* | ☐ *Closer*<br>☐ *Eyes Wide Shut* |
| ☐ *Eye of the Needle*<br>☐ *The Statement* | ☐ *Mulholland Drive*<br>☐ *Mulholland Falls* |

*Figure B.22. Multiple-Rank-Orders*

Indicate your preferences for each group of three movies below.
In the boxes to the left of each group of three movies, enter
1 next to your favorite movie
2 next to your second favorite movie
3 next to your least favorite movie

| | |
|---|---|
| ☐ *American Graffiti*<br>☐ *Harry and Tonto*<br>☐ *The Paper Chase* | ☐ *The Sting*<br>☐ *The Grifters*<br>☐ *The Italian Job* |
| ☐ *Octopussy*<br>☐ *For Your Eyes Only*<br>☐ *The Spy Who Loved Me* | ☐ *Ferris Bueller's Day Off*<br>☐ *The Blues Brothers*<br>☐ *Risky Business* |
| ☐ *Rocketeer*<br>☐ *Star Wars*<br>☐ *Indiana Jones* | ☐ *Ed TV*<br>☐ *Pleasantville*<br>☐ *The Truman Show* |
| ☐ *The Godfather*<br>☐ *Scarface*<br>☐ *Pulp Fiction* | ☐ *Dirty Harry*<br>☐ *Lethal Weapon*<br>☐ *The French Connection* |
| ☐ *Eight Men Out*<br>☐ *Silver City*<br>☐ *Sunshine State* | ☐ *Broadcast News*<br>☐ *The China Syndrome*<br>☐ *All the President's Men* |
| ☐ *Alice Doesn't Live Here Anymore*<br>☐ *Plain Dirty (Briar Patch)*<br>☐ *What's Eating Gilbert Grape?* | ☐ *Alien*<br>☐ *Men in Black*<br>☐ *Invasion of the Body Snatchers* |

*Figure B.23.* *Best-Worst Item Provides Partial Paired Comparisons*

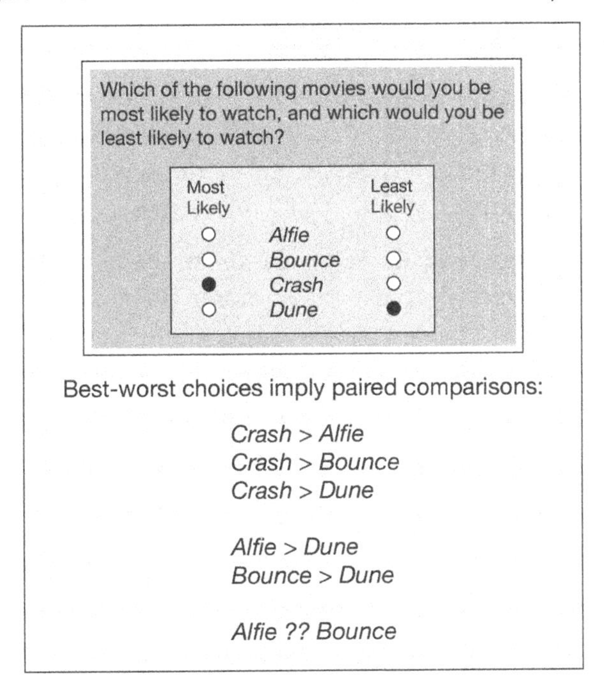

Obtaining interval measures from rank orders is a general way of thinking about scaling methods. We can use paired comparisons, rank orders, multiple-rank-orders, or best-worst responses from consumers to generate a paired-comparison preference matrix showing the proportion of times each object is ranked over each other object. And from a paired-comparison preference matrix, we can generate scale scores for the objects being compared. Interval measures convey more information than ranks. Seeing how consumer movie preferences fall along a real number line, for example, is much more informative than seeing the rank order of movies.

Many marketing analysts prefer choice tasks to assess consumer preferences and to make recommendations about product design and pricing. They argue that choice studies provide a straightforward approach to estimating market shares for products and brands. Furthermore, a well designed choice study bears a close resemblance to what consumers do in the marketplace. Consumers reveal their preferences in the choices they make.

The simplest of choice tasks is a paired comparison choice. Earlier we showed simple paired comparisons of movie titles. In figure B.24, we show a paired comparison choice task with more information about the movies to be compared. Paired comparisons are easy to make, even when there is extensive information about the objects being compared.[1]

Choice studies often present sets of three or four profiles, with each profile defined by a large number of attributes. Figure B.25 shows a choice task with three new movie ideas and profiles defined by six attributes. The consumer's task is to select one of the three hypothetical new movies. Choice tasks sometimes include a *none* option, permitting the consumer to move to the next item without making a choice.

Additional developments in preference measurement involve product configuration tasks and menu-based choice, as shown in figure B.26. When buying a computer or selecting meal items at a restaurant, we have many

[1] Paired comparisons are a convenient mechanism for obtaining choice and ratings data. Taste tests are commonly conducted as paired comparisons. One food is tasted. Water is drunk to clear the pallet. And then a second food is tasted. Asking which food tastes better or which food is preferred is a simple paired comparison. The optometrist asks us to compare two lenses in sequence. The patient looks through the first lens and then the second. Upon presenting the second lens, the optometrist asks a simple question, "Is this better or worse, or the same?" That is a paired comparison with three alternatives. Proximity in time is key to the comparison process, and using only one pair of lenses at a time simplifies the information processing task.

***Figure B.24.*** *Paired Comparison Choice Task*

Which of the following two movies would you prefer to view through your online service? Click on the circle to show your preference.

| *Mars Attacks* (1996)<br>Director: Tim Burton<br>Actors:<br>Jack Nicholson<br>Glenn Close<br>Annette Bening<br>Pierce Brosnan<br>Danny DeVito<br>Martin Short<br>○ | *Desperately Seeking Susan* (1985)<br>Director: Susan Seidelman<br>Actors:<br>Rosanna Arquette<br>Aidan Quinn<br>Madonna<br>Robert Joy<br>Mark Blum<br>○ |
|---|---|

***Figure B.25.*** *Choice Set with Three Product Profiles*

Suppose that your local movie theater is able to show only three movies at a time, and the movies have different prices. Assuming that the movie-going experience is alike in all other ways, which of these future movies would you most like to see? Show your choice by clicking on the circle to the right of the movie. Click the NEXT box when you are ready to move on.

|  |  |
|---|---|
| *French Curve*<br>Action/Adventure<br>Jennifer Lawrence<br>Director: Brian DePalma<br>Location: France<br>Length: 120 minutes | ○ |
| *Miss Conception*<br>Comedy<br>Jessica Chastain<br>Director: Woody Allen<br>Location: New York City<br>Length: 95 minutes | ○ |
| *Pluperfect Passion*<br>Drama<br>Caitlin Gerard<br>Director: Ben Affleck<br>Location: Mexico<br>Length: 105 minutes | ○ |

NEXT

options from which to choose. We build a customized product instead of choosing from a set of preconfigured product profiles.

It is only fitting that the measurement techniques we use to assess consumer preferences reflect the way consumers purchase products. In most product categories, choice tasks reflect consumer behavior in the marketplace in a way that rankings and ratings do not. And in certain product categories, menus reflect consumer behavior in the marketplace in a way that choice tasks do not.

We can also assess consumer preferences through a sequence of choices in an elimination pick list. Figure B.27 shows an elimination pick list for a set of seven movies. Elimination pick list instructions ask the respondent to choose one movie at a time. A computer program records each choice as it is made and updates the screen image by eliminating the movie that has just been chosen. The consumer continues to pick movies until there are no more movies she wants to watch. The elimination pick list represents an enhanced choice task with a *none* alternative. It may also be thought of as a computer-assisted ranking task.

The elimination pick list is designed to be a faithful representation of consumer choice in the marketplace. We often choose more than one product, buy more than one thing at at time. The elimination pick list is simultaneously a ranking and a choice task. Compared to a ranking task, note that picking one alternative at a time is easier than ranking a set of alternatives, yet it accomplishes the same purpose. The elimination pick list yields censored rank-order data with the possibility of tied ranks at the low end.

The pick list is consistent with the way consumers evaluate and choose products in the marketplace. When faced with a set of choices, consumers would be unlikely to ask themselves about their most preferred and least preferred products, as in a best-worst choice task. Rather, consumer buyers focus at the top of the list, on their most preferred products. By providing direct evidence about consumer consideration sets, the elimination pick list has an advantage over other choice and ranking tasks.

***Figure B.26.*** *Menu-based Choice Task*

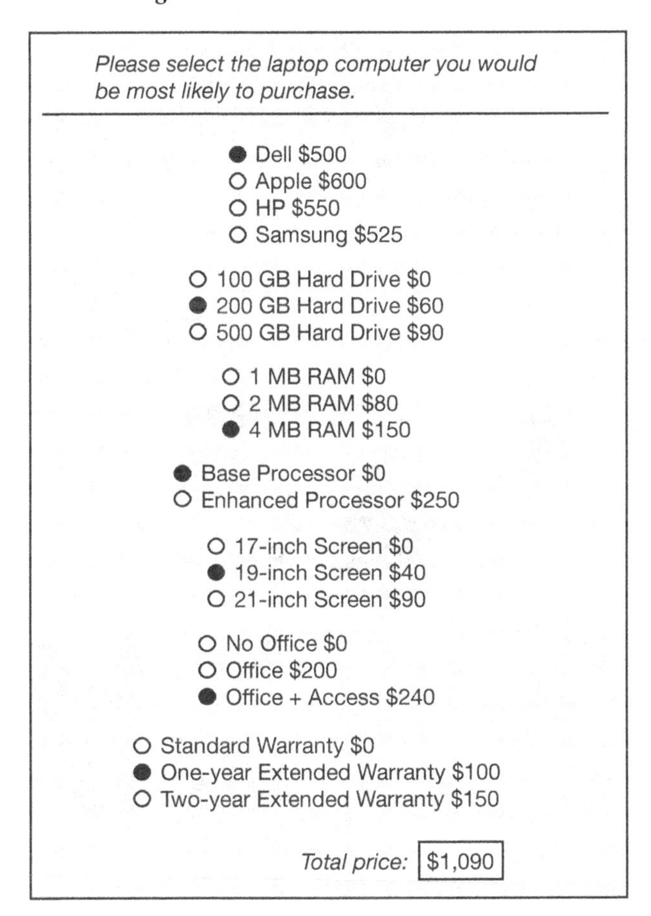


Adapted from Orme (2013) with permission of the publisher.

*Figure B.27.* *Elimination Pick List*

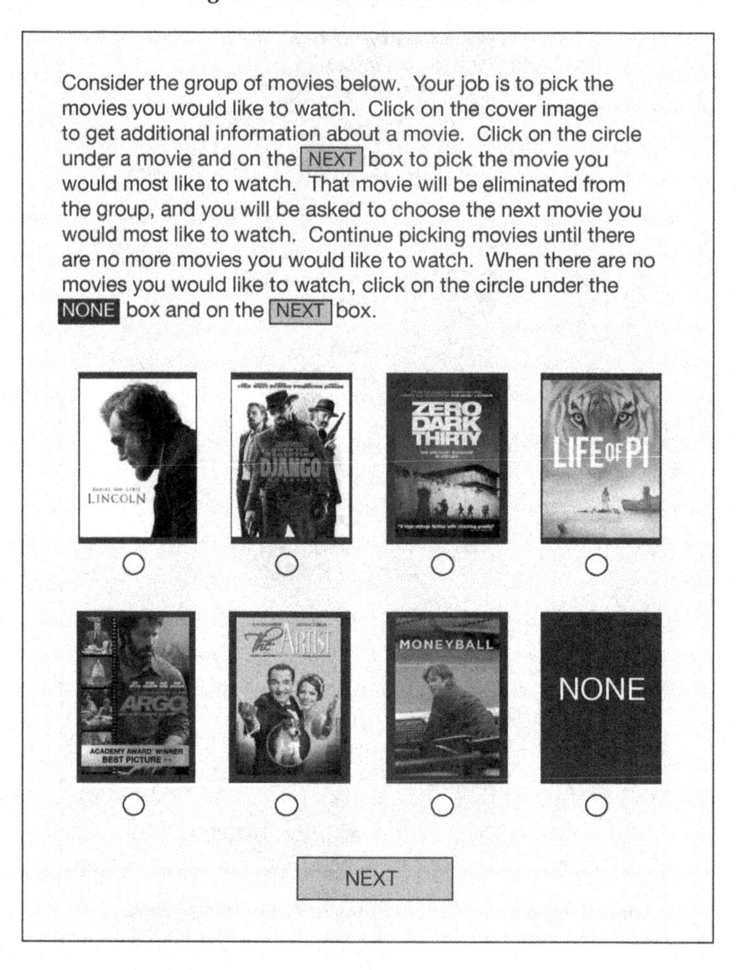

Obtaining meaningful and accurate measures is a prerequisite to building predictive models we can trust. When designing consumer surveys, we need to use wording that is easy to understand and unambiguous. We need to provide choices that are distinct and clearly specified. We try to make tasks as simple and straightforward as possible, while presenting as much detail as is needed for the measurement objective.

It is important to recognize the complexity of the task we present to consumer respondents. Paired comparisons are easier to make than multiple comparisons. Ranking a short list of alternatives is easier than ranking a long list of alternatives. Picking one alternative at a time is easier than ranking a set of alternatives. If we make a task similar to what consumers normally do in the marketplace, then consumers can more faithfully execute that task. This is the art and science of measurement.

Groves et al. (2009) provides a comprehensive review of survey research, including sampling, data collection modalities, data quality, nonresponse, and analysis issues. Miller (2008) provides discussion of measurement, item formats, and alternative scales of measurement for business research. Scaling methods for obtaining interval measures from paired comparisons and rank orders are documented in the psychometric literature (Stevens 1946; Guilford 1954; Torgerson 1958). Rounds, Miller, and Dawis (1978) review paired comparison and multiple-rank-order scaling, demonstrating strong agreement between measurements obtained from these two methods. Louviere (1993) introduces best-worst scaling methods.

For those who want to learn more about sampling and survey research, the American Statistical Association (ASA) Section on Survey Research Methods and the American Association of Public Opinion Researchers (AAPOR) are active in this area. Cochran (1977) provides a classic introduction to the field of sampling. Applied researchers can look to Levy and Lemeshow (1999) for an overview of conventional sampling techniques. Chan (2000) and Churchill and Iacobucci (2004) provide useful introductions for business and consumer research. Groves and Couper (1998), Miller (2001), and Miller and Dickson (2001) review issues in online survey research.

## B.8 Experiments

Two essential characteristics define experiments: control and randomization. When we run experiments, we do everything we can to control experimental conditions. We identify variables that could have an effect on consumer response—the explanatory variables. We try to keep these explanatory variables constant or in check. Control in experimentation is more easily accomplished in a laboratory setting, but some modicum of control or standardization is needed for field experiments as well.

A treatment represents the experimental manipulation of explanatory variables or factors. The treatment is itself a type of control: we employ the same treatment with each experimental unit assigned to that treatment. An experimental unit in consumer research could be an individual consumer or a group of consumers, such as a test market. Most experiments in business and marketing involve factorial designs. These are designs with treatments defined in terms of two or more factors.

What we cannot control in an experiment, we randomize. Randomization refers to the random assignment of experimental units to treatment combinations. What distinguishes one type of experiment from the next is the way experimental units are assigned to treatments. This is the domain of experimental design, a branch of statistical practice.

Along with control and randomization, a critical concept in experimental design is replication. This refers to repeated runs of an experimental treatment or repeated runs of sets of treatments. The more replications, the larger the sample size and the better the chance of observing a statistically significant effect. We seek designs that are efficient—that get the most possible information from each experimental unit.

Critical to experimental design is the enumeration of appropriate combinations of factors defining experimental treatments. We look for designs in which all levels of factors and combinations of factors are equally represented. And we look for designs in which equal numbers of experimental units are assigned to each level of each factor. In other words, we look for balanced designs. In many contexts, however, it is not possible to use a balanced design. Instead, we try to identify an optimal unbalanced design—this is a design that satisfies certain mathematical properties of optimality.

In designing new products, marketing managers need information about a wide variety of product features. Full factorial designs, completely crossed designs with as many cells as there are feature combinations, are impractical. Instead, marketing researchers can select feature combinations by referring to plans for fractional factorial designs, such as Latin squares, Graeco-Latin squares, lattice, or balanced incomplete block designs. Or they use computer programs to construct optimal unbalanced factorial designs.

The term "A/B testing" is used to describe online testing. This is an unfortunate term because it suggests that experiments must be simple comparisons of one treatment with another. Much more can be learned by employing factorial (muli-factor) designs.

Experimental research is sometimes called causal research. This is a misnomer because few (if any) experiments in business may be interpreted as providing evidence of cause and effect. Nonetheless, properly designed and executed experiments can provide strong evidence of relationships between explanatory and response variables.

Experimenters must be vigilant in designing and executing experiments because there are many potential sources of error. Figure B.28 provides a summary of validation issues. A distinction is made between internal and external validity. Internal validity relates to the design itself—proper control and randomization, as well as minimal confounding of factors. When researchers find an effect, they want the interpretation to be straightforward. External validity relates to the relevance of a study to other areas (generalizability), such as the relevance of laboratory findings to actual results in the field.

Experimental design is a large and well developed area of statistics. Classic statistical references include Kempthorne (1952), Cochran and Cox (1957), Cox (1958), Fisher (1971), and Box, Hunter, and Hunter (2005). Kirk (2013), Winer, Brown, and Michels (1991), and Keppel and Wickens (2004) provide introductions for social and behavioral scientists. Manly (1992) presents an overview of experimental and survey research designs with a few examples from business. Banks (1965), Patzer (1996), and the book of readings by Aaker (1971) provide examples of experiments in marketing. Rasch et al. (2011) provide tools for designing experiments in R.

*Figure B.28. Factors affecting the validity of experiments*

Validity of Experiments

Internal Validity

Can differences across experimental treatments be attributed to the treatments themselves?

History
Events outside the experiment

Maturation
Changes in subjects over time unrelated to the experiment

Testing
Process of experimentation itself (including experimenter effects)

Instrumentation
Changes in instruments over time or improper control across treatments

Regression
Statistical regression to the mean over time (between pretest and posttest)

Selection Bias
Nonrandom assignment of subjects to treatments

Mortality
Differential censoring or loss of subjects across treatments

External Validity

To what extent can we generalize from the experiment to the real world of business?

Measures
Measures irrelevant to actual variables of interest to management

Environment
Experimental laboratory or field setting unlike actual business environment

Subjects
Study subjects unrepresentative or unlike subjects of interest to management

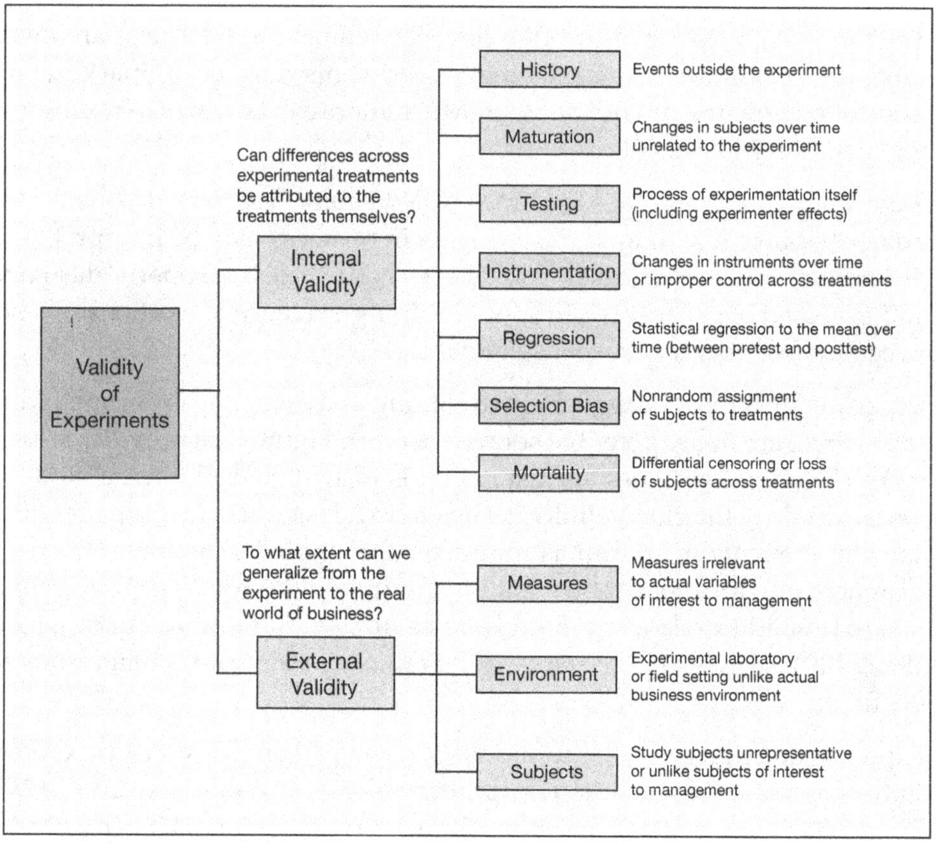

Source: Adapted from Miller (2008).

A classic reference for social scientists and business researchers is the book by Campbell and Stanley (1963), which reviewed practical considerations in the design and interpretation of experiments, as well as validity issues. Campbell and Stanley introduced the term *quasiexperiment* to refer to a non-standard experiment or an experiment that would not fit neatly into a framework for statistical analysis. They were most interested in field experiments that take place over time—time-series and pretest-posttest experiments.

Most quasiexperiments can be analyzed with statistical methods, but the analyses are more complex than those for completely randomized factorial designs. Cook and Campbell (1979) discussed the use of interrupted time-series methods. Repeated measures, nested or hierarchical designs, and statistical controls (adjusting for covariates) may be required.

Unlike laboratory experiments, field experiments are lacking in controls. It fact, it may be better to think of field experiments as observational studies. These require special care with regard to statistical analysis (Rosenbaum 1995). Longitudinal, mixed, and multi-level models are often useful for such designs (Diggle, Liang, and Zeger 1994; Pinheiro and Bates 2009; Gelman and Hill 2007).

## B.9 Interviews

Our understanding of the consumer would be incomplete if we were to rely solely on consumer surveys and designed experiments, and the numbers they provide. To achieve a holistic view of the consumer, we employ qualitative research, including in-depth interviews, focus groups, and ethnographic research.

Qualitative research is less structured than surveys or experiments. With qualitative research, we let the consumer take the lead. We know the purpose of our study. We know what we are trying to learn. But we conduct research in a way that encourages the consumer to tell us what is important.

Qualitative research is open-ended. We put ourselves in the consumer's shoes and try to see the world as she sees it. Variables are not defined in advance. As we encounter new facts, we modify our theories. We are open to new ways of thinking, ways suggested through interaction with the consumer.

Qualitative research is holistic. Instead of dividing the world of consumer experience into variables defined in advance, we consider the entire consumer experience. We look at consumer attitudes and behavior in context. What is the context? The consumer's life, the consumer's worldview.

An interviewer asks, "Why did you rent *Body Double* instead of *Body Heat*? How did you go about making that choice? What were you thinking at the time? How did you feel about the options? What influenced your decision?" The respondent's answers can take many forms.

A qualitative researcher would argue that it is not product choice itself that matters. What matters is the process of making the choice. If we can understand the process—the thinking behind the choice—then we are on our way to understanding the consumer. One way to understand the process is to conduct interviews, as reviewed in this section.

When we interview, we learn about the consumer from her own words. We listen to the voice of the consumer as she tells us about her feelings, needs, personal preferences, and experiences. Long interviews are sometimes called in-depth interviews or IDIs.

A semistructured or structured interview is a directed or guided conversation. The typical one-on-one interview begins by identifying the research topic and reviewing participant responsibilities and rights. Personal introductions follow with the interviewer describing herself and asking the participant to do the same. The bulk of the interview is comprised of open-ended questions with discussion. There is talking and listening. There is interaction between the interviewer and interviewee. When answers seem incomplete, there is probing and additional discussion.

In an unstructured interview, questions and ways of asking questions may vary considerably from one participant to the next. In a semistructured or structured interview, on the other hand, all participants are asked the same questions in the same way. An interview outline or guide, prepared in advance by the researcher, provides structure and guidance for the interviewer. Figure B.29, an interview guide for studying consumer attitudes about Alfred Hitchcock and his films, shows questions one might encounter in a semistructured or structured interview.

Most important to interviewing success is the preparation and talent of the interviewer. The interviewer is the instrument of data collection. Knowledge about the topic of the interview is important—knowledge of people more so. An interviewer must be adept at human interaction, able to make quick decisions about what, when, and how to ask questions. The task of the interviewer is to allow the interviewee's story to come through.

Projective tasks can be a good way to get people talking about their feelings without asking them direct questions. In keeping with the movie theme we have used in previous examples, we consider a projective task from Esipova et al. (2004), in which students were asked to describe characters in a movie. Figure B.30 shows the discussion guide for the scientist character.

Kvale (1996) and Stein and Paterno (2001) provide introductions to interviewing. A classic book by Hyman (1954), based on extensive interviews with interviewers, reviews methodological issues in interviewing, including interviewer bias. And the edited volume by Gubrium and Holstein (2002) provides discussion of interviewing in social research.

How is the researcher involved in the research process? To relate as an interviewer, to have a conversation, is to be involved with the interviewee.

But how do we talk with research participants without influencing what they say? How do we collect data in the field, working as participant observers, without affecting the nature of those data? It is not easy to address the problem of interviewer or researcher bias. There is no avoiding the fact that we are in the field with or in communication with the subjects of our research. There is a limit to what we can do with unobtrusive methods, as described in the classic book by Webb et al. (1966).

## B.10 Focus Groups

Focus groups are group interviews. A discussion guide provides the focus. An interviewer or moderator facilitates conversation among the participants. In consumer research it is common to have focus groups of six to twelve participants and to conduct focus groups that last less than two hours.

The focus group moderator, like the one-on-one interviewer, must be an excellent communicator. She should be outgoing and comfortable being the

*Figure B.29. Interview Guide*

Introduce topic and review participant responsibilities and rights.
Note that the interview will take no more than ninety minutes.
Do personal introductions with a focus upon movie-going habits.
Allow five to ten minutes to ask and discuss each of the following:

1. What do you know about Alfred Hitchcock and his movies?
2. When did you see your first Hitchcock movie? What was it like?
3. What are your feelings about Hitchcock as a director, and what contributed to those feelings?
4. Have your feelings about Hitchcock changed over the years?
5. As you look back on the Hitchcock movies you have seen, are there any particular scenes that stand out in your mind?
6. What is it about Hitchcock movies that makes them different from the movies of other directors?
7. What about the actors and their roles in Hitchcock movies? What do you think about them?
8. Describe the Hitchcock movie experience as you know it.
9. What advice would you give to someone new to Hitchcock?
10. Is there anything else you would like to add? Anything about Hitchcock, his movies, or the movie-going experience?

***Figure B.30.*** *Interview Projective Task*

Suppose you were making a movie. One of the characters in the movie is a scientist. What kind of personality would you like the character to have? How would the scientist look? What would the scientist wear? What kind of car would the scientist drive? Give me a picture of the kind of person this would be.

center of attention. She must be a good listener, able to put the group at ease and quick to react to the mood of the group. The moderator is a facilitator of conversation. She needs to be assertive, getting dominant group members to speak less and submissive group members to speak more. In addition, the moderator needs to be nondirective and unbiased regarding the topic of conversation. That is, she must avoid expressing her own or the client's opinions about products and services under study.

There are variations on traditional interviews and focus groups. A dyad involves an interviewer and two interviewees. A triad involves an interviewer and three interviewees. These small focus groups or minigroups are especially useful when investigating group purchase decisions. We have a dyad, for example, when interviewing couples about household purchases.

Another alternative to the traditional focus group is a structured sequence of focus groups. For example, a group of doctors could first act as observers of a focus group consisting only of patients and then become participants in a focus group consisting only of doctors.

As with one-on-one interviews, focus group interviews may be conducted face-to-face, by telephone, or online. Face-to-face groups are conducted with a moderator sitting in a focus group room with the participants. A one-way mirror often separates the group from observers. Audio tapes and videotapes are used to record the discussion. After the focus group has been conducted, audio and video tapes are transcribed, providing a text transcript of the discussion.

Telephone focus groups are usually conducted through commercial telephone conferencing facilities. The moderator and participants call into the conferencing service from their offices or homes. Their discussion is recorded on audio tapes and converted to text transcripts.

An online focus group is a moderated group interview conducted online either synchronously or asynchronously. The moderator, technical support staff, and participants log on to a common application on a server computer. Participants in online focus groups have the option of working from any computer with an Internet connection and a web browser. Text transcripts with participants are generated automatically from what the moderator and participants type.

When conducted synchronously, an online focus group is called a real-time focus group or chat. This is the online analogue of face-to-face and telephone focus groups. When conducted asynchronously, the online focus group is called a bulletin, message, or discussion board.

A primary difference between real-time and asynchronous groups is time frame for completion. A real-time focus group usually takes an hour or two with moderator and participants logged on at the same time. An asynchronous focus group, on the other hand, can take days or weeks to complete. Moderators and participants log on at different days and times. Another difference between real-time and asynchronous groups is the size of the group. While it is difficult to accommodate more than six to twelve real-time focus group members, asynchronous groups may be composed of groups of twelve to twenty, or even thirty, members. Many have noted the advantages of online bulletin board applications in obtaining information-rich exchanges among research participants (James 2002; Langer 2002; Wellner 2003).

In their study comparing focus group modalities, Esipova et al. (2004) found no statistically significant differences in the amount of social interaction within online, telephone, and face-to-face groups. This study showed that a wide variety of focus group tasks and question types could be used successfully online and that online recruitment and participation rates were acceptable. The student participants in the study indicated a preference for face-to-face and online modalities over the telephone modality. These researchers argued that the online modality offered a realistic alternative to telephone and face-to-face modalities.

Advances in technology foster the development of alternative methods and modalities for collecting consumer data. Facilities for web presentations and software demonstrations provide rich media enhancement for telephone and online focus groups. Online research participants can view images and videos while talking with one another.

Another method for conducting qualitative research is focused conversation, developed by the author and first described in Miller and Walkowski (2004). A focused conversation is an unmoderated conversation with participants working independently, generating qualitative data. Often conducted online with only two participants, a focused conversation is neither an interview nor a focus group. It is a small self-moderating group with a conversation guide to provide focus.

With similarities to autodriving in ethnographic research (Heisley and Levy 1991), focused conversation represents a useful technique for hearing the voice of the consumer. Because there is no interviewer or moderator to influence or direct the conversation, focused conversation is less susceptible to interviewer bias than traditional interviews and focus groups.

For additional reading, focus group methods and moderating skills are discussed by Morgan and Krueger (1998), Edmunds (1999), and Langer (2001). Miller and Walkowski (2004) and Miller (2015c) review methods of online qualitative research.

## B.11 Field Research

How do we learn about consumers? In addition to listening to what they say, we can observe how they behave in the field. With ethnographic research, we see how products are integrated into the lives of consumers.

Field research is naturalistic to the extent that it can be. As ethnographers or field researchers, we go to the places where consumers live, work, and shop. We observe consumer behavior in its natural habitat. Field research is process oriented. We want to observe the process of consumer choice as well as the choice itself.

What could be a better way to learn about consumers than to watch them? We follow them into stores to see how they shop. We visit their homes to see how they use products. We watch people as they go about living their lives, interacting with products and services. We observe and record information about product search, shopping behavior, and consumption.

Early ethnographers were academic researchers who went somewhere (often new, strange, and remote) to study cultures unlike their own. They were cultural anthropologists and sociologists. Their methods included behavioral observation, interviewing, and the examination of cultural artifacts.

Specialized methods of data collection and interpretation have emerged under the rubric of ethnography, including naturalistic inquiry, the long interview, life history and lifestyle research, humanistic research, hermeneutics, semiotics, and phenomenology. Among the coterie of academic researchers who describe themselves as ethnographers, there are, as well, individuals who promote philosophies such as feminism and postmodernism.

It is method, not political philosophy, that defines ethnography. Ethnography is fieldwork. It is behavioral observation *in situ.* Ethnography is the recording of observations. It is storytelling, with narrative flowing from the act of collecting and reviewing the raw data of field observation and cultural artifacts.

Paco Underhill (1999, 2004), who describes himself as an urban geographer and retail anthropologist, presents results from the observation of shopping behavior. Underhill conducts in-store observation, utilizing small video cameras and qualitative observation techniques, mapping programs, and attitudinal interviews with shoppers. Field research helps retailers identify store layout factors that influence purchase decisions.

From the observations of Underhill (1999), we learn that shoppers who walk fast have a narrower field of vision. Most shoppers slow down when they see reflective surfaces. When walking up and down aisles, shoppers in the United States are more likely to turn right than left. And, as we might suspect, there is great variability in the experience of shopping between men and women and across individuals from various ethnic groups.

Ethnographic interviews take many forms. There are casual conversations among researchers and participants. Life history and key informant interviews focus on the life experiences of individual subjects. Most interviews are semistructured or structured interviews. Surveys, questionnaires, and secondary research with regard to consumer artifacts are also part of the ethnographer's toolkit.

Pure observation assumes no contact between the researcher and the subjects of research. This is rare in ethnographic studies of consumers. Participant observation, as its name would imply, has the researcher participating in the miniculture, engaging in cultural activities along with research participants, and interacting with research participants.

To practice field research is to immerse oneself in the lives of consumers, to go where they live, work, and shop. It is hard to be unobtrusive when working as an interviewer or ethnographer (Wolcott 2001).

Keeping records of one's observation usually involves more than taking notes. Audio taping and videotaping are common. Today's ethnography is becoming digital ethnography, with cell phones, including those with integrated cameras and data collection software, being carried into the field. Aided by technology, consumer subjects can themselves provide raw data for studies of consumer cultures. The researcher need not go into the field as such. All she needs to do is make sure the cameras and recording equipment go into the field.

Heisley and Levy (1991) describe the technique called autodriving, in which consumer informants drive the process of data collection themselves. Each study participant records her own words and activities, listening to herself on audio tape, taking photographs of herself and her surroundings, seeing herself in pictures, watching herself on videotape, and reflecting on what she has recorded in additional recordings. Heisley and Levy (1991) stress the value of photographs in ethnographic research.

Another interesting development is online ethnography, virtual ethnography, or "netnography," as it has been called by Sherry and Kozinets (2001). This is the ethnographic study of online behavior itself, which draws on computer- or network-based data and artifacts. There have been numerous studies of computer-assisted artistic expression (Holtzman 1994; Aarseth 1997; Bolter and Grusin 1999; Ryan 1999; Gauntlett 2000; Lunenfeld 2000; Scholder and Crandall 2001; Weibel and Druckrey 2001).

Online technology and specialized software make virtual ethnography a realistic alternative to traditional ethnography or fieldwork, especially when studying the digital-savvy younger generation. There are commercial systems that provide a structure for conducting online ethnographic research, including data collection, archiving, and analysis. And various technologies have been employed with online communities.

Generally speaking, an online community is a group of individuals who interact online over an extended period of time. When used for primary consumer research, online communities may consist of groups of fifty to two hundred people who have been recruited to participate in research about related topics over a period of three to twelve months. Large online communities, composed of thousands of participants, are possible. Werry and Mowbray (2001) provide numerous examples of large corporate-sponsored and nonprofit online communities.

Research with online communities incorporates a variety of methods, including online surveys, real-time focus groups, and bulletin boards. Much of this work would be characterized as qualitative and ethnographic. Preece (2000), Powazak (2001), McArthur and Bruza (2001), and Hessan (2003) review critical success factors in developing online communities, including shared purpose, knowledge, and beliefs, participant commitment to repeated access over time, and technology for communication, shared resources, and data storage and retrieval.

A special concern of consumer research practitioners is the time it takes to conduct qualitative research. On this score there is less debate. Detractors and protagonists alike agree that thorough qualitative studies take weeks, months, or years to complete. Qualitative data collection is labor-intensive, requiring interviewers, moderators, trained observers, and ethnographers. And qualitative data analysis takes days to execute, as researchers review collections of artifacts, tapes, and text.

An introduction to ethnographic methods from an anthropological perspective may be found in Agar (1996), Ellen (1984), and Sanjek (1990). Arnould and Wallendorf (1994) review ethnographic methods for marketing research. The books by Wolcott (1994, 1999, 2001) provide an introduction to ethnography and the art and science of fieldwork.

Ethnography is no longer the sole province of academics. Spradley and McCurdy (1972) demonstrate how ethnographic methods can be used by almost anyone to study almost any culture (including one's own). Consumer minicultures have been the subject of numerous ethnographic research studies. Examples include the bar scene (Cavan 1966), garage sales, flea markets, and swap meets (Levinson 1986; Belk, Sherry, and Wallendorf 1988; Sherry 1999), skydiving (Celsi, Rose, and Leigh 1993), river rafting (Arnould and Price 1993), and collecting (Pearce 1999; Belk 2001).

Much has been written about online ethnography as it relates to communication, culture, and social theory (Walther 1996; Davis and Brewer 1997; Jones 1997; Johnson 1997; Lévy 1997; Holeton 1998; Jones 1999; Jonscher 1999; Smith and Kollock 1999; Herman and Swiss 2000; Mann and Stewart 2000; Miller and Slater 2000; Huberman 2001; Lévy 2001). There are numerous studies of virtual communities or aggregations of individuals that emerge through communication over the Internet (Porter 1997; Markham 1998; Cherny 1999; Smith and Kollock 1999; Rheingold 2000). Ellis, Oldridge, and Vasconcelos (2004) provide a useful review of this literature.

Ethnography is not without its detractors. Data collection and interpretation rely upon the talent and subjective opinion of the researcher. Field research often relies on small, specially selected consumer groups. Results may be criticized for not being standardized, reliable, or generalizable (Kvale 1994). Nonetheless, as data scientists, we can employ qualitative methods, collect data and use these data in models, much as we would use any other data.

# C Case Studies

Doing marketing data science means working with clients and understanding the business context of research. Case studies demonstrate the process. This appendix introduces case studies in marketing data science. Case study data and programs for analyzing these data are provided on the book's website: `http://www.ftpress/miller/`.

## C.1 AT&T Choice Study

Following its breakup in 1986, AT&T wanted to identify factors relating to customer choice of long-distance carriers. The firm collected respondent data from telephone interviews, household service and billing information from corporate databases, and census data linked to household addresses. Table C.1 shows variable names and definitions.

With data from one thousand long-distance telephone customers, we can develop models for predicting telephone customer choices. We can also examine issues of customer retention and churn and advise management on plans for target marketing.

---

The original data for this case were provided by James W. Watson and distributed as part of the S system from AT&T Bell Laboratories. S and later SPlus were precursors of R. The details of the AT&T Choice Study were discussed in Chambers and Hastie (1992).

---

*Table C.1. Variables for the AT&T Choice Study*

| *Variable* | *Description and Coding* |
|---|---|
| pick | Customer choice of long-distance service:<br>(AT&T or OCC = Other common carrier) |
| income | Income level of the household (in thousands of dollars) |
| moves | Number of times the household moved in the last five years |
| age | Age level of the respondent in years<br>(18–24, 25–34, 35–44, 45–54, 55–64, 65+) |
| education | <HS = Less than high school; HS = High school graduate<br>Voc = Vocational school; Coll = Some college<br>BA = College graduate; >BA = Graduate school |
| employment | D = Disabled; F = Full-time<br>H = Homemaker; P = Part-time<br>R = Retired; S = Student<br>U = Unemployed |
| usage | Telephone usage (average minutes per month) |
| nonpub | Does the household have an unlisted telephone number? |
| reachout | Has the household participated in the AT&T<br>"Reach Out America" telephone service plan? |
| card | Does the household have an AT&T Calling Card? |

## C.2 Anonymous Microsoft Web Data

The data for this case come from the Microsoft website `www.microsoft.com` as it existed for one week in February 1998. There are 37,711 users in the sample, and the case is called "anonymous" because the data files contain no personally identifiable information for these users.

While we often think of website nodes as being pages, in this study it is website areas that are the nodes. Page-view requests of users are categorized to reflect areas of the Microsoft website visited. There are 294 distinct areas, identified by name and number.

To provide an honest evaluation of predictive models, Microsoft user data are partitioned into training and test sets, with 32,711 users in the training file and 5,000 users in the test file. Data rows in these files shows a user identification number along with a website area visited during the one-week time frame of the study. A separate file shows area identification numbers, directory names for website areas, and area descriptions.

Analysis of the Microsoft data can begin by looking at areas visited and characterizing user behavior. Network area structure may be gleaned from user behavior by construing joint area usage as a link between area nodes.

A more ambitious goal, one consistent with published studies drawing on these data, would be to predict which areas of the website a user will visit based upon other areas visited. For predictive models of this type, we could utilize methods of association rule analysis and recommender systems.

---

The data for this case come from the University of California–Irvine Machine Learning Repository of the Center for Machine Learning and Intelligent Systems (Bache and Lichman 2013). The data sets and documentation are available at `http://archive.ics.uci.edu/ml/datasets/Anonymous+Microsoft+Web+Data`. The original data were created by Jack S. Breese, David Heckerman, and Carl M. Kadie of Microsoft Research in Redmond, Washington, and these data were used in testing models for predicting areas of the website a user would visit based on data about other areas that user had visited (Breese, Heckerman, and Kadie 1998).

## C.3 Bank Marketing Study

A Portuguese bank conducted seventeen telephone marketing campaigns between May 2008 and November 2010. The bank recorded client contact information for each telephone call. Table C.2 shows variable names and descriptions for the study.

Client characteristics include demographic factors: age, job type, marital status, and education. The client's previous use of banking services is also noted.

Current contact information shows the date of the telephone call and the duration of the call. There is also information about the call immediately preceding the current call, as well as summary information about all calls with the client.

The bank wants its clients to invest in term deposits. A term deposit is an investment such as a certificate of deposit. The interest rate and duration of the deposit are set in advance. A term deposit is distinct from a demand deposit.

The bank is interested in identifying factors that affect client responses to new term deposit offerings, which are the focus of the marketing campaigns. What kinds of clients are most likely to subscribe to new term deposits? What marketing approaches are most effective in encouraging clients to subscribe?

---

Data for this case come from the University of California–Irvine Machine Learning Repository of the Center for Machine Learning and Intelligent Systems at `http://archive.ics.uci.edu/ml/datasets/Bank+Marketing`. The original data were part of marketing studies documented in Moro, Laureano, and Cortez (2011) and Moro, Cortez, and Rita (2014).

---

***Table C.2.*** *Bank Marketing Study Variables*

| *Variable Name* | *Description (Possible Values)* |
|---|---|
| | *Client Demographics* |
| age | Age in years |
| job | Type of job (admin., unknown, unemployed, management, housemaid, entrepreneur, student, blue-collar, self-employed, retired, technician, services) |
| marital | Marital status (married, divorced, single) [Note: "divorced" means divorced or widowed] |
| education | Level of education (unknown, secondary, primary, tertiary) |
| | *Client Banking History* |
| default | Has credit in default? (yes, no) |
| balance | Average yearly balance (in Euros) |
| housing | Has housing loan? (yes, no) |
| loan | Has personal loan? (yes, no) |
| | *Data from Most Recent Marketing Contact/Call* |
| contact | Contact communication type (unknown, telephone, cellular) |
| day | Last contact day of the month |
| month | Last contact month of year (jan, feb, mar, . . ., nov, dec) |
| duration | Last contact duration (in seconds) |
| | *Data from All Marketing Contacts/Calls* |
| campaign | Number of contacts performed during this campaign for this client (includes last contact) |
| pdays | Number of days that passed since the client was last contacted from a previous campaign (-1 means client was not previously contacted) |
| previous | Number of contacts performed before this campaign for this client |
| poutcome | Outcome of the previous marketing campaign (unknown, other, failure, success) |
| | *Response to Most Recent Marketing Contact/Call* |
| response | Has the client subscribed to a term deposit? (yes, no) |

## C.4 Boston Housing Study

The Boston Housing Study is a market response study of sorts, with the market being 506 census tracts in the Boston metropolitan area. The objective of the study was to examine the effect of air pollution on housing prices, controlling for the effects of other explanatory variables. The response variable is the median price of homes (in 1970 dollars) in the census track. Table C.3 shows variables included in the case. Short variable names correspond to those used in previously published studies.

---

The original data from the Boston Housing Study (Harrison and Rubinfeld 1978) were published by Belsley, Kuh, and Welsch (1980) in their book about regression diagnostics. In subsequent years, versions of these data have been used by statisticians to introduce and evaluate regression methods, including classification and regression trees (Breiman et al. 1984), treed regression (Alexander and Grimshaw 1996), and monotone regression (Dole 1999). Miller (1999) used the Boston Housing Study data to explore sample size requirements for a number of modern data-adaptive regression methods. Data provided for this case represent an updated version of the original data, following the suggested revisions of Gilley and Pace (1996).

---

*Table C.3.* *Boston Housing Study Variables*

| *Variable Name* | *Description* |
|---|---|
| neighborhood | Name of the Boston neighborhood (location of the census tract) |
| mv | Median value of homes in dollars |
| nox | Air pollution (nitrogen oxide concentration) |
| crim | Crime rate |
| zn | Percent of land zoned for lots |
| indus | Percent of business that is industrial or nonretail |
| chas | On the Charles River (1) or not (0) |
| rooms | Average number of rooms per home |
| age | Percentage of homes built before 1940 |
| dis | Weighted distance to employment centers |
| rad | Accessibility to radial highways |
| tax | Tax rate |
| ptratio | Pupil/teacher ratio in public schools |
| lstat | Percentage of population of lower socio-economic status |

## C.5 Computer Choice Study

In 1998 Microsoft introduced a new operating system. Computer manufacturers were interested in making predictions about the personal computer marketplace.

To help manufacturers understand the market for personal computers, we conducted a computer choice study involving eight computer brands, price, and four other attributes of interest: compatibility, performance, reliability, and learning time. Table C.4 provides a description of attribute levels used in the study.

The computer choice study was a nationwide study. We identified people who expressed an interest in buying a new personal computer within the next year. Consumers volunteering for the study were sent questionnaire booklets, answer sheets, and postage-paid return-mail envelopes. Each respondent received $25 for participating in the study. The survey consisted of sixteen pages, with each page showing a choice set of four product profiles. For each choice set, survey participants were asked first to select the computer they most preferred, and second, to indicate whether or not they would actually buy that computer. For some analyses it may be sufficient to focus on the initial choice or most preferred computer in each set. Figure C.1 shows the first page of the survey (the first choice set).

Being diligent data scientists, we might want to define a training-and-test regimen. One approach in this context would be to build predictive models on twelve choice sets and test on four sets. We can arbitrarily select sets 3, 7, 11, and 15 as hold-out choice sets, for example, and let the remaining item sets 1, 2, 4, 5, 6, 8, 9, 10, 12, 13, 14, and 16 serve as training sets. With sixteen choice sets of four, we have 64 product profiles for each individual in the study. The training data would include 48 rows of product profiles for each individual and the test data would include 16 rows of product profiles for each individual. The data for one individual are shown in table C.5.

---

This is a retrospective study, as many of the companies involved have changed their roles in the computer industry or have left the industry entirely. Using a study more than ten years old has its advantages. None of the companies in question will care what our analysis shows. Like most of the examples in the book, these are real data, and at one time they had real meaning. The study is based on research supported by Sharon Chamberlain.

---

*Table C.4. Computer Choice Study: Product Attributes*

| Attribute | Level Code | Level Description |
|---|---|---|
| Brand | Apple | Manufacturer: Apple |
| | Compaq | Manufacturer: Compaq |
| | Dell | Manufacturer: Dell |
| | Gateway | Manufacturer: Gateway |
| | HP | Manufacturer: HP |
| | IBM | Manufacturer: IBM |
| | Sony | Manufacturer: Sony |
| | Sun | Manufacturer: Sun Microsystems |
| Compatibility | 1 | 65% Compatible |
| | 2 | 70% Compatible |
| | 3 | 75% Compatible |
| | 4 | 80% Compatible |
| | 5 | 85% Compatible |
| | 6 | 90% Compatible |
| | 7 | 95% Compatible |
| | 8 | 100% Compatible |
| Performance | 1 | Just as fast |
| | 2 | Twice as fast |
| | 3 | Three times as fast |
| | 4 | Four times as fast |
| Reliability | 1 | As likely to fail |
| | 2 | Less likely to fail |
| Learn | 1 | 4 hours to learn |
| | 2 | 8 hours to learn |
| | 3 | 12 hours to learn |
| | 4 | 16 hours to learn |
| | 5 | 20 hours to learn |
| | 6 | 24 hours to learn |
| | 7 | 28 hours to learn |
| | 8 | 32 hours to learn |
| Price | 1 | $1,000 |
| | 2 | $1,250 |
| | 3 | $1,500 |
| | 4 | $1,750 |
| | 5 | $2,000 |
| | 6 | $2,250 |
| | 7 | $2,500 |
| | 8 | $2,750 |

*Figure C.1. Computer Choice Study: One Choice Set*

Computer System 1.A

- Manufacturer: Sun Microsystems
- 65% of all applications that currently run on Windows 95 will run on this system
- Four times as fast as a similarly configured Windows 95 system
- Less likely to have a system failure than a Windows 95 system
- Takes 32 hours to learn how to use
- Base system price: $2750

Computer System 1.B

- Manufacturer: Apple
- 95% of all applications that currently run on Windows 95 will run on this system
- Three times as fast as a similarly configured Windows 95 system
- Just as likely to have a system failure as a Windows 95 system
- Takes 24 hours to learn how to use
- Base system price: $1500

Computer System 1.C

- Manufacturer: IBM
- 90% of all applications that currently run on Windows 95 will run on this system
- Just as fast as a similarly configured Windows 95 system
- Less likely to have a system failure than a Windows 95 system
- Takes 16 hours to learn how to use
- Base system price: $1750

Computer System 1.D

- Manufacturer: Compaq
- 100% of all applications that currently run on Windows 95 will run on this system
- Three times as fast as a similarly configured Windows 95 system
- Less likely to have a system failure than a Windows 95 system
- Takes 8 hours to learn how to use
- Base system price: $2500

*Table C.5. Computer Choice Study: Data for One Individual*

| id | profile | setid | position | brand | compat | perform | reliab | learn | price | choice | buy |
|---|---|---|---|---|---|---|---|---|---|---|---|
| 1 | 1 | 1 | Top-Left | Sun | 1 | 4 | 2 | 8 | 8 | 0 | 0 |
| 1 | 2 | 1 | Top-Right | Apple | 7 | 3 | 1 | 6 | 3 | 0 | 0 |
| 1 | 3 | 1 | Bottom-Left | IBM | 6 | 1 | 2 | 4 | 4 | 1 | 0 |
| 1 | 4 | 1 | Bottom-Right | Compaq | 8 | 3 | 2 | 2 | 7 | 0 | 0 |
| 1 | 5 | 2 | Top-Left | Gateway | 4 | 2 | 2 | 2 | 2 | 0 | 0 |
| 1 | 6 | 2 | Top-Right | Sony | 8 | 4 | 1 | 5 | 8 | 0 | 0 |
| 1 | 7 | 2 | Bottom-Left | Sun | 1 | 1 | 2 | 7 | 3 | 0 | 0 |
| 1 | 8 | 2 | Bottom-Right | Dell | 5 | 3 | 1 | 4 | 5 | 1 | 0 |
| 1 | 9 | 3 | Top-Left | Dell | 4 | 4 | 1 | 6 | 7 | 1 | 0 |
| 1 | 10 | 3 | Top-Right | Sony | 7 | 2 | 2 | 2 | 6 | 0 | 0 |
| 1 | 11 | 3 | Bottom-Left | Apple | 3 | 3 | 2 | 8 | 2 | 0 | 0 |
| 1 | 12 | 3 | Bottom-Right | HP | 8 | 1 | 1 | 2 | 5 | 0 | 0 |
| 1 | 13 | 4 | Top-Left | Compaq | 6 | 2 | 1 | 6 | 5 | 0 | 0 |
| 1 | 14 | 4 | Top-Right | IBM | 5 | 4 | 1 | 2 | 3 | 1 | 1 |
| 1 | 15 | 4 | Bottom-Left | HP | 7 | 3 | 2 | 7 | 1 | 0 | 0 |
| 1 | 16 | 4 | Bottom-Right | Apple | 2 | 1 | 1 | 4 | 7 | 0 | 0 |
| 1 | 17 | 5 | Top-Left | Sony | 6 | 1 | 1 | 8 | 3 | 0 | 0 |
| 1 | 18 | 5 | Top-Right | Compaq | 5 | 4 | 2 | 3 | 1 | 1 | 1 |
| 1 | 19 | 5 | Bottom-Left | Dell | 1 | 3 | 2 | 2 | 8 | 0 | 0 |
| 1 | 20 | 5 | Bottom-Right | Gateway | 8 | 2 | 1 | 1 | 1 | 0 | 0 |
| 1 | 21 | 6 | Top-Left | IBM | 1 | 4 | 2 | 4 | 2 | 1 | 1 |
| 1 | 22 | 6 | Top-Right | Dell | 8 | 4 | 2 | 8 | 6 | 0 | 0 |
| 1 | 23 | 6 | Bottom-Left | Sun | 4 | 2 | 1 | 3 | 4 | 0 | 0 |
| 1 | 24 | 6 | Bottom-Right | Gateway | 6 | 3 | 1 | 1 | 8 | 0 | 0 |
| 1 | 25 | 7 | Top-Left | HP | 2 | 4 | 2 | 1 | 3 | 1 | 1 |
| 1 | 26 | 7 | Top-Right | Compaq | 7 | 2 | 1 | 2 | 5 | 0 | 0 |
| 1 | 27 | 7 | Bottom-Left | Sony | 1 | 3 | 1 | 3 | 6 | 0 | 0 |
| 1 | 28 | 7 | Bottom-Right | IBM | 6 | 2 | 2 | 7 | 7 | 0 | 0 |
| 1 | 29 | 8 | Top-Left | Dell | 6 | 1 | 2 | 5 | 1 | 0 | 0 |
| 1 | 30 | 8 | Top-Right | HP | 5 | 2 | 2 | 6 | 6 | 0 | 0 |
| 1 | 31 | 8 | Bottom-Left | Gateway | 2 | 3 | 2 | 3 | 5 | 1 | 0 |
| 1 | 32 | 8 | Bottom-Right | Compaq | 1 | 4 | 1 | 1 | 4 | 0 | 0 |
| 1 | 33 | 9 | Top-Left | Apple | 1 | 2 | 2 | 5 | 5 | 0 | 0 |
| 1 | 34 | 9 | Top-Right | Gateway | 3 | 4 | 1 | 7 | 6 | 1 | 0 |
| 1 | 35 | 9 | Bottom-Left | IBM | 7 | 1 | 1 | 3 | 8 | 0 | 0 |
| 1 | 36 | 9 | Bottom-Right | Sony | 5 | 3 | 2 | 1 | 7 | 0 | 0 |
| 1 | 37 | 10 | Top-Left | HP | 4 | 1 | 2 | 4 | 8 | 0 | 0 |
| 1 | 38 | 10 | Top-Right | Dell | 8 | 3 | 2 | 1 | 2 | 1 | 1 |
| 1 | 39 | 10 | Bottom-Left | IBM | 2 | 2 | 1 | 5 | 6 | 0 | 0 |
| 1 | 40 | 10 | Bottom-Right | Sun | 7 | 4 | 1 | 8 | 5 | 0 | 0 |
| 1 | 41 | 11 | Top-Left | Compaq | 4 | 3 | 2 | 5 | 3 | 0 | 0 |
| 1 | 42 | 11 | Top-Right | Sun | 3 | 4 | 2 | 6 | 8 | 0 | 0 |
| 1 | 43 | 11 | Bottom-Left | Gateway | 5 | 1 | 2 | 8 | 4 | 0 | 0 |
| 1 | 44 | 11 | Bottom-Right | HP | 6 | 4 | 1 | 3 | 2 | 1 | 1 |
| 1 | 45 | 12 | Top-Left | IBM | 8 | 3 | 2 | 6 | 4 | 1 | 1 |
| 1 | 46 | 12 | Top-Right | Gateway | 7 | 4 | 2 | 5 | 7 | 0 | 0 |
| 1 | 47 | 12 | Bottom-Left | Sony | 3 | 2 | 1 | 2 | 1 | 0 | 0 |
| 1 | 48 | 12 | Bottom-Right | Apple | 4 | 1 | 1 | 1 | 6 | 0 | 0 |
| 1 | 49 | 13 | Top-Left | Sun | 5 | 1 | 1 | 5 | 2 | 0 | 0 |
| 1 | 50 | 13 | Top-Right | Apple | 6 | 4 | 2 | 2 | 4 | 0 | 0 |
| 1 | 51 | 13 | Bottom-Left | Dell | 3 | 2 | 2 | 3 | 3 | 0 | 0 |
| 1 | 52 | 13 | Bottom-Right | IBM | 4 | 3 | 1 | 8 | 1 | 1 | 1 |
| 1 | 53 | 14 | Top-Left | Sony | 4 | 4 | 2 | 7 | 5 | 0 | 0 |
| 1 | 54 | 14 | Top-Right | HP | 3 | 3 | 1 | 5 | 4 | 1 | 0 |
| 1 | 55 | 14 | Bottom-Left | Compaq | 2 | 2 | 2 | 8 | 8 | 0 | 0 |
| 1 | 56 | 14 | Bottom-Right | Gateway | 1 | 1 | 1 | 6 | 1 | 0 | 0 |
| 1 | 57 | 15 | Top-Left | Apple | 5 | 2 | 1 | 7 | 8 | 0 | 0 |
| 1 | 58 | 15 | Top-Right | Sun | 6 | 3 | 2 | 4 | 6 | 1 | 0 |
| 1 | 59 | 15 | Bottom-Left | HP | 1 | 2 | 1 | 8 | 7 | 0 | 0 |
| 1 | 60 | 15 | Bottom-Right | Sony | 2 | 1 | 2 | 6 | 2 | 0 | 0 |
| 1 | 61 | 16 | Top-Left | Gateway | 8 | 2 | 1 | 4 | 3 | 0 | 0 |
| 1 | 62 | 16 | Top-Right | IBM | 3 | 1 | 2 | 1 | 5 | 0 | 0 |
| 1 | 63 | 16 | Bottom-Left | Compaq | 5 | 4 | 2 | 3 | 1 | 1 | 1 |
| 1 | 64 | 16 | Bottom-Right | Dell | 2 | 1 | 1 | 7 | 4 | 0 | 0 |

Training: SET1, SET2, SET4, SET5, SET6, SET8, SET9, SET10, SET12, SET13, SET14, SET16
Test: SET3, SET7, SET11, SET15

## C.6 DriveTime Sedans

DriveTime in 2001 is an automobile dealership and financing firm with seventy-six dealerships in eight states. In a typical month the firm sells about four thousand used vehicles and processes about ten thousand credit applications. Virtually all sales are financed. The firm's stated mission is: "To be the auto dealership and finance company for people with less than perfect credit."

DriveTime generates traffic at its dealerships through television and radio advertising, referrals from other dealerships, and through its website. Customers who need financing to purchase vehicles are run through a custom credit risk scorecard, which uses both credit bureau and application information to determine credit worthiness. A generated risk score is used to determine the appropriate deal structure and credit policy.

DriveTime purchases most of its vehicles at auctions and from wholesalers. Vehicles include many makes and models of cars and trucks. The firm uses an information service known as Experian Autocheck to ensure that vehicles have correct odometer readings, have not been previously "totaled" (that is, evaluated as having no value after an accident), and have no other significant negative history. Vehicles that fail the Experian check are rejected and sent back to sellers. Those that pass are sent to a DriveTime reconditioning and inspection center, where they are put through additional checks and repaired as necessary. Vehicles are then delivered to the dealerships for sale.

Normal dealer sales occur within ninety days of delivery to the dealership. If a vehicle does not sell within ninety days, it is called an overage vehicle, meaning that it has been on the lot too long to generate normal dealer profits. Each overage vehicle has its sales price reduced in order to encourage a sale within the ensuing 91- to 119-day period. Profits on vehicles sold within the 91- to 119-day period are much lower than profits on vehicles sold within the normal 90-day period. Furthermore, if an overage vehicle fails to sell within 120 days, the vehicle is taken off the lot and sold at auction. DriveTime takes a loss on vehicles sold at auction.

---

Written by Thomas W. Miller and Steve Zemaitis. Data provided by DriveTime.


---

Table C.6 provides a hypothetical example, showing how normal and overage sales translate into business profits or losses for DriveTime. This example demonstrates the value of using a statistical model to select vehicles for sale. Profit contributions in the example represent gross rather than net profits. They do not account for operating costs, overhead costs, or taxes.

Table C.7 describes variables from the DriveTime vehicles database. The data, which represent 17,506 sedans sold and financed in the second half of 2001, are divided into three data sets for modeling work: 8,753 sedans comprise the training set, 4,377 the validation set, and 4,376 the test set.

Table C.8 shows how researchers use eight color categories to represent twenty-seven colors in the vehicles database. Color categories are defined so that each category has a sufficiently large frequency to warrant its use in modeling work. Gold becomes a catch-all or other color category, including gold, tan, cream, yellow, and brown tones.

Certain variables may be useful in developing vehicle selection models. Newer, lower mileage vehicles, for example, may be expected to sell faster than older, higher mileage vehicles. Sales prices are not included in the vehicles database, but we can assume that prices for vehicles sold within ninety days (normal dealer sales) are marked up, so that the firm recovers costs associated with purchasing, repairs, operations, and interest, and makes an appropriate profit.

DriveTime managers wonder whether it is possible to develop selection models for sedans using data from the vehicles database. Is a single model sufficient, or should separate models be built for the states in which DriveTime operated in 2001 (Arizona, California, Florida, Georgia, Nevada, New Mexico, Texas, and Virginia)? What would the models look like, and how much profit improvement would result from using the models?

***Table C.6.*** *Hypothetical Profits from Model-guided Vehicle Selection*

The table below reflects hypothetical profits associated with DriveTime vehicle sales, given an average total cost per vehicle of $5,000, a 20 percent markup for normal dealer sales, 10 percent markup for overage dealer sales, and 20 percent loss for overage vehicles sold at auction. This example assumes that, of the approximately four thousand vehicles sold each month, about 85 percent are normal dealer sales, 10 percent overage dealer sales (within the 91- to 119-day period), and 5 percent overage auction sales.

| | *Type of Sale* | | | |
|---|---|---|---|---|
| | *Normal Dealer* | *Overage Dealer* | *Overage Auction* | *Monthly Totals* |
| Unit total cost | $5,000 | $5,000 | $5,000 | |
| Unit price | $6,000 | $5,500 | $4,000 | |
| Unit margin profit (loss) | $1,000 | $ 500 | ($1,000) | |
| Units sold | 3,400 | 400 | 200 | 4,000 |
| | (85%) | (10%) | (5%) | |
| Profit (loss) | $3,400,000 | $200,000 | ($200,000) | $3,400,000 |

Suppose that researchers are able to develop a model that is reasonably accurate in predicting how long it takes to sell a vehicle. Suppose further that, using this time-to-sale model to guide inventory decisions, DriveTime is able to increase normal dealer sales from 85 to 90 percent, with corresponding declines in overage vehicle sales. Assuming no change in vehicle costs or prices, what would be the effect upon profits? The following table suggests that monthly profits would increase by $220,000. Twelve months of sales of this type would contribute more than $2.6 million in profit a year. This demonstrates the value of using statistical models to guide business decisions.

| | *Type of Sale* | | | |
|---|---|---|---|---|
| | *Normal Dealer* | *Overage Dealer* | *Overage Auction* | *Monthly Totals* |
| Unit total cost | $5,000 | $5,000 | $5,000 | |
| Unit price | $6,000 | $5,500 | $4,000 | |
| Unit margin profit (loss) | $1,000 | $ 500 | ($1,000) | |
| Units sold | 3,600 | 280 | 120 | 4,000 |
| | (90%) | (7%) | (3%) | |
| Profit (loss) | $3,600,000 | $140,000 | ($120,000) | $3,620,000 |

*Table C.7. DriveTime Data for Sedans*

| *Variable Name* | *Description* |
| --- | --- |
| data.set | Data set for modeling (TRAIN, VALIDATE, or TEST) |
| total.cost | Total cost of vehicle (purchase cost + repair cost + other costs) |
| lot.sale.days | Days from vehicle delivery to dealership to sale (days on lot) |
| overage | Overage vehicle (NO = 0–90 days on lot, YES = 91+ days on lot) |
| vehicle.type | Type of sedan (ECONOMY, FAMILY.SMALL, FAMILY.MEDIUM, FAMILY.LARGE, or LUXURY) |
| domestic.import | Type of manufacturer (Domestic or Import) |
| vehicle.age | Age of vehicle in years (year of sale minus model year) |
| vehicle.age.group | Age group of vehicle (ONE-THREE, FOUR, FIVE, SIX, or SEVEN+) |
| color.set | Color category (BLACK, WHITE, ..., GOLD) |
| makex | Make/manufacturer of vehicle (BUICK, CADILLAC, ..., TOYOTA) |
| state | State location of dealership where vehicle sold (AZ = Arizona, CA = California, ..., VA = Virginia) |
| make.model | Make and model of sedan (ACURA.INTEGRA, BUICK.CENTURY, ..., TOYOTA.TERCEL) |

***Table C.8.*** *DriveTime Sedan Color Map with Frequency Counts*

| *Color in Database* | *Color Category Defined by Researchers* | | | | | | | | *Count* |
|---|---|---|---|---|---|---|---|---|---|
| | Black | White | Blue | Green | Red | Purple | Silver | Gold | |
| Aluminum/ Silver | 0 | 0 | 0 | 0 | 0 | 0 | 1234 | 0 | 1234 |
| Beige | 0 | 0 | 0 | 0 | 0 | 0 | 0 | 123 | 123 |
| Black | 1216 | 0 | 0 | 0 | 0 | 0 | 0 | 0 | 1216 |
| Blue | 0 | 0 | 2149 | 0 | 0 | 0 | 0 | 0 | 2149 |
| Blue - Dark | 0 | 0 | 16 | 0 | 0 | 0 | 0 | 0 | 16 |
| Blue - Light | 0 | 0 | 53 | 0 | 0 | 0 | 0 | 0 | 53 |
| Bronze | 0 | 0 | 0 | 0 | 0 | 0 | 0 | 15 | 15 |
| Brown | 0 | 0 | 0 | 0 | 0 | 0 | 0 | 64 | 64 |
| Burgundy/ Maroon | 0 | 0 | 0 | 0 | 0 | 1410 | 0 | 0 | 1410 |
| Cream | 0 | 0 | 0 | 0 | 0 | 0 | 0 | 76 | 76 |
| Chrome/ Stainless Steel | 0 | 0 | 0 | 0 | 0 | 0 | 1 | 0 | 1 |
| Copper | 0 | 0 | 0 | 0 | 0 | 0 | 0 | 9 | 9 |
| Gray | 0 | 0 | 0 | 0 | 0 | 0 | 618 | 0 | 618 |
| Gold | 0 | 0 | 0 | 0 | 0 | 0 | 0 | 1003 | 1003 |
| Green | 0 | 0 | 0 | 3309 | 0 | 0 | 0 | 0 | 3309 |
| Green - Dark | 0 | 0 | 0 | 59 | 0 | 0 | 0 | 0 | 59 |
| Green - Light | 0 | 0 | 0 | 20 | 0 | 0 | 0 | 0 | 20 |
| Lavender | 0 | 0 | 0 | 0 | 0 | 8 | 0 | 0 | 8 |
| Mauve | 0 | 0 | 0 | 0 | 12 | 0 | 0 | 0 | 12 |
| Orange | 0 | 0 | 0 | 0 | 9 | 0 | 0 | 0 | 9 |
| Pink | 0 | 0 | 0 | 0 | 7 | 0 | 0 | 0 | 7 |
| Purple | 0 | 0 | 0 | 0 | 0 | 366 | 0 | 0 | 366 |
| Red | 0 | 0 | 0 | 0 | 1406 | 0 | 0 | 0 | 1406 |
| Tan | 0 | 0 | 0 | 0 | 0 | 0 | 0 | 414 | 414 |
| Taupe | 0 | 0 | 0 | 0 | 0 | 0 | 0 | 11 | 11 |
| Teal | 0 | 0 | 289 | 0 | 0 | 0 | 0 | 0 | 289 |
| Turquoise | 0 | 0 | 2 | 0 | 0 | 0 | 0 | 0 | 2 |
| White | 0 | 3603 | 0 | 0 | 0 | 0 | 0 | 0 | 3603 |
| Yellow | 0 | 0 | 0 | 0 | 0 | 0 | 0 | 4 | 4 |
| *Count* | 1216 | 3603 | 2509 | 3388 | 1434 | 1784 | 1853 | 1719 | 17506 |

## C.7 Lydia E. Pinkham Medicine Company

Lydia E. Pinkham (1819–1883) was an advocate for women's health and a developer of herbal medicines for women. After her death, family members began the mass marketing of a product known as Lydia E. Pinkham's Vegetable Compound. The product was heavily advertised for many years, and historical data on sales and advertising were made available to economic researchers.

The data come in two files. The first file provides complete annual data for sales revenue, advertising expenses, and income in thousands of dollars. The file covers the years 1907–1960 and includes binary indicator variables for years prior to, during, and after Prohibition. The medicine, which contained 40 percent alcohol, continued to be sold during Prohibition, which extended from January 17, 1920 through December 5, 1933.

The second file contains monthly sales revenue and advertising expenses in dollars for the period from January 1907 through December 1926 and from January 1937 through June 1960. There are missing monthly data for the middle time period.

For more than forty years, the Lydia E. Pinkham Medicine Company case has been used to demonstrate sales forecasting, time series, and econometric methods. Exemplary studies include Caines, Sethi, and Brotherton (1977), Helmer and Johansson (1977), Winer (1979), Bhattacharyya (1982), Heyse and Wei 1985, and Baghestani (1991). More recently, Kim (2005) used bootstrap methods in a new analysis of the case.

---

This classic data set from the econometric literature was distributed with a textbook by Berndt (1991). The story of Lydia E. Pinkham has been documented by Washburn (1931) and Stage (1979) and in Lydia's own book (Pinkham 1900). An advertisement for Lydia E. Pinkham's original formula appeared in the *Saint John Daily Evening News* on April 17, 1883. Here is a link to an image of that advertisement: `https://news.google.com/newspapers?id=N9kIAAAAIBAJ&sjid=5TcDAAAAIBAJ&dq=montreal+hackett&pg=5642,125153&hl=en`

---

## C.8 Procter & Gamble Laundry Soaps

The Procter & Gamble Company developed a new formula for one of its laundry soaps. Before introducing the new formula to the marketplace, the company wanted to know whether consumers would prefer the new formula, called *X*, to the original formula, called *M*.

Consumers in 1,008 households, some of whom were previous users of the original formula *M*, were given the opportunity to try formulas *X* and *M* in blind preference tests. At the end of the tests, consumers were asked to indicate their soap preferences by choosing either *X* or *M*. Water temperature (cold or hot) and type (hard, medium, soft) were noted for each household. Data were coded as shown in Table C.9. Results from the field test represent cross-classified categorical data. See table C.10.

***Table C.9.*** *Variables for the Laundry Soap Experiment*

| *Variable* | *Description and Coding* |
|---|---|
| choice | *X* = consumer prefers new formula *X*<br>*M* = consumer prefers original formula *M* |
| muser | NO = not a previous user of formula *M*<br>YES = previous user of formula *M* |
| wtemp | Household water temperature<br>LOW = 101 to 115 degrees Fahrenheit<br>HIGH = over 115 degrees Fahrenheit |
| wtype | Household water type<br>HARD = hard water<br>MEDIUM = neither hard nor soft water<br>SOFT = soft water |

Source data for this study came from Reis and Smith (1963).

***Table C.10.*** *Cross-Classified Categorical Data for the Laundry Soap Experiment*

| | | Previous User of M | | Previous Non-user of M | |
|---|---|---|---|---|---|
| Water Softness | Brand Preference | High Temperature | Low Temperature | High Temperature | Low Temperature |
| Soft | X | 19 | 57 | 29 | 63 |
| | M | 29 | 49 | 27 | 53 |
| Medium | X | 23 | 47 | 33 | 66 |
| | M | 47 | 55 | 23 | 50 |
| Hard | X | 24 | 37 | 42 | 68 |
| | M | 43 | 52 | 30 | 42 |

## C.9 Return of the Bobbleheads

The Dodgers are one of thirty Major League Baseball teams using promotions to increase attendance. Reports suggest that bobblehead promotions in particular are on the rise, with 2.27 million dolls distributed in 2012 (Broughton 2012) and an estimated 2.7 million dolls in 2013 (Foster 2013).

We provide complete promotion and attendance data for all teams for the 2012 season on the website for the book. These data have a format similar to the Dodgers data in table 8.1, except that there are extra columns for the year and home team. Having data for all teams allows us to explore alternative modeling approaches, such as building a model for each team, aggregate models for groups of teams, or hierarchical models for game-day observations within teams. When predicting attendance at Major League Baseball parks, we would need to consider the fact that ballparks are often filled to capacity. Special models may be required to accommodate this high-end censoring (Lemke, Leonard, and Tlhokwane 2010).

---

Major League Baseball data for promotions and attendance were collected by Erica Costello in December 2012. She graciously contributed these data so students could learn from them.

## C.10 Studenmund's Restaurants

Managers of a nationwide restaurant chain, which we will call Studenmund's Restaurants, want to find new restaurant locations. Gross restaurant sales and the number of competitors within a two-mile radius are noted at existing restaurant locations. Census data for population and income are also collected for these locations. Table C.11 shows the variable names and definitions, and Table C.12 shows the observed data from thirty-three restaurants.

Researchers at Studenmund's wonder if it is possible to define a model for predicting restaurant sales. Could such a model be trusted to yield accurate predictions? Could the model be used to pick future restaurant locations?

***Table C.11.*** *Variables for Studenmund's Restaurants*

| *Variable* | *Description* |
|---|---|
| sales | Gross sales volume at the restaurant location (This is likely to be the number of customers/sales receipts in a year. Or it could be average monthly sales dollars or some other measure relating to sales revenue.) |
| competition | Number of direct competitors within a two-mile radius of the restaurant location |
| population | Number of people living within a three-mile radius of the restaurant location |
| income | Average household income of people living within a three-mile radius of the restaurant location |

The original data for this case were given in Studenmund (1992), an econometrics textbook now in its sixth edition (Studenmund 2010).

*Table C.12. Data for Studenmund's Restaurants*

| sales | competition | population | income |
|---|---|---|---|
| 107,919 | 3 | 65,044 | 13,240 |
| 118,866 | 5 | 101,376 | 22,554 |
| 98,579 | 7 | 124,989 | 16,916 |
| 122,015 | 2 | 55,249 | 20,967 |
| 152,827 | 3 | 73,775 | 19,576 |
| 91,259 | 5 | 48,484 | 15,039 |
| 123,550 | 8 | 138,809 | 21,857 |
| 160,931 | 2 | 50,244 | 26,435 |
| 98,496 | 6 | 104,300 | 24,024 |
| 108,052 | 2 | 37,852 | 14,987 |
| 144,788 | 3 | 66,921 | 30,902 |
| 164,571 | 4 | 166,332 | 31,573 |
| 105,564 | 3 | 61,951 | 19,001 |
| 102,568 | 5 | 100,441 | 20,058 |
| 103,342 | 2 | 39,462 | 16,194 |
| 127,030 | 5 | 139,900 | 21,384 |
| 166,755 | 6 | 171,740 | 18,800 |
| 125,343 | 6 | 149,894 | 15,289 |
| 121,886 | 3 | 57,386 | 16,702 |
| 134,594 | 6 | 185,105 | 19,093 |
| 152,937 | 3 | 114,520 | 26,502 |
| 109,622 | 3 | 52,933 | 18,760 |
| 149,884 | 5 | 203,500 | 33,242 |
| 98,388 | 4 | 39,334 | 14,988 |
| 140,791 | 3 | 95,120 | 18,505 |
| 101,260 | 3 | 49,200 | 16,839 |
| 139,517 | 4 | 113,566 | 28,915 |
| 115,236 | 9 | 194,125 | 19,033 |
| 136,749 | 7 | 233,844 | 19,200 |
| 105,067 | 7 | 83,416 | 22,833 |
| 136,872 | 6 | 183,953 | 14,409 |
| 117,146 | 3 | 60,457 | 20,307 |
| 163,538 | 2 | 65,065 | 20,111 |

## C.11 Sydney Transportation Study

Residents of the north suburbs of Sydney, Australia can commute to downtown Sydney by car or train. Their choice of transportation will be due, in part, to the time and cost of commuting by car and train. On the day of the *Journey to Work Survey*, Sydney commuters indicate their primary method of transportation (car or train) and their best alternative method of transportation (car or train). For both the chosen and alternative methods, 333 commuters provide time and cost estimates for all trip components by car, train, bus, walking, and other modes of transportation.

Table C.13 shows names and descriptions of selected variables from the *Journey to Work Survey*. Time measurements reflect the total commute time by car or train, summing across all components of the trip. Cost measurements reflect total costs by car or train, summing across all components of the trip. Car costs are adjusted for the number of persons in the car and include parking charges. Using these data, we build a models for predicting the transportation choices of Sydney commuters.

**Table C.13.** *Variables for the Sydney Transportation Study*

| *Variable* | *Description* |
|---|---|
| cartime | Time of the commute by car (in minutes) |
| carcost | Cost of the commute by car (in cents) |
| traintime | Time of the commute by train (in minutes) |
| traincost | Cost of the commute by train (in cents) |
| choice | Choice of transportation mode (CAR or TRAIN) on the day of the *Journey to Work Survey* |

Source data for this case come from Hensher and Johnson (1981).

## C.12 ToutBay Begins Again

In March 2015, ToutBay remains in start-up mode awaiting the release of its first products. In what is becoming an increasingly data-driven world, ToutBay owner Tom Miller sees opportunities for *data science as a service (DSaaS)*, a term he uses to describe ToutBay's business model. The goal for the ToutBay division of Research Publishers LLC is to be a market maker in the data science space, publishing and distributing time-sensitive information and competitive intelligence.

The ToutBay website `www.toutbay.com` tells the story of a company founded in December 2013 to provide access to applications developed by analysts, modelers, researchers, and data scientists across the world. These subject matter experts—touts—work with data and develop models that are of use to many people. As a two-minute video on the website claims, *ToutBay gets people together—people who have answers and people who need answers.* The video introduces the firm and explains why information from ToutBay can be more valuable than information freely obtained from search engines.

ToutBay products are expected to fall under sports, finance, marketing, and health and fitness. Sports touts go beyond raw data about players and teams to build models that predict future performance. ToutBay works with sports touts to make their predictive models available to players, owners, managers, and sports enthusiasts.

Finance touts help individuals and firms make informed decisions about when and where to make investments. These touts have expertise in econometrics and time series analysis. They understand markets and predictive models. They detect trends in the past and make forecasts about the future.

One of ToutBay's first products is expected to be a stock portfolio constructor. This is a financial model designed by Dr. Ernest P. Chan, a recognized expert in the area of quantitative finance and author of two books on the subject (Chan 2009, 2013). The idea behind this product is to allow a stock investor to specify his/her investment objectives and time horizon, as well as the domain of stocks being considered and the number of stocks de-

---

This case draws from information at the ToutBay website `http://www.toutbay.com` and from Google Analytics reports, including reports summarizing Scroll Depth plug-in data.

---

sired in a portfolio. Then, using current information about stock prices and performance, as well as selected economic factors, the Stock Portfolio Constructor creates a customized stock portfolio for the investor. It lists the selected stocks and shows their expected future return over the investor's time horizon, assuming an equal level of investment in each stock. The Stock Portfolio Constructor also shows what would have been the historical performance of that portfolio in recent years.

Marketing touts play a similar expert role, going beyond raw sales data to provide consumer and marketplace insights. They have formal training in measurement, statistics, or machine learning, as well as extensive business consulting experience. The results of their models for site selection, product positioning, segmentation, or target marketing are of special interest to business managers.

Health and fitness, a fourth product area, involves scientists with expertise in nutrition, physiology, and molecular mechanisms of health and disease. These touts provide relevant information based on scientific research. They deliver personalized plans developed from real-world models that predict future health and fitness. ToutBay intends to make health and fitness plans available to individual consumers, personal trainers and medical practitioners.

ToutBay's major public event to date has been the R User Conference, also known as UseR!, June 30 through July 3, 2014. The conference was held on the UCLA campus in Los Angeles, California, and attracted around 700 scientists and software engineers, people who write programs (scripts) in the open-source language R (a widely used language in statistics and data science). ToutBay was one of the sponsors of UseR!, along with major software developers and publishers.

ToutBay's goal at UseR! was to introduce itself to potential touts. The company's message was simple: *You do the research and modeling, and we do the rest. We turn scripts into products.* The idea is that, by working with ToutBay, data scientists can focus on data science and ToutBay will take care of marketing, communications, sales, order processing, distribution, and customer support. The ToutBay website has a *For Touts* page that provides the details.

Because ToutBay operates entirely online, its business depends on having a website that conveys a clear message to visitors or guests. Success means converting website guests into ToutBay account holders. And after information products become available, success will mean converting account holders into subscribers to information products.

Revenues will come from customer subscriptions, with touts setting prices for their information products and ToutBay charging a fee for online sales and distribution of those products. In recruiting future touts, ToutBay has a simple message: *If you were the author of a book, you would look for a publisher, and you would hope that the publisher would work with bookstores to sell your book. But what if you are the author of a predictive model? Where do you go to publish your model? Where do you go to sell the results of your model? ToutBay—that's where.*

Since opening its website in April 2014, ToutBay has been tracking user traffic with Google Analytics. Recently, the firm has been reviewing data relating to visits, page views, and time on the site. There may have been a slight increase in traffic around the time of the UseR! conference. Otherwise, traffic has been limited, which is a source of concern for the company.

The ToutBay website employs a single-page design, with extensive information on the home page, including the two-minute video introduction to the company. A single-page approach to website design provides better overall performance than a multi-page approach because a single-page approach requires fewer data transmissions between the client browser and the website server.

One difficulty in employing a single-page approach, however, is that standard page-view statistics provide an incomplete picture of website usage. Recognizing this, ToutBay website developers employed JavaScript code to detect how far down users were scrolling on the home page. These scrolling data are included in user traffic information for the site. Table C.14 shows variables and variable definitions for website data under review.

The ToutBay's owner hopes that a detailed analysis of website content and structure, as well as data about website usage, will provide guidance in developing future versions of the website, coinciding with the introduction of the company's first products.

**Table C.14.** *ToutBay Begins: Website Data*

| *Variable* | *Description* |
|---|---|
| date | Date coded as mm/dd/yy |
| sessions | Number of sessions (a session is a defined period of time, such as 30 minutes, that a user is actively engaged with the website) |
| users | Number of users who have at least one session |
| new_sessions | Estimate of the percentage of first-time visits |
| pageviews | Number of pages viewed (repeated views of the same page are counted) |
| pages_per_session | Average number of pages viewed during a session (repeated views of the same page are counted) |
| ave_session_duration | Average length of a session in hours, minutes, and seconds (hh:mm:ss) |
| bounce_rate | Percentage of single-page visits (user enters and leaves from the same page, usually the home page) |
| scroll_videopromo | Number of users who scroll down as far as the video introducing ToutBay on the home page |
| scroll_whatstoutbay | Number of users who scroll down as far as the *What's ToutBay* section of the home page |
| scroll_howitworks | Number of users who scroll down as far as the *How It Works* section of the home page |
| scroll_faq | Number of users who scroll down as far as the *FAQ* section of the home page |
| scroll_latestfeeds | Number of users who scroll down as far as the *Press Releases* section of the home page |
| chrome | Number of sessions from the Google Chrome browser |
| safari | Number of sessions from the Apple Safari browser |
| firefox | Number of sessions from the Mozilla Firefox browser |
| internet_explorer | Number of sessions from the Microsoft Internet Explorer browser |
| windows | Number of sessions from Microsoft Windows operating systems |
| macintosh | Number of sessions from Apple Macintosh operating systems |
| ios | Number of sessions from Apple iOS operating systems |
| android | Number of sessions from Android operating systems |

In preparing these data, we first created an external traffic reporting segment by filtering out traffic coming from website developers and ToutBay principals. The variables in the data set include data gathered from Google Analytics reports for `www.toutbay.com` for the period from April 12, 2014 through September 19, 2014. Also included are counts from Scroll Depth, a Google Analytics plug-in that tracks how far users scroll down a page. Scroll Depth is especially useful for a website that puts a lot of information on individual pages such as the home page (a single-page approach). Documentation for Scroll Depth is available at `http://scrolldepth.parsnip.io/`. When using Google Analytics, we do not have access to the original data that have been collected. Rather, we use the variables and reporting aggregates that Google Analytics defines. Documentation for Google Analytics measures (dimensions and metrics) is available at `https://developers.google.com/analytics/devguides/reporting/core/dimsmets`.

## C.13 Two Month's Salary

I never understood why giving a diamond was the social norm when proposing marriage. As I began searching for an engagement ring, two thoughts kept racing through my mind: "How will I be able to find the right diamond?" and "What is this thing going to cost me?" It goes without saying that my fiancée-to-be is worth the expense, but very seldom in our lives do we spend two month's salary on a product we know so little about.

Most guys are like me. They do not want to spend a lot of time talking to jewelers, doing extensive research, and comparing prices. So for the sake of my male cohort, I took my statistical education to the streets to find out what goes into diamond pricing and value.

I visited ten brick-and-mortar jewelers where I talked with salespeople, tracked data, and viewed more than one hundred diamonds. Then I visited seven online jewelers, gathering information on more than three hundred additional diamonds from two active stores. All observations in my data set represented round-cut diamonds. Although prices of alternative shapes or cuts might be comparable, I only looked at round-cut stones because that shape was the most common, held the most value, and was the only one my girlfriend wanted.

Shortly after beginning my research, I realized why a diamond is the perfect gift to represent an engagement. A diamond symbolizes your choice in a mate because a perfect one is very rare and all of them are unique, complete with imperfections and positive aspects that make them sparkle.

Uniqueness in diamonds is measured using four characteristics called the four Cs: color, clarity, carat, and cut. These traits combine to give a diamond its brilliance and fire. A low level of any one of these attributes can significantly decrease a diamond's value. Here is what I learned about the four Cs.

**Carat.** Carat is the standard unit of weight used for gemstones (one carat equals 0.200 grams or 200 milligrams). Diamonds are rounded to the nearest hundredth of a carat or point. A 1.27-carat diamond is said to be "one

---

Written by Brian A. Pope.

hundred and twenty-seven points." Typical diamond sizes vary from one-quarter to three carats. Diamonds are sized in one-quarter-carat increments, and jewelers typically carry stock of diamonds at each one-quarter-carat increment. According to jewelry store personnel, not only does price increase with the weight of a stone, but, as a diamond passes each one-quarter-carat threshold, its price jumps correspondingly.

**Color.** Because diamonds are formed through heat and pressure, the presence of various gases can cause them to take on various tints. Some diamonds are clear. Others have a yellow or brown tint. The Gemological Institute of America (GIA) has established a standard color scale for grading diamonds from D to Z based on tint or color. This scale was used by all twelve of the jewelers I visited. It breaks color grades down into categories like "colorless" and "near colorless." Jewelers indicate that the price of a diamond decreases as you move away from a D grade, which is considered perfectly colorless. In most cases, however, differences in color grade can only be seen when diamonds are compared with one another.

**Clarity.** The clarity of a diamond measures the purity of the stone. There are often carbon pockets that form imperfections in diamonds called inclusions. Clarity summarizes the number and size of inclusions. The GIA has created a scale that rates inclusions by their visibility to the naked eye. From a flawless (FL) diamond to one that has slight inclusions (SI1 and SI2), salespeople will tell you that the price and value of a diamond decreases as the number of noticeable inclusions increases. But when you shop, you will rarely see a perfectly flawless diamond, and most often you cannot visually detect inclusions at the VVS or VS levels.

**Cut.** As you go from one jeweler to the next, carat, color, and clarity are defined and measured in a generally universal way. A grade D diamond is perfectly colorless. A diamond with I2 clarity will have plainly visible flaws. And a 1.03-carat diamond has the same weight anywhere you shop. That leaves the type and quality of a stone's cut to differentiate diamond products. The type of cut determines the shape of the diamond, but I limited my study to round-cut diamonds. Determining the quality of cut was more problematic.

I often felt like I was being deceived when salespeople explained why their cut scale was the only appropriate way to measure the quality of cut. A few jewelers used three criteria that the GIA says make an ideally cut stone: depth, symmetry, and polish. Variations in depth and symmetry can cause a diamond to lose its brilliance. In addition to these two qualities, the overall finish or polish of the stone can have a substantial effect on how well it shines. In the end, I simplified my definition of the cut variable based on my shopping experiences. Regardless of what was said about cut, most jewelers would show two levels of cut. One of the levels would be described as ideal and the other non-ideal. The difference between ideally and non-ideally cut diamonds is not likely to be noticeable to the naked eye, but a diamond will undoubtedly cost more if a jeweler describes it as ideally cut. In addition to the four Cs, I wanted to see if price varied across sales channels. I gathered data from three separate types of jewelers.

**Independent Jewelers.** These businesses are usually not in an enclosed mall. They are limited to a single community rather than chain stores. Many of the independent jewelers I visited operated at only one location. At independent jewelers I would be given a selection of seven to ten round-cut stones, and store personnel took a non-pressured approach to the sales process.

**Mall Jewelers.** Located within enclosed malls, many of these jewelers were local branches of national chains. I found the selection of stones to be higher in number but lower in quality. The main factor that I did not like here was the pushy nature of the sales force. I often felt like I was buying a used car.

**Internet Jewelers.** I looked for online jewelers to complete my analysis. I found two stores with a vast selection of stones. I took a sample of more than three hundred stones from the over four thousand round-cut diamonds available at these two stores. Although online jewelers provided pictures of about half their stones, I would find it difficult to buy a diamond I could not see in person.

Now that the data have been gathered and coded according to the rules summarized in table C.15, I need to figure out which diamond to buy my girlfriend. Furthermore, some of the jewelers are asking questions about why I am collecting this information. One of the independent jewelers is interested in my study. He thinks he might be able to use the results to guide his own diamond buying.

*Table C.15.* *Diamonds Data: Variable Names and Coding Rules*

| *Variable* | *Description and Coding* | |
|---|---|---|
| carat | Weight of the diamond in carats (1 carat = 200 milligrams) | |
| color | D = 1 | I = 6 |
| | E = 2 | J = 7 |
| | F = 3 | K = 8 |
| | G = 4 | L = 9 |
| | H = 5 | M = 10 |
| clarity | FL = 1 | SI1 = 7 |
| | IF = 2 | SI2 = 8 |
| | VVS1 = 3 | I1 = 9 |
| | VVS2 = 4 | I2 =10 |
| | VS1 = 5 | I3 = 11 |
| | VS2 = 6 | |
| cut | Not Ideal = 0 | |
| | Ideal = 1 | |
| channel | Mall = 0 | |
| | Independent = 1 | |
| | Internet = 2 | |
| store | Goodman's = 1 | Kay = 7 |
| | Chalmer's = 2 | Zales = 8 |
| | Fred Meyer = 3 | Danford = 9 |
| | R. Holland = 4 | Blue Nile = 10 |
| | Ausman's = 5 | Ashford = 11 |
| | University = 6 | Riddle's = 12 |
| price | Price in U.S. dollars (April 2001) | |

## C.14 Wisconsin Dells

Wisconsin Dells, a sprawling resort and entertainment center in south central Wisconsin, is one of the Midwest's favorite vacation destinations. The Dells area is a mixture of beautiful valleys, canyons, hills, forests, and recreational businesses nestled around an interlocking series of lakes and rivers. Wisconsin Dells is an hour north of Madison, Wisconsin (the state capital), three to four hours from Chicago, and four hours from Minneapolis/St. Paul. The Dells offers a wide variety of activities. In summer, people come for its water parks and amphibious tours. In winter, people come for cross-country skiing and snowmobiling. Indoor attractions are open year-round.

In the summer of 1995 Wisconsin Dells business owners were developing plans for drawing visitors to their attractions. They had many questions about their customers and potential customers. To answer these questions, business owners, represented by the Wisconsin Dells Visitor and Convention Bureau, enlisted the aid of Chamberlain Research Consultants, a marketing research firm headquartered in Madison. The firm conducted 1,698 in-person interviews with visitors to Wisconsin Dells. These interviews took place on the main street of Wisconsin Dells and at water parks, hotels, restaurants, and other area attractions. Interviewers obtained demographic and vacation trip information from visitors. The Wisconsin Dells area offers many popular tourist activities and attractions. Let us review some of the more popular attractions.

**Tommy Bartlett's Thrill Show.** Started in 1952, this is one of the most famous Dells attractions. The show is a combination of on-stage performances (including juggling, tumbling, and music) and a water-skiing show. The water show has highly choreographed stunts, including a three-tier human pyramid on water skis. The Thrill Show auditorium holds five thousand people, and there are three performances daily between Memorial Day and Labor Day.

**Water Parks.** The Dells area is home to several water parks, including Noah's Ark, which is reportedly the largest water park in the nation.

---

Written by Jonathan C. Harrington. Based on research supported by Sharon Chamberlain.

**The Ducks.** When people talk about "The Ducks in the Dells," they are not talking about waterfowl. These Ducks are amphibious vehicles built by the U.S. Army during World War II as a means of transporting soldiers over land and water. The Ducks are used to give tours of the natural wonders of the area. Duck Tours take visitors up hills, down into valleys, across rivers, and through lakes. Along the way, visitors see all manner of intriguing rock formations and beautiful scenery. Duck Tours run from March through October, weather permitting.

**Circus World Museum.** Wisconsin Dells is located just north of Baraboo, Wisconsin, former home of the famous Ringling brothers, founders of the Ringling Brothers and Barnum & Bailey Circus. Owned by the State Historical Society of Wisconsin, Circus World Museum celebrates the history of the circus with exhibits, circus performances, variety shows, clown shows, animal shows, and a petting menagerie. The museum is open year-round with extended hours during the summer.

**Boat Tours.** The Dells area stretches along the Wisconsin River and includes several lakes. An alternative to Duck Tours are the boat tours, which stick to the waterways and attractions along the shorelines.

**Stand Rock.** The Dells has fascinating natural rock formations because the upper layers of rock are more resistant to erosion than are the underlying layers. Stand Rock is an unusual formation, with a large, round, table-like rock supported by a far narrower column. This formation is near another tall rock formation with a gap in between. To commemorate a famous leap across the gap, the tour of this site includes a dog leaping from rock to rock. Stand Rock is accessible by boat.

**Gambling.** Ho-Chunk Casino is located one mile south of downtown Wisconsin Dells. This Indian casino features slots, video poker, blackjack, and various forms of entertainment.

Additional area attractions include a wax museum, numerous campgrounds, many shopping opportunities, go-carts, a fifties revival show, golf courses, nature walks, a UFO and science fiction museum, a motor speedway, fishing trips, riding stables, laser tag facilities, movie theaters, and various other museums and shows.

Exhibit C.16 shows visitor variables and their coding. Interviewers asked visitors whether they had participated in or were likely to participate in any of a number of activities around the Wisconsin Dells. Exhibit C.17 shows variables relating to participation in these activities.

Taking the role of a Dells business owner or a representative of the Wisconsin Dells Visitor and Convention Bureau, we have many questions to answer. What can we learn about the people who visit the Dells? Are there discernible patterns in visitor activities? Is it possible to identify consumer segments among the visitors? What kinds of activities would we recommend for visitor groups identified by demographics or type of visiting party?

A majority of current Dells advertising takes the form of brochures and pamphlets placed at various attractions in the Dells. Business owners would like to target advertising to those people most likely to visit attractions. What can we learn from the Dells data to help business owners in their advertising and marketing activities?

***Table C.16.*** *Dells Survey Data: Visitor Characteristics*

| *Variable* | *Interview Item and Coding* |
|---|---|
| nnights | Length of stay (number of nights)<br>Coded as five ordered categories: 0, 1, 2, 3, 4+ |
| nadults | Number of adults in the party, including the respondent<br>Coded as five ordered categories: 1, 2, 3, 4, 5+ |
| nchildren | Number of children under 18 in the party<br>Coded as six ordered categories: None, 1, 2, 3, 4, 5+ |
| planning | How far in advance the vacation was planned<br>One month or more ago<br>This month (between two and four weeks ago)<br>This week |
| sex | Sex of respondent, coded by sight and sound of voice |
| age | Age category of respondent in years<br>Less than 25, 25–34, 35–44, 45–54 , 55–64, 65+ |
| education | Highest level of education completed by respondent<br>High school graduate or less<br>Some college<br>College graduate<br>Attended or completed graduate school |
| income | Level of total household annual income<br>Lower income ($50,000 or less)<br>Middle income (between $50,000 and $100,000)<br>Upper income ($100,000 or more) |
| region | ZIP code recoded into one of six regions<br>Chicago, Minneapolis/St. Paul<br>Milwaukee, Other Wisconsin<br>Madison, Other |

*Table C.17. Dells Survey Data: Visitor Activities*

| *Variable* | *Activity (Participation coded yes or no)* |
|---|---|
| shopping | Shopping |
| antiquing | Antiquing |
| scenery | Driving to look at scenery |
| eatfine | Eating out at fine restaurants |
| eatcasual | Casual theme restaurants |
| eatfamstyle | Family-style restaurants/buffets |
| eatfastfood | Fast-food restaurants |
| museums | Going to museums/indoor activities |
| indoorpool | Relaxing in indoor pool areas where staying |
| outdoorpool | Relaxing in outdoor pool areas where staying |
| hiking | Hiking |
| gambling | Gambling |
| boatswim | Boating/swimming/outdoor sports |
| fishing | Fishing |
| golfing | Golfing |
| boattours | Boat tours |
| rideducks | Riding The Ducks |
| amusepark | Amusement or theme park |
| minigolf | Miniature golf |
| gocarting | Go-carting |
| waterpark | Water park |
| circusworld | Circus World Museum |
| tbskishow | Tommy Bartlett's ski show |
| helicopter | Helicopter rides |
| horseride | Horseback riding |
| standrock | Stand Rock Ceremony |
| outattract | Outdoor attractions (not an amusement park) |
| nearbyattract | Nearby area attractions |
| movietheater | Movie theater |
| concerttheater | Concert/theater/evening entertainment |
| barpubdance | Bars, pubs, dancing |
| shopbroadway | Shop, browse on Broadway |
| bungeejumping | Bungee jumping |

## C.15 Wisconsin Lottery Sales

It is January 1999, and Wisconsin Lottery administrators have basic questions about the market for lottery tickets. Who are the Wisconsin Lottery's customers, and what makes them buy lottery tickets? Wisconsin Lottery sales contribute to State of Wisconsin revenues. Wisconsin Lottery and Department of Revenue administrators want to have accurate ways of predicting these sales. If they knew who their customers were and how to find them, administrators could do a better job in selecting new lottery ticket retailers. Administrators are concerned about what appears to be a drop in demand for instant lottery tickets. They are also concerned that both online and instant ticket sales could be affected by the opening of new Indian casinos.

There are two general classes of lottery games: online and instant. Online lottery tickets, which are sold at selected retail establishments in Wisconsin, require the buyer to pick numbers to be entered at an online lottery terminal. These tickets are sold to individual customers every day of the year. Some tickets are sold for $0.50, $2.00, and $5.00, but the great majority of lottery tickets are sold for $1.00. Odds of winning are extremely low, but jackpots can be huge. Jackpots for the online game PowerBall sometimes exceed $100 million.

Instant lottery tickets, also called "scratch tickets," come in many varieties. These have smaller jackpots and better odds of winning than online lottery tickets. The Wisconsin Lottery sells bundles of instant tickets to legitimate for-profit and nonprofit organizations in Wisconsin, and these organizations, in turn, sell individual tickets to consumers. Most online ticket retailers also sell instant lottery tickets. We might assume that, on average, online ticket retailers place orders for instant lottery ticket bundles about once every four weeks.

Competing lottery games are offered by the neighboring states of Illinois, Iowa, Michigan, and Minnesota. States sometimes cooperate with one an-

---

Wisconsin State administrators provided data for lottery sales and Wisconsin Indian casinos. And David R. Blough provided geographical measurements for Wisconsin ZIP codes. Frees and Miller (2004) used the Wisconsin Lottery data to demonstrate forecasting methods for panel/longitudinal data.

other. In 1998 the popular game PowerBall, for example, derived its large jackpots by pooling ticket sales from eighteen states and the District of Columbia.

Substitute or competing (and legal) gaming products include bingo and slots at Indian casinos in Wisconsin and neighboring states. We identified fourteen Wisconsin casinos operational at the time of the study. We also learned that new casinos were planned for Madison in 1999 and for Milwaukee in 2000. The Potawatomi Nation plan for Milwaukee (ZIP code 53233) included a 256,000 square foot casino complex with 1,000 slot machines.

In developing models for lottery sales, we can draw upon observations of people familiar with lottery activities. We can also draw upon our intuition and anecdotal evidence. There are a number of hypotheses to consider:

- Ticket sales are higher shortly after new lottery games are introduced with television or radio advertising.
- Higher lottery jackpots lead to higher online ticket sales. There may also be some carry-over effect on instant lottery ticket sales.
- Ticket sales are higher in those areas that are better served by online ticket retailers. That is, higher numbers of retailers should lead to higher sales.
- Ticket sales are lower in areas served by substitute gaming facilities, such as Indian casinos.
- Lower income, less educated people buy more lottery tickets per capita than higher income, more educated people.
- On average, senior citizens buy more lottery tickets than people in other age groups. The thinking here is that senior citizens have more free time to engage in recreational and gaming activities.
- Ticket sales are higher during the first week of the month because many people get paid or receive government support checks, such as Social Security checks, on the first day of the month.

Although we might expect advertising to affect sales, State of Wisconsin law restricts the use of extensive advertising by the Wisconsin Lottery. The only time that the Wisconsin Lottery is allowed to advertise is when a new lottery game is introduced. New lottery games are usually instant games, and only a small proportion of these games receives television or radio advertising. For example, in the forty-week period for this study, twenty-seven new instant lottery games were introduced. Six of these games received television advertising, and one received radio advertising. We might assume that each new instant lottery game that received advertising received it for one month (for the week of new product launch and for three weeks thereafter).

Sales data for the Wisconsin Lottery are like the sales data of many organizations. These are hierarchical or panel data, having both a cross-sectional and a time-series organization. For each retail establishment selling online lottery tickets, the State has a record of the number of lottery tickets sold, their cost, and the time of the sale. Retail establishments fall within sales territories or areas. For the Wisconsin Lottery we might think of ZIP codes as sales territories.

We organized Wisconsin Lottery sales data by ZIP code and time. We aggregated instant ticket sales across retail establishments within ZIP codes, and we also obtained instant ticket sales within ZIP codes. We used weeks as our unit of aggregation across time. Weeks began on Sundays and ended on Saturdays; we obtained data for 40 consecutive weeks (the weeks ending April 4, 1998 through January 2, 1999). Table C.18 provides names and descriptions for the relevant variables. Sales data were not available for Wisconsin Indian casinos, but we did obtain measures of gaming capacity (casino size and the number of slot machines). Table C.19 shows names and descriptions for information fields in the casino data set.

We can link lottery sales data, casino data, and demographic data using ZIP codes. We derived ZIP code demographics from 1990 United States Census data, with revised Census estimates from 1995. We also recorded the centroid of each ZIP code region in East-West and North-South coordinates. Table C.20 shows names and descriptions for the ZIP code demographic and location variables.

A geographer helped us to locate the East-West and North-South coordinates for ZIP code centroids. He explained that ZIP regions are highly irregular polygons and that the centroid of a ZIP code is at best an approximate center of the ZIP code region. To get coordinates for Wisconsin ZIP code centriods, the geographer used the Wisconsin Transverse Mercator

Geo-referencing System, which measures coordinate axes in meters, with the origin set as an arbitrary point in Iowa, southwest of all Wisconsin ZIP codes. Centriod coordinates should not be thought of as centers of population because it is unlikely that population would be evenly distributed across ZIP code regions.

When fitting linear models to the lottery sales data, we should note that some explanatory variables, such as the size of lottery jackpots, vary across time, but are constant across ZIP code locations. Other explanatory variables, such as population, vary across ZIP code locations, but are treated as constant across time (for the 40 weeks that we are considering). Still other variables, such as lottery sales response variables, vary across time and locations. In fitting models to these data, we need to identify appropriate error structures, noting which variables vary with time and which vary with location.

We do not have to make a distinction between sales dollars and sales volume because most lottery tickets are sold for $1.00. Just the same, we need to define appropriate response variables. In testing certain research hypotheses, we may want to use per capita measures rather than original measures. And, given the characteristics of online and instant sales (online being sales to consumers and instant being sales to retailers), we may choose to develop separate models for the online and instant sales responses. Alternatively, we could try to synchronize online and instant sales information by shifting or lagging one sales time-series relative to the other.

State of Wisconsin administrators want to predict online and instant ticket sales and to identify future potentially productive online ticket sales locations. In the process of fitting models, we might think about providing meaningful tests of hypotheses about what affects lottery sales. We have sufficient data to fit a variety of models, including time-series, panel, and spatial data models. Where shall we begin?

*Table C.18.* Wisconsin Lottery Data

| Variable | Description |
|---|---|
| zip | ZIP code |
| weekindx | Index for the week of the study (number between 1 and 40) |
| weekend | End of week date |
| weekofmo | Index of the week of the month. The week containing the first day of the month is given the index number 1. Subsequent weeks in the month are numbered in sequence. |
| wjackpot | Wednesday online PowerBall jackpot in millions of dollars |
| sjackpot | Saturday online PowerBall jackpot in millions of dollars |
| nads | Number of new instant lottery games receiving either television or radio advertising (assuming a four-week advertising period) |
| zolsales | Online lottery sales to individual consumers in dollars summed across retail locations |
| zinsales | Estimated instant lottery sales to for-profit retailers in dollars summed across retail locations |
| nretail | Number of listed retailers in the ZIP code |

*Table C.19.* Wisconsin Casino Data

| Variable | Description |
|---|---|
| zip | ZIP code |
| casino | Casino name |
| city | City name |
| owner | Owner/tribe name |
| size | Building size (thousands of square feet) |
| slots | Number of slot machines |
| yearest | Year established |

*Table C.20.* *Wisconsin ZIP Code Data*

| *Variable* | *Description* |
|---|---|
| zip | ZIP code |
| countycd | County code number |
| county | County name |
| perperhh | Mean persons per household times 10 |
| mschool | Median years of schooling times 10 |
| mhomeval | Median home value in $100s for owner-occupied homes |
| prent | Percent of housing that is renter occupied |
| polder | Percent of population that is 55 and older |
| mage | Median age of persons in household |
| mincome | Median household income in $100s |
| populat | Population |
| ewlocate | East-West coordinate for the centroid of the ZIP code |
| nslocate | North-South coordinate for the centroid of the ZIP code |

## C.16 Wikipedia Votes

The Wikipedia online encyclopedia is a collaborative writing project open to all. Jimmy Wales and Larry Sanger started Wikipedia January 15, 2001 using wiki software from Ward Cunningham.

The Wikipedia website grew slowly during its first four years. Between its fifth and sixth years of operation, however, the website doubled in size, growing from 500 thousand articles to more than one million articles. By September 2014, Wikipedia consisted of more than 33 million articles in 287 languages, with more than 48 million contributors.

Wikipedia is maintained by a set of elected administrators. Votes are cast by existing administrators and by non-administrator users. A set of votes over any selected period of time may be used to define a social network. The act of voting defines a link in a directed network, with a user/voter linked to another user/candidate.

Wikipedia Votes represents a network data set of 7,115 nodes (users, voters, candidates) and 103,689 links (votes). The data span the first seven years of Wikipedia, January 2001 through January 2008, documenting the early growth of the site and user collaboration in building the site.

Table C.21 lists the top ten websites worldwide in September 2014 according to Alexa Internet, Inc., a subsidiary of Amazon.com. The ranking is based on page view and daily visitor counts. Wikipedia ranks sixth on the list and is the only member not maintained by a corporation.

---

Data showing the from-node and to-node structure of this social network are drawn from Lestovec, Huttenlocher, and Kleinbert (2010a, 2010b) and are available as part of the Stanford Large Network Dataset Collection at `https://snap.stanford.edu/data/wiki-Vote.html`. Background information about Wikipedia was obtained from Wikipedia (2014b).

---

***Table C.21.*** *Top Sites on the Web, September 2014*

| *Rank* | *Name* | *Description* |
|---|---|---|
| 1 | Google.com | Search engine |
| 2 | Facebook.com | Social network, photo sharing |
| 3 | Youtube.com | Online videos |
| 4 | Yahoo.com | Internet portal |
| 5 | Baidu.com | Chinese search engine |
| 6 | Wikipedia.org | Online encyclopedia |
| 7 | Twitter.com | Social network, microblogging |
| 8 | Amazon.com | Online retailer |
| 9 | Qq.com | Chinese Internet portal |
| 10 | Linkedin.com | Social network for professionals |

Adapted from Alexa Internet (2014).

# D Code and Utilities

Doing marketing data science with R means looking for task views posted with the Comprehensive R Archive Network (CRAN). We go to RForge and GitHub. We read package vignettes and papers in *The R Journal* and the *Journal of Statistical Software*. At the time of this writing, the R programming environment consists of more than 5,000 packages, many of them focused on modeling methods. Useful general references for learning R include Fox and Weisberg (2011), Matloff (2011) and Wickham (2015). Chambers and Hastie (1992) and Venables and Ripley (2002), although written with S/SPlus in mind, remain critical references in the statistical programming community.

Doing marketing data science with Python means gathering programs and documentation from GitHub and staying in touch with organizations such as PyCon, SciPy and PyData. At the time of this writing, the Python Package Index (PyPI) consists of more than 57,000 packages. There are large open-source developer communities working on scientific programming packages like NumPy, SciPy, and SciKit-Learn. The Python Software Foundation supports code development and education. Useful general references for learning Python include Chun (2007), Beazley (2009), Beazley and Jones (2013), and Lubanovic (2015).

This appendix provides code and utilities associated with marketing data science and modeling techniques in predictive analytics discussed in this book. The code and accompanying data sets are open source, downloadable from the book's website at `http://www.ftpress.com/miller`.

The R code for the spine chart function in exhibit D.1 starting on page 400 shows how to construct a customized data visualization for conjoint studies. Using standard R graphics functions, the spine chart is built one point, line, and text string at a time. The precise placement of points, lines, and text is under control of the programmer. This utility is used in the chapter about preference and choice.

Market simulation utilities are provided in exhibit D.2 on page 408. These utilities are called by the programs in the chapter about brand and price.

R utilities for grid and ggplot2 graphics are provided in exhibit D.3. The split-plotting utilities in exhibit D.3 (pages 409 through 411) are used in a number of chapters to render R ggplot2 graphics objects with proper margins and multiple-plot layout.

R utilities for computing distances between points defined by longitude and latitude are useful in spatial data analysis. These are provided in exhibit D.4 on page 413.

An R correlation heat map utility is provided in exhibit D.5 on page 413. This utility draws on the lattice package documented by Sarkar (2008, 2014).

A Python utility for evaluating binary classifiers is provided in exhibit D.6 on page 414.

Graphical user interfaces and integrated development environments can be of great assistance to programmers when building end-user applications. Tools for building solutions with R are provided with open-source systems such as KNIME and RStudio and in commercial systems from Alteryx, IBM, Microsoft, and SAS.

Tools for building solutions with Python are provided with open-source systems such as IPython, KNIME (Berthold et al. 2007), and Orange (Demšar and Zupan 2013) and in commercial systems from Continuum Analytics, Enthought, Microsoft, and IBM.

Programs in data science do not have to be perfect or beautiful. We prefer programs that are efficient, versatile, and readable, as well-constructed programs often are. We want code that produces correct answers, although rarely will we be asked to prove a program correct. Programs in data science do not have to be all-Python or all-R. Many routines in both languages call on underlying executables from C, C++, or Fortran. What do programs in data science need to do? They need to get the job done.

For the practicing data scientist, there are considerable advantages to being multilingual. One of the good things about knowing both R and Python is that we can test scripts developed in one language against scripts developed in the other. Thus we end the programming portion of this dual-language guide. Keep coding.

**Exhibit D.1.** *Conjoint Analysis Spine Chart (R)*

```
# Conjoint Analysis Spine Chart (R)

# spine chart accommodates up to 45 part-worths on one page
# |part-worth| <= 40 can be plotted directly on the spine chart
# |part-worths| > 40 can be accommodated through standardization

print.digits <- 2  # set number of digits on print and spine chart

# user-defined function for printing conjoint measures
if (print.digits == 2)
  pretty.print <- function(x) {sprintf("%1.2f",round(x,digits = 2))}
if (print.digits == 3)
  pretty.print <- function(x) {sprintf("%1.3f",round(x,digits = 3))}

# --------------------------------------------------
# user-defined function for spine chart
# --------------------------------------------------
spine.chart <- function(conjoint.results,
  color.for.part.worth.point = "blue",
  color.for.part.worth.line = "blue",
  left.side.symbol.to.print.around.part.worths = "(",
  right.side.symbol.to.print.around.part.worths = ")",
  left.side.symbol.to.print.around.importance = "",
  right.side.symbol.to.print.around.importance = "",
  color.for.printing.importance.text = "dark red",
  color.for.printing.part.worth.text = "black",
  draw.gray.background = TRUE,
  draw.optional.grid.lines = TRUE,
  print.internal.consistency = TRUE,
  fix.max.to.4 = FALSE,
  put.title.on.spine.chart = FALSE,
  title.on.spine.chart = paste("TITLE GOES HERE IF WE ASK FOR ONE",sep=""),
  plot.framing.box = TRUE,
  do.standardization = TRUE,
  do.ordered.attributes = TRUE) {

  # fix.max.to.4  option to override the range for part-worth plotting

  if(!do.ordered.attributes) effect.names <- conjoint.results$attributes
  if(do.ordered.attributes) effect.names <-
    conjoint.results$ordered.attributes

  number.of.levels.of.attribute <- NULL
  for(index.for.factor in seq(along=effect.names))
    number.of.levels.of.attribute <- c(number.of.levels.of.attribute,
      length(conjoint.results$xlevels[[effect.names[index.for.factor]]]))

  # total number of levels needed for vertical length of spine the spine plot
  total.number.of.levels <- sum(number.of.levels.of.attribute)
```

```
 # define size of spaces based upon the number of part-worth levels to plot
  if(total.number.of.levels <= 20) {
    smaller.space <- 0.01
    small.space <- 0.02
    medium.space <- 0.03
    large.space <- 0.04
    }
  if(total.number.of.levels > 20) {
    smaller.space <- 0.01 * 0.9
    small.space <- 0.02 * 0.9
    medium.space <- 0.03 * 0.9
    large.space <- 0.04 * 0.9
    }
  if(total.number.of.levels > 22) {
    smaller.space <- 0.01 * 0.85
    small.space <- 0.02 * 0.85
    medium.space <- 0.03 * 0.825
    large.space <- 0.04 * 0.8
    }
  if(total.number.of.levels > 25) {
    smaller.space <- 0.01 * 0.8
    small.space <- 0.02 * 0.8
    medium.space <- 0.03 * 0.75
    large.space <- 0.04 * 0.75
    }
  if(total.number.of.levels > 35) {
    smaller.space <- 0.01 * 0.65
    small.space <- 0.02 * 0.65
    medium.space <- 0.03 * 0.6
    large.space <- 0.04 * 0.6
    }

# of course there is a limit to how much we can plot on one page
if (total.number.of.levels > 45)
  stop("\n\nTERMINATED: More than 45 part-worths on spine chart\n")

if(!do.standardization)
  part.worth.plotting.list <- conjoint.results$part.worths

if(do.standardization)
  part.worth.plotting.list <- conjoint.results$standardized.part.worths

# check the range of part-worths to see which path to go down for plotting
# initialize these toggles to start

max.is.less.than.40 <- FALSE
max.is.less.than.20 <- FALSE
max.is.less.than.10 <- FALSE
max.is.less.than.4 <- FALSE
max.is.less.than.2 <- FALSE
max.is.less.than.1 <- FALSE
```

```
if (max(abs(min(unlist(part.worth.plotting.list),na.rm=TRUE)),
  max(unlist(part.worth.plotting.list),na.rm=TRUE)) <= 40) {
  max.is.less.than.40 <- TRUE
  max.is.less.than.20 <- FALSE
  max.is.less.than.10 <- FALSE
  max.is.less.than.4 <- FALSE
  max.is.less.than.2 <- FALSE
  max.is.less.than.1 <- FALSE
  }
if (max(abs(min(unlist(part.worth.plotting.list),na.rm=TRUE)),
  max(unlist(part.worth.plotting.list),na.rm=TRUE)) <= 20) {
  max.is.less.than.40 <- FALSE
  max.is.less.than.20 <- TRUE
  max.is.less.than.10 <- FALSE
  max.is.less.than.4 <- FALSE
  max.is.less.than.2 <- FALSE
  max.is.less.than.1 <- FALSE
  }
if(max(abs(min(unlist(part.worth.plotting.list),na.rm=TRUE)),
  max(unlist(part.worth.plotting.list),na.rm=TRUE)) <= 10) {
  max.is.less.than.40 <- FALSE
  max.is.less.than.20 <- FALSE
  max.is.less.than.10 <- TRUE
  max.is.less.than.4 <- FALSE
  max.is.less.than.2 <- FALSE
  max.is.less.than.1 <- FALSE
  }
if (max(abs(min(unlist(part.worth.plotting.list),na.rm=TRUE)),
  max(unlist(part.worth.plotting.list),na.rm=TRUE)) <= 4) {
  max.is.less.than.40 <- FALSE
  max.is.less.than.20 <- FALSE
  max.is.less.than.4 <- TRUE
  max.is.less.than.10 <- FALSE
  max.is.less.than.2 <- FALSE
  max.is.less.than.1 <- FALSE
  }
if(max(abs(min(unlist(part.worth.plotting.list),na.rm=TRUE)),
  max(unlist(part.worth.plotting.list),na.rm=TRUE)) <= 2) {
  max.is.less.than.40 <- FALSE
  max.is.less.than.20 <- FALSE
  max.is.less.than.4 <- FALSE
  max.is.less.than.10 <- FALSE
  max.is.less.than.2 <- TRUE
  max.is.less.than.1 <- FALSE
  }
 if(max(abs(min(unlist(part.worth.plotting.list),na.rm=TRUE)),
  max(unlist(part.worth.plotting.list),na.rm=TRUE)) <= 1) {
  max.is.less.than.40 <- FALSE
  max.is.less.than.20 <- FALSE
  max.is.less.than.4 <- FALSE
  max.is.less.than.10 <- FALSE
  max.is.less.than.2 <- FALSE
  max.is.less.than.1 <- TRUE
  }
```

```
# sometimes we override the range for part-worth plotting
# this is not usually done... but it is an option
if (fix.max.to.4) {
  max.is.less.than.40 <- FALSE
  max.is.less.than.20 <- FALSE
  max.is.less.than.10 <- FALSE
  max.is.less.than.4 <- TRUE
  max.is.less.than.2 <- FALSE
  max.is.less.than.1 <- FALSE
  }

if (!max.is.less.than.1 & !max.is.less.than.2 & !max.is.less.than.4 &
  !max.is.less.than.10 & !max.is.less.than.20 & !max.is.less.than.40)
    stop("\n\nTERMINATED: Spine chart cannot plot |part-worth| > 40")

# determine point positions for plotting part-worths on spine chart
if (max.is.less.than.1 | max.is.less.than.2 | max.is.less.than.4 |
  max.is.less.than.10 | max.is.less.than.20 | max.is.less.than.40) {
# begin if-block plotting when all part-worths in absolute value
# are less than one of the tested range values
# part-worth positions for plotting
# end if-block plotting when all part-worths in absolute value
# are less than one of the tested range values
# offsets for plotting vary with the max.is.less.than setting
  if(max.is.less.than.1) {
    list.scaling <- function(x) {0.75 + x/5}
    part.worth.point.position <-
      lapply(part.worth.plotting.list,list.scaling)
    }
  if(max.is.less.than.2) {
    list.scaling <- function(x) {0.75 + x/10}
    part.worth.point.position <-
      lapply(part.worth.plotting.list,list.scaling)
    }

  if(max.is.less.than.4) {
    list.scaling <- function(x) {0.75 + x/20}
    part.worth.point.position <-
      lapply(part.worth.plotting.list,list.scaling)
    }

   if(max.is.less.than.10) {
    list.scaling <- function(x) {0.75 + x/50}
    part.worth.point.position <-
      lapply(part.worth.plotting.list,list.scaling)
    }

  if(max.is.less.than.20) {
    list.scaling <- function(x) {0.75 + x/100}
    part.worth.point.position <-
      lapply(part.worth.plotting.list,list.scaling)
    }
```

```
    if(max.is.less.than.40) {
      list.scaling <- function(x) {0.75 + x/200}
      part.worth.point.position <-
        lapply(part.worth.plotting.list,list.scaling)
      }

    part.worth.point.position <- lapply(part.worth.plotting.list,list.scaling)
    }

  if (plot.framing.box) plot(c(0,0,1,1),c(0,1,0,1),xlab="",ylab="",
    type="n",xaxt="n",yaxt="n")

  if (!plot.framing.box) plot(c(0,0,1,1),c(0,1,0,1),xlab="",ylab="",
    type="n",xaxt="n",yaxt="n", bty="n")

  if (put.title.on.spine.chart) {
    text(c(0.50),c(0.975),pos=3,labels=title.on.spine.chart,cex=01.5)
    y.location <- 0.925  # starting position with title
    }

  if (!put.title.on.spine.chart) y.location <- 0.975  # no-title start

  # store top of vertical line for later plotting needs
  y.top.of.vertical.line <- y.location

  x.center.position <- 0.75  # horizontal position of spine

  # begin primary plotting loop
  # think of a plot as a collection of text and symbols on screen or paper
  # we are going to construct a plot one text string and symbol at a time
  # (note that we may have to repeat this process at the end of the program)
  for(k in seq(along=effect.names)) {
    y.location <- y.location - large.space
    text(c(0.4),c(y.location),pos=2,
      labels=paste(effect.name.map(effect.names[k])," ",sep=""),cex=01.0)
    text(c(0.525),c(y.location),pos=2,col=color.for.printing.importance.text,
    labels=paste(" ",left.side.symbol.to.print.around.importance,
    pretty.print(
      unlist(conjoint.results$attribute.importance[effect.names[k]])),"%",
      right.side.symbol.to.print.around.importance,sep=""),cex=01.0)

  # begin loop for printing part-worths
    for(m in seq(1:number.of.levels.of.attribute[k])) {
      y.location <- y.location - medium.space
      text(c(0.4),c(y.location),pos=2,
      conjoint.results$xlevel[[effect.names[k]]][m],cex=01.0)
   #   part.worth.label.data.frame[k,m],cex=01.0)

      text(c(0.525),c(y.location),pos=2,
      col=color.for.printing.part.worth.text,
      labels=paste(" ",left.side.symbol.to.print.around.part.worths,
      pretty.print(part.worth.plotting.list[[effect.names[k]]][m]),
      right.side.symbol.to.print.around.part.worths,sep=""),cex=01.0)
```

```
    points(part.worth.point.position[[effect.names[k]]][m],y.location,
      type = "p", pch = 20, col = color.for.part.worth.point, cex = 2)
    segments(x.center.position, y.location,
    part.worth.point.position[[effect.names[k]]][m], y.location,
      col = color.for.part.worth.line, lty = 1, lwd = 2)
    }
  }
y.location <- y.location - medium.space

# begin center axis and bottom plotting
y.bottom.of.vertical.line <- y.location  # store top of vertical line

below.y.bottom.of.vertical.line <- y.bottom.of.vertical.line - small.space/2

if (!draw.gray.background) {
# four optional grid lines may be drawn on the plot parallel to the spine
  if (draw.optional.grid.lines) {
    segments(0.55, y.top.of.vertical.line, 0.55,
      y.bottom.of.vertical.line, col = "black", lty = "solid", lwd = 1)
    segments(0.65, y.top.of.vertical.line, 0.65,
      y.bottom.of.vertical.line, col = "gray", lty = "solid", lwd = 1)
    segments(0.85, y.top.of.vertical.line, 0.85,
      y.bottom.of.vertical.line, col = "gray", lty = "solid", lwd = 1)
    segments(0.95, y.top.of.vertical.line, 0.95,
      y.bottom.of.vertical.line, col = "black", lty = "solid", lwd = 1)
    }
  }

# gray background for plotting area of the points
if (draw.gray.background) {
  rect(xleft = 0.55, ybottom = y.bottom.of.vertical.line,
    xright = 0.95, ytop = y.top.of.vertical.line, density = -1, angle = 45,
    col = "light gray", border = NULL, lty = "solid", lwd = 1)

# four optional grid lines may be drawn on the plot parallel to the spine
  if (draw.optional.grid.lines) {
    segments(0.55, y.top.of.vertical.line, 0.55,
      y.bottom.of.vertical.line, col = "black", lty = "solid", lwd = 1)

    segments(0.65, y.top.of.vertical.line, 0.65,
      y.bottom.of.vertical.line, col = "white", lty = "solid", lwd = 1)

    segments(0.85, y.top.of.vertical.line, 0.85,
      y.bottom.of.vertical.line, col = "white", lty = "solid", lwd = 1)

    segments(0.95, y.top.of.vertical.line, 0.95,
      y.bottom.of.vertical.line, col = "black", lty = "solid", lwd = 1)
    }
  }

# draw the all-important spine on the plot
segments(x.center.position, y.top.of.vertical.line, x.center.position,
  y.bottom.of.vertical.line, col = "black", lty = "dashed", lwd = 1)
```

```
# horizontal line at top
segments(0.55, y.top.of.vertical.line, 0.95, y.top.of.vertical.line,
     col = "black", lty = 1, lwd = 1)

# horizontal line at bottom
segments(0.55, y.bottom.of.vertical.line, 0.95, y.bottom.of.vertical.line,
       col = "black", lty = 1, lwd = 1)

# plot for ticks and labels
segments(0.55, y.bottom.of.vertical.line,
   0.55, below.y.bottom.of.vertical.line,
   col = "black", lty = 1, lwd = 1)  # tick line at bottom
segments(0.65, y.bottom.of.vertical.line,
   0.65, below.y.bottom.of.vertical.line,
   col = "black", lty = 1, lwd = 1)  # tick line at bottom
segments(0.75, y.bottom.of.vertical.line,
   0.75, below.y.bottom.of.vertical.line,
   col = "black", lty = 1, lwd = 1)  # tick line at bottom
segments(0.85, y.bottom.of.vertical.line,
   0.85, below.y.bottom.of.vertical.line,
   col = "black", lty = 1, lwd = 1)  # tick line at bottom
segments(0.95, y.bottom.of.vertical.line,
   0.95, below.y.bottom.of.vertical.line,
   col = "black", lty = 1, lwd = 1)  # tick line at bottom

# axis labels vary with the max.is.less.than range being used
if (max.is.less.than.1) text(c(0.55,0.65,0.75,0.85,0.95),
  rep(below.y.bottom.of.vertical.line,times=5),
  pos=1,labels=c("-1","-0.5","0","+0.5","+1"),cex=0.75)
if (max.is.less.than.2) text(c(0.55,0.65,0.75,0.85,0.95),
  rep(below.y.bottom.of.vertical.line,times=5),
  pos=1,labels=c("-2","-1","0","+1","+2"),cex=0.75)
if (max.is.less.than.4) text(c(0.55,0.65,0.75,0.85,0.95),
  rep(below.y.bottom.of.vertical.line,times=5),
  pos=1,labels=c("-4","-2","0","+2","+4"),cex=0.75)
if (max.is.less.than.10) text(c(0.55,0.65,0.75,0.85,0.95),
  rep(below.y.bottom.of.vertical.line,times=5),
  pos=1,labels=c("-10","-5","0","+5","+10"),cex=0.75)
if (max.is.less.than.20) text(c(0.55,0.65,0.75,0.85,0.95),
  rep(below.y.bottom.of.vertical.line,times=5),
  pos=1,labels=c("-20","-10","0","+10","+20"),cex=0.75)
if (max.is.less.than.40) text(c(0.55,0.65,0.75,0.85,0.95),
  rep(below.y.bottom.of.vertical.line,times=5),
  pos=1,labels=c("-40","-20","0","+20","+40"),cex=0.75)
y.location <- below.y.bottom.of.vertical.line - small.space

if(do.standardization)
  text(.75,y.location,pos=1,labels=c("Standardized Part-Worth"),cex=0.95)

if(!do.standardization) text(.75,y.location,pos=1,labels=c("Part-Worth"),
  cex=0.95)

y.location <- below.y.bottom.of.vertical.line - small.space
```

```
  if(do.standardization)
    text(0.75,y.location,pos=1,labels=c("Standardized Part-Worth"),cex=0.95)

  if(!do.standardization) text(0.75,y.location,pos=1,labels=c("Part-Worth"),
    cex=0.95)

  if(print.internal.consistency) {
    y.location <- y.location - medium.space
    text(c(0.525),c(y.location),pos=2,labels=paste("Internal consistency: ",
    pretty.print(conjoint.results$internal.consistency),
    sep=""))
    }

  # if we have grid lines we may have plotted over part-worth points
  # if we have a gray background then we have plotted over part-worth points
  # so let us plot those all-important part-worth points and lines once again
  if(draw.gray.background || draw.optional.grid.lines) {
    y.location <- y.top.of.vertical.line  # retreive the starting value

  # repeat the primary plotting loop
  for(k in seq(along=effect.names)) {
    y.location <- y.location - large.space
    text(c(0.4),c(y.location),pos=2,
      labels=paste(effect.name.map(effect.names[k])," ",sep=""),cex=01.0)
    text(c(0.525),c(y.location),pos=2,col=color.for.printing.importance.text,
      labels=paste(" ",left.side.symbol.to.print.around.importance,
      pretty.print(
      unlist(conjoint.results$attribute.importance[effect.names[k]])),"%",
      right.side.symbol.to.print.around.importance,sep=""),cex=01.0)

 # begin loop for printing part-worths
      for(m in seq(1:number.of.levels.of.attribute[k])) {
         y.location <- y.location - medium.space
         text(c(0.4),c(y.location),pos=2,
         conjoint.results$xlevel[[effect.names[k]]][m],cex=01.0)
         text(c(0.525),c(y.location),
           pos=2,col=color.for.printing.part.worth.text,
           labels=paste(" ",left.side.symbol.to.print.around.part.worths,
           pretty.print(part.worth.plotting.list[[effect.names[k]]][m]),
           right.side.symbol.to.print.around.part.worths,sep=""),cex=01.0)

      points(part.worth.point.position[[effect.names[k]]][m],y.location,
         type = "p", pch = 20, col = color.for.part.worth.point, cex = 2)
      segments(x.center.position, y.location,
      part.worth.point.position[[effect.names[k]]][m], y.location,
         col = color.for.part.worth.line, lty = 1, lwd = 2)
      }
    }
  }
}
# save spine.chart function for future work
save(spine.chart,file="mtpa_spine_chart.Rdata")
```

*Exhibit D.2.* *Market Simulation Utilities (R)*

```
# Market Simulation Utilities (R)

# user-defined function for first-choice simulation rule
first.choice.simulation.rule <- function(response, alpha = 1) {
  # begin function for first-choice rule
  # returns binary vector or response vector with equal division
  # of 1 across all locations at the maximum
  # use alpha for desired sum across respondents
  # alpha useful when the set of tested profiles is not expected to be one
  if(alpha < 0 || alpha > 1) stop("alpha must be between zero and one")
  response.vector <- numeric(length(response))
  for(k in seq(along=response))
    if(response[k] == max(response)) response.vector[k] <- 1
  alpha*(response.vector/sum(response.vector))
  } # end first-choice rule function

# user-defined function for predicted choices from four-profile choice sets
choice.set.predictor <- function(predicted.probability) {
  predicted.choice <- length(predicted.probability)  # initialize
  index.fourth <- 0  # initialize block-of-four choice set indices
  while (index.fourth < length(predicted.probability)) {
    index.first  <- index.fourth + 1
    index.second <- index.fourth + 2
    index.third  <- index.fourth + 3
    index.fourth <- index.fourth + 4
    this.choice.set.probability.vector <-
      c(predicted.probability[index.first],
      predicted.probability[index.second],
      predicted.probability[index.third],
      predicted.probability[index.fourth])
    predicted.choice[index.first:index.fourth] <-
      first.choice.simulation.rule(this.choice.set.probability.vector)
    }
  predicted.choice <- factor(predicted.choice, levels = c(0,1),
    labels = c("NO","YES"))
  predicted.choice
  } # end choice.set.predictor function

# save market simulation utilities for future work
save(first.choice.simulation.rule,
  choice.set.predictor,
  file="mtpa_market_simulation_utilities.Rdata")
```

***Exhibit D.3.*** *Split-plotting Utilities (R)*

```
# Split-Plotting Utilities with grid Graphics (R)

library(grid)  # grid graphics foundation of split-plotting utilities

# functions used with ggplot2 graphics to split the plotting region
# to set margins and to plot more than one ggplot object on one page/screen

vplayout <- function(x, y)
viewport(layout.pos.row=x, layout.pos.col=y)


# grid graphics utility plots one plot with margins
ggplot.print.with.margins <- function(ggplot.object.name,left.margin.pct=10,
  right.margin.pct=10,top.margin.pct=10,bottom.margin.pct=10)
{ # begin function for printing ggplot objects with margins
  # margins expressed as percentages of total... use integers
 grid.newpage()
pushViewport(viewport(layout=grid.layout(100,100)))
print(ggplot.object.name,
  vp=vplayout((0 + top.margin.pct):(100 - bottom.margin.pct),
  (0 + left.margin.pct):(100 - right.margin.pct)))
} # end function for printing ggplot objects with margins



# grid graphics utility plots two ggplot plotting objects in one column
special.top.bottom.ggplot.print.with.margins <-
  function(ggplot.object.name,ggplot.text.tagging.object.name,
  left.margin.pct=5,right.margin.pct=5,top.margin.pct=5,
  bottom.margin.pct=5,plot.pct=80,text.tagging.pct=10) {
# begin function for printing ggplot objects with margins
# and text tagging at bottom of plot
# margins expressed as percentages of total... use integers
  if((top.margin.pct + bottom.margin.pct + plot.pct + text.tagging.pct) != 100)
    stop(paste("function special.top.bottom.ggplot.print.with.margins()",
    "execution terminated:\n   top.margin.pct + bottom.margin.pct + ",
    "plot.pct + text.tagging.pct not equal to 100 percent",sep=""))
  grid.newpage()
  pushViewport(viewport(layout=grid.layout(100,100)))
  print(ggplot.object.name,
  vp=vplayout((0 + top.margin.pct):
    (100 - (bottom.margin.pct + text.tagging.pct)),
  (0 + left.margin.pct):(100 - right.margin.pct)))

  print(ggplot.text.tagging.object.name,
    vp=vplayout((0 + (top.margin.pct + plot.pct)):(100 - bottom.margin.pct),
    (0 + left.margin.pct):(100 - right.margin.pct)))
} # end function for printing ggplot objects with margins and text tagging
```

```
# grid graphics utility plots three ggplot plotting objects in one column
three.part.ggplot.print.with.margins <- function(ggfirstplot.object.name,
ggsecondplot.object.name,
ggthirdplot.object.name,
left.margin.pct=5,right.margin.pct=5,
top.margin.pct=10,bottom.margin.pct=10,
first.plot.pct=25,second.plot.pct=25,
third.plot.pct=30) {
# function for printing ggplot objects with margins and top and bottom plots
# margins expressed as percentages of total... use integers
if((top.margin.pct + bottom.margin.pct + first.plot.pct +
  second.plot.pct  + third.plot.pct) != 100)
    stop(paste("function special.top.bottom.ggplot.print.with.margins()",
         "execution terminated:\n   top.margin.pct + bottom.margin.pct",
         "+ first.plot.pct + second.plot.pct  + third.plot.pct not equal",
         "to 100 percent",sep=""))
grid.newpage()
pushViewport(viewport(layout=grid.layout(100,100)))

print(ggfirstplot.object.name, vp=vplayout((0 + top.margin.pct):
  (100 - (second.plot.pct  + third.plot.pct + bottom.margin.pct)),
  (0 + left.margin.pct):(100 - right.margin.pct)))

print(ggsecondplot.object.name,
  vp=vplayout((0 + top.margin.pct + first.plot.pct):
  (100 - (third.plot.pct + bottom.margin.pct)),
  (0 + left.margin.pct):(100 - right.margin.pct)))

print(ggthirdplot.object.name,
  vp=vplayout((0 + top.margin.pct + first.plot.pct + second.plot.pct):
  (100 - (bottom.margin.pct)),(0 + left.margin.pct):
  (100 - right.margin.pct)))
}

# grid graphics utility plots two ggplot plotting objects in one row
# primary plot graph at left... legend at right
special.left.right.ggplot.print.with.margins <-
  function(ggplot.object.name, ggplot.text.legend.object.name,
  left.margin.pct=5, right.margin.pct=5, top.margin.pct=5,
  bottom.margin.pct=5, plot.pct=85, text.legend.pct=5) {
# begin function for printing ggplot objects with margins
# and text legend at bottom of plot
# margins expressed as percentages of total... use integers
  if((left.margin.pct + right.margin.pct + plot.pct + text.legend.pct) != 100)
    stop(paste("function special.left.right.ggplot.print.with.margins()",
    "execution terminated:\n   left.margin.pct + right.margin.pct + ",
    "plot.pct + text.legend.pct not equal to 100 percent",sep=""))
  grid.newpage()
  pushViewport(viewport(layout=grid.layout(100,100)))
  print(ggplot.object.name,
  vp=vplayout((0 + top.margin.pct):(100 - (bottom.margin.pct)),
  (0 + left.margin.pct + text.legend.pct):(100 - right.margin.pct)))
```

```
  print(ggplot.text.legend.object.name,
    vp=vplayout((0 + (top.margin.pct)):(100 - bottom.margin.pct),
    (0 + left.margin.pct + plot.pct):(100 - right.margin.pct)))
} # end function for printing ggplot objects with margins and text legend

# save split-plotting utilities for future work
save(vplayout,
  ggplot.print.with.margins,
  special.top.bottom.ggplot.print.with.margins,
  three.part.ggplot.print.with.margins,
  special.left.right.ggplot.print.with.margins,
  file="mtpa_split_plotting_utilities.Rdata")
```

**Exhibit D.4.** *Utilities for Spatial Data Analysis (R)*

```
# Utilities for Spatial Data Analysis (R)

# user-defined function to convert degrees to radians
# needed for lat.long.distance function
degrees.to.radians <- function(x) {
  (pi/180)*x
  } # end degrees.to.radians function

# user-defined function to convert distance between two points in miles
# when the two points (a and b) are defined by longitude and latitude
lat.long.distance <- function(longitude.a,latitude.a,longitude.b,latitude.b) {
  radius.of.earth <- 24872/(2*pi)
  c <- sin((degrees.to.radians(latitude.a) -
    degrees.to.radians(latitude.b))/2)^2 +
    cos(degrees.to.radians(latitude.a)) *
    cos(degrees.to.radians(latitude.b)) *
    sin((degrees.to.radians(longitude.a) -
    degrees.to.radians(longitude.b))/2)^2
  2 * radius.of.earth * (asin(sqrt(c)))
  } # end lat.long.distance function

save(degrees.to.radians,
  lat.long.distance,
  file = "mtpa_spatial_distance_utilities.R")
```

***Exhibit D.5.*** *Correlation Heat Map Utility (R)*

```
# Correlation Heat Map Utility (R)
#
# Input correlation matrix. Output heat map of correlation matrix.
# Requires R lattice package.

correlation_heat_map <- function(cormat, order_variable = NULL) {
    if (is.null(order_variable)) order_variable = rownames(cormat)[1]
    cormat_line <- cormat[order_variable, ]
    ordered_cormat <-
        cormat[names(sort(cormat_line, decreasing=TRUE)),
            names(sort(cormat_line, decreasing=FALSE))]
    x <- rep(1:nrow(ordered_cormat), times=ncol(ordered_cormat))
    y <- NULL
    for (i in 1:ncol(ordered_cormat))
        y <- c(y,rep(i,times=nrow(ordered_cormat)))
    # use fixed format 0.XXX in cells of correlation matrix
    cortext <- sprintf("%0.3f", as.numeric(ordered_cormat))
    text.data.frame <- data.frame(x, y, cortext)
    text.data.frame$cortext <- as.character(text.data.frame$cortext)
    text.data.frame$cortext <- ifelse((text.data.frame$cortext == "1.000"),
    NA,text.data.frame$cortext)  # define diagonal cells as missing
    text.data.frame <- na.omit(text.data.frame)  # diagonal cells have no text
    # determine range of correlations all positive or positive and negative
    if (min(cormat) > 0)
        setcolor_palette <- colorRampPalette(c("white", "#00BFC4"))
    if (min(cormat) < 0)
        setcolor_palette <- colorRampPalette(c("#F8766D", "white", "#00BFC4"))
    # use larger sized type for small matrices
    set_cex = 1.0
    if (nrow(ordered_cormat) <= 4) set_cex = 1.5
    print(levelplot(ordered_cormat, cuts = 25, tick.number = 9,
        col.regions = setcolor_palette,
        scales=list(tck = 0, x = list(rot=45), cex = set_cex),
        xlab = "",
        ylab = "",
        panel = function(...) {
            panel.levelplot(...)
            panel.text(text.data.frame$x, text.data.frame$y,
            labels = text.data.frame$cortext, cex = set_cex)
            }))
    }

save(correlation_heat_map, file = "correlation_heat_map.RData")
```

***Exhibit D.6.*** *Evaluating Predictive Accuracy of a Binary Classifier (Python)*

```
# Evaluating Predictive Accuracy of a Binary Classifier (Python)

def evaluate_classifier(predicted, observed):
    import pandas as pd
    if(len(predicted) != len(observed)):
        print('\nevaluate_classifier error:',\
            ' predicted and observed must be the same length\n')
        return(None)
    if(len(set(predicted)) != 2):
        print('\nevaluate_classifier error:',\
            ' predicted must be binary\n')
        return(None)
    if(len(set(observed)) != 2):
        print('\nevaluate_classifier error:',\
            ' observed must be binary\n')
        return(None)

    predicted_data = predicted
    observed_data = observed
    input_data = {'predicted': predicted_data,'observed':observed_data}
    input_data_frame = pd.DataFrame(input_data)

    cmat = pd.crosstab(input_data_frame['predicted'],\
        input_data_frame['observed'])
    a = float(cmat.ix[0,0])
    b = float(cmat.ix[0,1])
    c = float(cmat.ix[1,0])
    d = float(cmat.ix[1,1])
    n = a + b + c + d
    predictive_accuracy = (a + d)/n
    true_positive_rate = a / (a + c)
    false_positive_rate = b / (b + d)
    precision = a / (a + b)
    specificity = 1 - false_positive_rate
    expected_accuracy = (((a + b)*(a + c)) + ((b + d)*(c + d)))/(n * n)
    kappa = (predictive_accuracy - expected_accuracy)\
       /(1 - expected_accuracy)
    return(a, b, c, d, predictive_accuracy, true_positive_rate, specificity,\
        false_positive_rate, precision, expected_accuracy, kappa)
```

# Bibliography


Aaker, D. A. (ed.) 1971. *Multivariate Analysis in Marketing*. Belmont, Calif.: Wadsworth.

Aaker, D. A. 2001. *Strategic Marketing Management* (sixth ed.). New York: Wiley.

Aarseth, E. J. 1997. *Cybertext: Perspectives on Ergodic Literature*. Baltimore: Johns Hopkins University Press. 349

Ackland, R. 2013. *Web Social Science: Concepts, Data and Tools for Social Scientists in the Digital Age*. Los Angeles: Sage.

Agar, M. H. 1996. *The Professional Stranger: An Informal Introduction to Ethnography* (second ed.). New York: Academic Press.

Agrawal, R., H. Mannila, R. Srikant, H. Toivonen, and A. I. Verkamo 1996. Fast discovery of association rules. In U. M. Fayyad, G. Piatetsky-Shapiro, P. Smyth, and R. Uthurusamy (eds.), *Handbook of Data Mining and Knowledge Discovery*, Chapter 12, pp. 307–328. Menlo Park, Calif. and Cambridge, Mass.: American Association for Artificial Intelligence and MIT Press. 153

Agresti, A. 2013. *Categorical Data Analysis* (third ed.). New York: Wiley.

Airoldi, E. M., D. Blei, E. A. Erosheva, and S. E. Fienberg (eds.) 2014. *Handbook of Mixed Membership Models and Their Applications*. Boca Raton, Fla.: CRC Press.

Aizaki, H. 2012, September 22. Basic functions for supporting an implementation of choice experiments in R. *Journal of Statistical Software, Code Snippets* 50(2):1–24. `http://www.jstatsoft.org/v50/c02`.

Aizaki, H. 2014. *support.CEs: Basic Functions for Supporting an Implementation of Choice Experiments*. Comprehensive R Archive Network. 2014. `http://cran.r-project.org/web/packages/support.CEs/support.CEs.pdf`.

Akaike, H. 1973. Information theory and an extension of the maximum likelihood principle. In B. N. Petrov and F. Csaki (eds.), *Second International Symposium on Information Theory*, pp. 267–281. Budapest: Akademiai Kiado. 21, 273

Albert, J. 2009. *Bayesian Computation with R*. New York: Springer. 264

Albert, R. and A.-L. Barabási 2002, January. Statistical mechanics of complex networks. *Reviews of Modern Physics* 74(1):47–97. 194

Alexa Internet, I. 2014. The top 500 sites on the web. Retrieved from the World Wide Web, October 21, 2014, at `http://www.alexa.com/topsites`.

Alexander, W. and S. Grimshaw 1996. Treed regression. *Journal of Computational and Graphical Statistics* 5(2):156–175. 358

Alfons, A. 2014a. *cvTools: Cross-Validation Tools for Regression Models*. Comprehensive R Archive Network. 2014. `http://cran.r-project.org/web/packages/cvTools/cvTools.pdf`.

Alfons, A. 2014b. *simFrame: Simulation Framework*. Comprehensive R Archive Network. 2014. `http://cran.r-project.org/web/packages/simFrame/simFrame.pdf`. 274

Alfons, A., M. Templ, and P. Filzmoser 2014. *An Object-Oriented Framework for Statistical Simulation: The R Package simFrame*. Comprehensive R Archive Network. 2014. `http://cran.r-project.org/web/packages/simFrame/vignettes/simFrame-intro.pdf`. 274

Allemang, D. and J. Hendler 2007. *Semantic Web for the Working Ontologist: Effective Modeling in RDFS and OWL* (second ed.). Boston: Morgan Kaufmann.

Allen, M. J. and W. M. Yen 2002. *Introduction to Measurement Theory*. Prospect Heights, Ill.: Waveland Press.

Allison, P. D. 2010. *Survival Analysis Using SAS: A Practical Guide* (second ed.). Cary, N.C.: SAS Institute Inc.

Anand, S. S. and A. G. Büchner 2002. Database marketing and web mining. In W. Klösgen and J. M. Żytkow (eds.), *Handbook of Data Mining and Knowledge Discovery*, Chapter 46.1, pp. 843–849. Oxford: Oxford University Press.

Andersen, P. K., Ø. Borgan, R. D. Gill, and N. Keiding 1993. *Statistical Models Based on Counting Processes*. New York: Springer.

Armstrong, J. S. (ed.) 2001. *Principles of Forecasting: A Handbook for Researchers and Practitioners*. Boston: Kluwer.

Arnould, E. J. and L. L. Price 1993, June. River magic: Extraordinary experience and the extended service encounter. *Journal of Consumer Research* 20:24–45. 351

Arnould, E. J. and M. Wallendorf 1994, November. Market-oriented ethnography: Interpretation building and marketing strategy formulation. *Journal of Marketing Research* 31(4):484–504.

Asur, S. and B. A. Huberman 2010. Predicting the future with social media. In *Proceedings of the 2010 IEEE/WIC/ACM International Conference on Web Intelligence and Intelligent Agent Technology - Volume 01*, WI-IAT '10, pp. 492–499. Washington, DC, USA: IEEE Computer Society. http://dx.doi.org/10.1109/WI-IAT.2010.63. 284

Athanasopoulos, G., R. A. Ahmed, and R. J. Hyndman 2009. Hierarchical forecasts for Australian domestic tourism. *International Journal of Forecasting* 25:146–166.

Bache, K. and M. Lichman 2013. UCI machine learning repository. `http://archive.ics.uci.edu/ml`. 355

Bacon, L. D. 2002. Marketing. In W. Klösgen and J. M. Żytkow (eds.), *Handbook of Data Mining and Knowledge Discovery*, Chapter 34, pp. 715–725. Oxford: Oxford University Press.

Baeza-Yates, R. and B. Ribeiro-Neto 1999. *Modern Information Retrieval*. New York: ACM Press.

Baghestani, H. 1991. Cointegration analysis of the advertising-sales relationship. *Journal of Industrial Economics* 39:671–681.

Bagozzi, R. P. and Y. Yi 1991. Multitrait-multimethod matrices in consumer research. *Journal of Consumer Research* 17:426–439. 295

Bahga, A. and V. Madisetti 2015a. *Cloud Computing: A Hands-On Approach*. Atlanta: Bahga and Madisetti. `www.cloudcomputingbook.info`.

Bahga, A. and V. Madisetti 2015b. *Internet of Things: A Hands-On Approach*. Atlanta: Bahga and Madisetti. `www.internet-of-things-book.com`.

Baker, B. O., C. D. Hardyck, and L. F. Petrinovich 1966. Weak measurements vs. strong statistics: An empirical critique of S. S. Stevens' proscriptions on statistics. *Educational and Psychological Measurement* 26:291–309. Reprinted in Mehrens and Ebel (1967).

Banks, S. 1965. *Experimentation in Marketing*. New York: McGraw-Hill.

Barabási, A.-L. 2003. *Linked: How Everything is Connected to Everything Else*. New York: Penguin/Plume.

Barabási, A.-L. 2010. *Bursts: The Hidden Pattern Behind Everything We Do*. New York: Penguin/Dutton.

Barabási, A.-L. and R. Albert 1999. Emergence of scaling in random networks. *Science* 286 (5439):509–512.

Barney, J. B. 1997. *Gaining and Sustaining Competitive Advantage*. Reading, Mass.: Addison-Wesley.

Bass, F. M. 1969, January. A new product growth model for consumer durables. *Management Science* 15:215–227.

Bates, D. M. and D. G. Watts 2007. *Nonlinear Regression Analysis and Its Applications*. New York: Wiley.

Baumohl, B. 2008. *The Secrets of Economic Indicators: Hidden Clues to Future Economic Trends and Investment Opportunities* (second ed.). Upper Saddle River, N.J.: Pearson.

Beazley, D. M. 2009. *Python Essential Reference* (fourth ed.). Upper Saddle River, N.J.: Pearson Education.

Beazley, D. M. and B. K. Jones 2013. *Python Cookbook* (third ed.). Sebastopol, Calif.: O'Reilly.

Becker, R. A. and W. S. Cleveland 1996. *S-Plus Trellis™ Graphics User's Manual*. Seattle: MathSoft, Inc. 276

Behnel, S. 2014, September 10. lxml. Supported at `http://lxml.de/`. Documentation retrieved from the World Wide Web, October 26, 2014, at `http://lxml.de/3.4/lxmldoc-3.4.0.pdf`.

Belew, R. K. 2000. *Finding Out About: A Cognitive Perspective on Search Engine Technology and the WWW*. Cambridge: Cambridge University Press.

Belk, R. W. 2001. *Collecting in a Consumer Society*. New York: Routledge. 351

Belk, R. W., J. F. Sherry, Jr., and M. Wallendorf 1988. A naturalistic inquiry into buyer and seller behavior at a swap meet. *Journal of Consumer Research* 14(4):449–470. 351

Belsley, D. A., E. Kuh, and R. E. Welsch 1980. *Regression Diagnostics: Identifying Influential Data and Sources of Collinearity*. New York: Wiley.

Ben-Akiva, M. and S. R. Lerman 1985. *Discrete Choice Analysis: Theory and Application to Travel Demand*. Cambridge: MIT Press.

Berk, R. A. 2008. *Statistical Learning from a Regression Perspective*. New York: Springer.

Berndt, E. R. 1991. *The Practice of Econometrics*. Reading, Mass.: Addison-Wesley.

Berners-Lee, T. 1999. *Weaving the Web: The Original Design and Ultimate Destiny of the World Wide Web*. San Francisco: HarperCollins.

Berry, M. and M. Browne 2005, October. Email surveillance using non-negative matrix factorization. *Computational and Mathematical Organization Theory* 11(3):249–264.

Berry, M. W. and M. Browne 1999. *Understanding Search Engines: Mathematical Modeling and Text Retrieval*. Philadelphia: Society for Industrial and Applied Mathematics.

Berthold, M. R., N. Cebron, F. Dill, T. R. Gabriel, T. Kötter, T. Meinl, P. Ohl, C. Sieb, K. Thiel, and B. Wiswedel 2007. KNIME: The Konstanz Information Miner. In *Studies in Classification, Data Analysis, and Knowledge Organization (GfKL 2007)*. New York: Springer. ISSN 1431-8814. ISBN 978-3-540-78239-1.

Besanko, D., D. Dranove, and M. Shanley 2006. *Economics of Strategy* (fourth ed.). New York: Wiley.

Bessen, J. 1993, September–October. Riding the marketing information wave. *Harvard Business Review*:150–160.

Best, R. J. 2002. *Market-Based Management: Strategies for Growing Customer Value and Profitability* (third ed.). Upper Saddle River, N.J.: Prentice Hall.

Betz, N. E. and D. J. Weiss 2001. Validity. In B. Bolton (ed.), *Handbook of Measurement and Evaluation in Rehabilitation* (third ed.)., pp. 49–73. Gaithersburg, Md.: Aspen Publishers.

Bhattacharyya, M. N. 1982. Lydia Pinkham data remodeled. *Journal of Time Series Analysis* 3:81–102.

Bickart, B. and D. Schmittlein 1999, May. The distribution of survey contact and participation in the United States: Constructing a survey-based estimate. *Journal of Marketing Research* 26:286–294.

Bird, S., E. Klein, and E. Loper 2009. *Natural Language Processing with Python: Analyzing Text with the Natural Language Toolkit*. Sebastopol, Calif.: O'Reilly. `http://www.nltk.org/book/`. 297

Bishop, C. M. 1995. *Neural Networks for Pattern Recognition*. Oxford: Oxford University Press.

Bishop, Y. M. M., S. E. Fienberg, and P. W. Holland 1975. *Discrete Multivariate Analysis: Theory and Practice*. Cambridge: MIT Press.

Bivand, R. A., E. J. Pebesma, and V. Gómez-Rubio 2008. *Applied Spatial Data Analysis with R*. New York: Springer. 242

Bizer, C., J. Lehmann, G. Kobilarov, S. Auer, C. Becker, R. Cyganiak, and S. Hellmann 2009. DBpedia—A crystallization point for the web of data. *Web Semantics* 7:154–165.

Blankenship, A. B., G. E. Breen, and A. Dutka 1998. *State of the Art Marketing Research* (second ed.). Chicago: American Marketing Association.

Blattberg, R. C., B.-D. Kim, and S. A. Neslin 2008. *Database Marketing: Analyzing and Managing Customers*. New York: Springer.

Blei, D., A. Ng, and M. Jordan 2003. Latent Dirichlet allocation. *Journal of Machine Learning Research* 3:993–1022. 275

Bock, R. D. and L. V. Jones 1968. *The Measurement and Prediction of Judgment and Choice*. San Francisco: Holden-Day.

Bogdanov, P. and A. Singh 2013. Accurate and scalable nearest neighbors in large networks based on effective importance. *CIKM '13*:1–10. ACM CIKM '13 conference paper retrieved from the World Wide Web at `http://www.cs.ucsb.edu/~petko/papers/bscikm13.pdf`.

Bojanowski, M. 2014. *Package intergraph: Coercion Routines for Network Data Objects in R*. Comprehensive R Archive Network. 2014. `http://cran.r-project.org/web/packages/intergraph/intergraph.pdf`.

Bollen, J., H. Mao, and X. Zeng 2011. Twitter mood predicts the stock market. *Journal of Computational Science* 2:1–8.

Bolter, J. D. and R. Grusin 1999. *Remediation: Understanding New Media*. Cambridge: MIT Press. 349

Borg, I. and P. J. F. Groenen 2010. *Modern Multidimensional Scaling: Theory and Applications* (second ed.). New York: Springer. 96

Borgatti, S. P., M. G. Everett, and J. C. Johnson 2013. *Analyzing Social Networks*. Thousand Oaks, Calif.: Sage. 199

Boser, B. E., I. M. Guyon, and V. N. Vapnik 1992. A training algorithm for optimal margin classifiers. In *Proceedings of the Fifth Conference on Computational Learning Theory*, pp. 144–152. Association for Computing Machinery Press. 78

Box, G. E. P. and D. R. Cox 1964. An analysis of transformations. *Journal of the Royal Statistical Society, Series B (Methodological)* 26(2):211–252.

Box, G. E. P., W. G. Hunter, and J. S. Hunter 2005. *Statistics for Experimenters: Design, Innovation, and Discovery* (second ed.). New York: Wiley.

Box, G. E. P., G. M. Jenkins, and G. C. Reinsel 2008. *Time Series Analysis: Forecasting and Control* (fourth ed.). New York: Wiley. 287

Boyd, D. 2014. *It's Complicated: The Social Lives of Networked Teens*. New Haven, Conn.: Yale University Press.

Boztug, Y. and T. Reutterer 2008. A combined approach for segment-specific market basket analysis. *European Journal of Operational Research* 187(1):294–312. 153

Brandes, U. and T. Erlebach (eds.) 2005. *Network Analysis: Methodological Foundations*. New York: Springer.

Brandes, U., L. C. Freeman, and D. Wagner 2014. Social networks. In R. Tamassia (ed.), *Handbook of Graph Drawing and Visualization*, Chapter 26, pp. 805–840. Boca Raton, Fla.: CRC Press/Chapman & Hall.

Brandt, D. R., K. Gupta, and J. Roberts 2004, Fall. Burden of proof: customer-centric measures must be linked to financial performance. *Marketing Research* 16(3):14–20.

Brealey, R., S. Myers, and F. Allen 2013. *Principles of Corporate Finance*. New York: McGraw-Hill/Irwin.

Breese, J. S., D. Heckerman, and C. M. Kadie 1998, May. Empirical analysis of predictive algorithms for collaborative filtering. Technical Report MSR-TR-98-12, Microsoft Research. 18 pp. `http://research.microsoft.com/apps/pubs/default.aspx?id=69656`. 355

Breiman, L. 2001. Statistical modeling: The two cultures. *Statistical Science* 16(3):199–215.

Breiman, L., J. H. Friedman, R. A. Olshen, and C. J. Stone 1984. *Classification and Regression Trees*. New York: Chapman & Hall.

Brin, S. and L. Page 1998. The anatomy of a large-scale hypertextual web search engine. *Computer Networks and ISDN Systems* 30:107–117.

Broughton, D. 2012, November 12–18. Everybody loves bobbleheads. *Sports Business Journal*:9. Retrieved from the World Wide Web at `http://www.sportsbusinessdaily.com/Journal/Issues/2012/11/12/Research-and-Ratings/Bobbleheads.aspx`. 372

Brown, F. G. 1976. *Principles of Educational and Psychological Testing* (Second ed.). New York: Holt, Rinehart, and Winston. 297

Brownlee, J. 2011. *Clever Algorithms: Nature-Inspired Programming Recipies*. Melbourne, Australia: Creative Commons. `http://www.CleverAlgorithms.com`. 275

Bruzzese, D. and C. Davino 2008. Visual mining of association rules. In S. Simoff, M. H. Böhlen, and A. Mazeika (eds.), *Visual Data Mining: Theory, Techniques and Tools for Visual Analytics*, pp. 103–122. New York: Springer.

Buchanan, M. 2002. *Nexus: Small Worlds and the Groundbreaking Science of Networks*. New York: W.W. Norton and Company.

Bühlmann, P. and S. van de Geer 2011. *Statistics for High-Dimensional Data: Methods, Theory and Applications*. New York: Springer. 241, 272

Burnham, K. P. and D. R. Anderson 2002. *Model Selection and Multimodel Inference: A Practical Information-Theoretic Approach* (second ed.). New York: Springer-Verlag.

Butts, C. T. 2008a, May 5. network: A package for managing relational data in R. *Journal of Statistical Software* 24(2):1–36. `http://www.jstatsoft.org/v24/i02`. Updated version available at `http://cran.r-project.org/web/packages/network/vignettes/networkVignette.pdf`.

Butts, C. T. 2008b, May 8. Social network analysis with sna. *Journal of Statistical Software* 24(6):1–51. `http://www.jstatsoft.org/v24/i06`.

Butts, C. T. 2014a. *Package network: Classes for Relational Data*. Comprehensive R Archive Network. 2014. `http://cran.r-project.org/web/packages/network/network.pdf`.

Butts, C. T. 2014b. *Package snad: Statistical Analysis of Network Data with R*. Comprehensive R Archive Network. 2014. `http://cran.r-project.org/web/packages/sand/sand.pdf`.

Butts, C. T. 2014c. *Package sna: Tools for Social Network Analysis*. Comprehensive R Archive Network. 2014. `http://cran.r-project.org/web/packages/sna/sna.pdf`. Also see the statnet website at `http://www.statnetproject.org/`.

Buvač, V. and P. J. Stone 2001, April 2. The General Inquirer user's guide. Software developed with the support of Harvard University and The Gallup Organization.

Caines, P. E., S. P. Sethi, and T. W. Brotherton 1977. Impulse response identification and causality detection for the Lydia-Pinkham data. *Annals of Economic and Social Measurements* 6:147–163.

Cairo, A. 2013. *The Functional Art: An Introduction to Information Graphics and Visualization*. Berkeley, Calif: New Riders.

Cameron, A. C. and P. K. Trivedi 1998. *Regression Analysis of Count Data*. Cambridge: Cambridge University Press.

Cami, A. and N. Deo 2008. Techniques for analyzing dynamic random graph models of web-like networks: An overview. *Networks* 51(4):211–255.

Campbell, D. T. and D. W. Fiske 1959. Convergent validity and discriminant validity by the multitrait-multimethod matrix. *Psychological Bulletin* 56:81–105.

Campbell, D. T. and J. C. Stanley 1963. *Experimental and Quasi-Experimental Designs for Research*. Skokie, IL: Rand McNally.

Campbell, S. and S. Swigart 2014. *Going beyond Google: Gathering Internet Intelligence* (5 ed.). Oregon City, Oreg.: Cascade Insights. `http://www.cascadeinsights.com/gbgdownload`. 225

Cantelon, M., M. Harter, T. J. Holowaychuk, and N. Rajlich 2014. *Node.js in Action*. Shelter Island, N.Y.: Manning.

Carley, K. M. and D. Skillicorn 2005, October. Special issue on analyzing large scale networks: The Enron corpus. *Computational and Mathematical Organization Theory* 11(3): 179–181.

Carlin, B. P. and T. A. Louis 1996. *Bayes and Empirical Bayes Methods for Data Analysis*. London: Chapman & Hall. 264

Carr, D., N. Lewin-Koh, and M. Maechler 2014. *hexbin: Hexagonal Binning Routines*. Comprehensive R Archive Network. 2014. `http://cran.r-project.org/web/packages/hexbin/hexbin.pdf`. 276

Carrington, P. J., J. Scott, and S. Wasserman (eds.) 2005. *Models and Methods in Social Network Analysis*. Cambridge, UK: Cambridge University Press.

Carroll, J. D. and P. E. Green 1995. Psychometric methods in marketing research: Part I, conjoint analysis. *Journal of Marketing Research* 32:385–391.

Carroll, J. D. and P. E. Green 1997. Psychometric methods in marketing research: Part II, multidimensional scaling. *Journal of Marketing Research* 34:193–204. 96

Cavan, S. 1966. *Liquor License: An Ethnography of Bar Behavior*. Chicago: Aldine. 351

Celsi, R. L., R. L. Rose, and T. W. Leigh 1993. An exploration of high-risk leisure consumption through skydiving. *Journal of Consumer Research* 20:1–23. 351

Cespedes, F. V., J. P. Dougherty, and B. S. Skinner, III 2013, Winter. How to identify the best customers for your business. *MIT Sloan Management Review* 54(2):53–59.

Chakrabarti, S. 2003. *Mining the Web: Discovering Knowledge from Hypertext Data*. San Francisco: Morgan Kaufmann.

Chakrapani, C. (ed.) 2000. *Marketing Research: State of the Art Perspectives*. Chicago: American Marketing Association.

Chambers, J. M. and T. J. Hastie (eds.) 1992. *Statistical Models in S*. Pacific Grove, Calif.: Wadsworth & Brooks/Cole. Champions of S, S-Plus, and R call this "the white book." It introduced statistical modeling syntax using S3 classes.

Chan, E. P. 2009. *Quantitative Trading: How to Build Your Own Algorithmic Trading Business*. New York: Wiley.

Chan, E. P. 2013. *Algorithmic Trading: Winning Strategies and Their Rationale*. New York: Wiley.

Chan, P. 2000. Sampling techniques. In C. Chakrapani (ed.), *Marketing Research: State-of-the-Art Perspectives*, Chapter 1, pp. 156–184. American Marketing Association.

Chang, W. 2013. *R Graphics Cookbook*. Sebastopol, Calif.: O'Reilly.

Chapanond, A., M. S. Krishnamoorthy, and B. Yener 2005, October. Graph theoretic and spectral analysis of Enron email data. *Computational and Mathematical Organization Theory* 11(3):265–281.

Charniak, E. 1993. *Statistical Language Learning*. Cambridge: MIT Press.

Chatterjee, S. and A. S. Hadi 2012. *Regression Analysis by Example* (fifth ed.). New York: Wiley.

Chaturvedi, A. and J. D. Carroll 1998. A perceptual mapping procedure for analysis of proximity data to determine common and unique product-market structures. *European Journal of Operational Research* 111:268–284.

Chau, M. and H. Chen 2003. Personalized and focused web spiders. In N. Zhong, J. Liu, and Y. Yao (eds.), *Web Intelligence*, Chapter 10, pp. 198–217. New York: Springer. 225

Cherny, L. 1999. *Conversation and Community: Chat in a Virtual World*. Stanford, Calif.: CSLI Publications. 351

Chodorow, K. 2013. *MongoDB: The Definitive Guide* (second ed.). Sebastopol, Calif.: O'Reilly.

Christakis, N. A. and J. H. Fowler 2009. *Connected: The Surprising Power of Our Social Networks and How They Shape Our Lives*. New York: Little Brown and Company.

Christensen, R. 1997. *Log-Linear Models and Logistic-Regression* (second ed.). New York: Springer.

Chun, W. J. 2007. *Core Python Programming* (second ed.). Upper Saddle River, N.J.: Pearson Education.

Churchill, Jr., G. A. and D. Iacobucci 2004. *Marketing Research: Methodological Foundations* (ninth ed.). Mason, Ohio: South-Western.

Cleveland, W. S. 1993. *Visualizing Data*. Murray Hill, N.J.: AT&T Bell Laboratories. Initial documentation for trellis graphics in S-Plus. 276

Cochran, W. G. 1977. *Sampling Techniques*. New York: Wiley.

Cochran, W. G. and G. M. Cox 1957. *Experimental Designs* (Second ed.). New York: Wiley.

Cohen, J. 1960, April. A coefficient of agreement for nominal data. *Educational and Psychological Measurement* 20(1):37–46. 270

Commandeur, J. J. F. and S. J. Koopman (eds.) 2007. *An Introduction to State Space Time Series Analysis*. Oxford: Oxford University Press. 288

Connolly, T. and C. Begg 2015. *Database Systems: A Practical Approach to Design, Implementation, and Management* (sixth ed.). New York: Pearson.

Conway, D. and J. M. White 2012. *Machine Learning for Hackers*. Sebastopol, Calif.: O'Reilly.

Cook, R. D. 1998. *Regression Graphics: Ideas for Studying Regressions through Graphics*. New York: Wiley.

Cook, R. D. 2007. Fisher lecture: Dimension reduction in regression. *Statistical Science* 22: 1–26.

Cook, R. D. and S. Weisberg 1999. *Applied Regression Including Computing and Graphics*. New York: Wiley.

Cook, T. D. and D. T. Campbell 1979. *Quasi-Experimentation: Design & Analysis Issues for Field Settings*. Boston: Houghton Mifflin.

Cooper, L. G. 1999. Market share models. In J. Eliashberg and G. L. Lilien (eds.), *Handbook of Operations Research and Management Science: Vol. 5, Marketing*, Chapter 1, pp. 259–314. New York: Elsevier North Holland. 174

Cooper, L. G. and M. Nakanishi 1988. *Market-Share Analysis*. Norwell, Mass.: Kluwer. 174

Cooper, R. G. 2011. *Winning at New Products: Creating Value Through Innovation* (fourth ed.). New York: Basic Books/Perseus.

Copeland, R. 2013. *MongoDB Applied Design Patterns* (second ed.). Sebastopol, Calif.: O'Reilly.

Cowpertwait, P. S. P. and A. V. Metcalfe 2009. *Introductory Time Series with R*. New York: Springer.

Cox, D. R. 1958. *Planning of Experiments*. New York: Wiley.

Cox, T. F. and M. A. A. Cox 1994. *Multidimensional Scaling*. London: Chapman & Hall. 96

Cressie, N. 1993. *Statistics for Spatial Data* (revised ed.). New York: Wiley. 242

Cristianini, N. and J. Shawe-Taylor 2000. *An Introduction to Support Vector Machines and Other Kernel-Based Learning Methods*. Cambridge: Cambridge University Press.

Crnovrsanin, T., C. W. Mueldera, R. Faris, D. Felmlee, and K.-L. Ma 2014. Visualization techniques for categorical analysis of social networks with multiple edge sets. *Social Networks* 37:56–64.

Crockford, D. 2008. *JavaScript: The Good Parts*. Sebastopol, Calif.: O'Reilly.

Croissant, Y. 2014. *mlogit: Multinomial Logit Model*. Comprehensive R Archive Network. 2014. `http://cran.r-project.org/web/packages/mlogit/mlogit.pdf`. 174

Croll, A. and S. Power 2009. *Complete Web Monitoring: Watching Your Visitors, Performance, Communities & Competitors*. Sebastopol, Calif.: O'Reilly.

Cronbach, L. J. 1951. Coefficient alpha and the internal structure of tests. *Psychometrika* 16:297–334. 297

Cronbach, L. J. 1995. Giving method variance its due. In P. E. Shrout and S. T. Fiske (eds.), *Personality Research, Methods, and Theory: A Festschrift Honoring Donald W. Fiske*, Chapter 10, pp. 145–157. Hillsdale, N.J.: Lawrence Erlbaum Associates.

Csardi, G. 2014a. *Package igraphdata: A Collection of Network Data Sets for the igraph Package*. Comprehensive R Archive Network. 2014. `http://cran.r-project.org/web/packages/igraphdata/igraphdata.pdf`.

Csardi, G. 2014b. *Package igraph: Network Analysis and Visualization*. Comprehensive R Archive Network. 2014. `http://cran.r-project.org/web/packages/igraph/igraph.pdf`.

Datta, A., S. Shulman, B. Zheng, S.-D. Lin, A. Sum, and E.-P. Lim (eds.) 2011. *Social Informatics: Proceedings of the Third International Conference SocInfo 2011*. New York: Springer.

Davenport, T. H. and J. G. Harris 2007. *Competing on Analytics: The New Science of Winning*. Boston: Harvard Business School Press. 259

Davenport, T. H., J. G. Harris, and R. Morison 2010. *Analytics at Work: Smarter Decisions, Better Results*. Boston: Harvard Business School Press. 259

Davies, R. and D. Rogers (eds.) 1984. *Store Location and Store Assessment Research*. New York: Wiley.

Davis, B. H. and J. P. Brewer 1997. *Electronic Discourse: Linguistic Individuals in Virtual Space*. Albany, N.Y.: State University of New York Press. 351

Davis, J. J. 2011. *Advertising Research: Theory & Practice* (second ed.). Upper Saddle River, N.J.: Prentice Hall.

Davison, M. L. 1992. *Multidimensional Scaling*. Melbourne, Fla.: Krieger. 96

Day, G. S. 1986. *Analysis for Strategic Marketing Decisions*. St. Paul, Minn.: West Publishing Company. 253

Day, G. S. 1999. *Market Driven Strategy: Processes for Creating Value*. New York: Free Press.

de Nooy, W., A. Mrvar, and V. Batagelj 2011. *Exploratory Social Network Analysis with Pajek*. Cambridge, U.K.: Cambridge University Press.

Dean, J. and S. Ghemawat 2004. MapReduce: Simplifed Data Processing on Large Clusters. Retrieved from the World Wide Web at `http://static.usenix.org/event/osdi04/tech/full_papers/dean/dean.pdf`.

Dehuri, S., M. Patra, B. B. Misra, and A. K. Jagadev (eds.) 2012. *Intelligent Techniques in Recommendation Systems*. Hershey, Pa.: IGI Global.

Delen, D., R. Sharda, and P. Kumar 2007. Movie forecast guru: A Web-based DSS for Hollywood managers. *Decision Support Systems* 43(4):1151–1170. 284

Demšar, J. and B. Zupan 2013. Orange: Data mining fruitful and fun—A historical perspective. *Informatica* 37:55–60. 398

Di Battista, G., P. Eades, R. Tomassia, and I. G. Tollis 1999. *Graph Drawing: Algorithms for the Visualization of Graphs*. Upper Saddle River, N.J.: Prentice Hall.

Di Battista, G. and M. Rimondini 2014. Computer networks. In R. Tamassia (ed.), *Handbook of Graph Drawing and Visualization*, Chapter 25, pp. 763–804. Boca Raton, Fla.: CRC Press/Chapman & Hall.

Dickson, P. R. 1997. *Marketing Management* (second ed.). Orlando, Fla.: Harcourt Brace & Company. 131

Diesner, J., T. L. Frantz, and K. M. Carley 2005, October. Communication networks from the Enron email corpus. *Computational and Mathematical Organization Theory* 11(3): 201–228.

Diggle, P. J., K.-Y. Liang, and S. L. Zeger 1994. *Analysis of Longitudinal Data*. Oxford: Oxford University Press. 341

Dippold, K. and H. Hruschka 2013. Variable selection for market basket analysis. *Computational Statistics* 28(2):519–539. `http://dx.doi.org/10.1007/s00180-012-0315-3`. ISSN 0943-4062. 153

Doctorow, C. 2014. *Information Doesn't Want to Be Free: Laws of the Internet Age*. San Francisco: McSweeney's.

Dole, D. 1999. CoSmo: A constrained scatterplot smoother for estimating convex, monotonic transformations. *Journal of Business and Economic Statistics* 17(4):444–455. 358

Dover, D. 2011. *Search Engine Optimization Secrets: Do What You Never Thought Possible with SEO*. New York: Wiley.

Draper, N. R. and H. Smith 1998. *Applied Regression Analysis* (third ed.). New York: Wiley.

DuCharme, B. 2013. *Learning SPARQL*. Sebastopol, Calif.: O'Reilly.

Duda, R. O., P. E. Hart, and D. G. Stork 2001. *Pattern Classification* (second ed.). New York: Wiley.

Dumais, S. T. 2004. Latent semantic analysis. In B. Cronin (ed.), *Annual Review of Information Science and Technology*, Volume 38, Chapter 4, pp. 189–230. Medford, N.J.: Information Today.

Durbin, J. and S. J. Koopman 2012. *Time Series Analysis by State Space Methods* (second ed.). New York: Oxford University Press. 288

Easley, D. and J. Kleinberg 2010. *Networks, Crowds, and Markets: Reasoning about a Highly Connected World*. Cambridge, UK: Cambridge University Press.

Edmunds, H. 1999. *The Focus Group Research Handbook*. Chicago: NTC Business Books.

Efron, B. 1986. Why isn't everyone a Bayesian (with commentary). *The American Statistician* 40(1):1–11.

Ehrenberg, A. and J. Bound 2000. Turning data into knowledge. In C. Chakrapani (ed.), *Marketing Research: State-of-the-Art Perspectives*, Chapter 2, pp. 23–46. Chicago: American Marketing Association. 209

Ehrenberg, A. S. C. 1988. *Repeat Buying: Facts, Theory, and Data* (second ed.). New York: Oxford University Press.

Eichenwald, K. 2005. *Conspiracy of Fools: A True Story*. New York: Broadway Books/Random House.

Elff, M. 2014. *mclogit: Mixed Conditional Logit*. Comprehensive R Archive Network. 2014. `http://cran.r-project.org/web/packages/mclogit/mclogit.pdf`. 174

Ellen, R. F. (ed.) 1984. *Ethnographic Research: A Guide to General Conduct*. London: Academic Press.

Ellis, D., R. Oldridge, and A. Vasconcelos 2004. Community and virtual community. In B. Cronin (ed.), *Annual Review of Information Science and Technology*, Volume 38, Chapter 3, pp. 145–186. Medford, N.J.: Information Today.

Enders, W. 2010. *Applied Econometric Time Series* (third ed.). New York: Wiley.

Engelbrecht, A. P. 2007. *Computational Intelligence: An Introduction* (second ed.). New York: Wiley. 275

Erdős, P. and A. Rényi 1959. On random graphs. *Publicationes Mathematicae* 6:290–297.

Erdős, P. and A. Rényi 1960. On the evolution of random graphs. *Publications of the Mathematical Institute of the Hungarian Academy of Sciences* 5:17–61.

Erdős, P. and A. Rényi 1961. On the strength of connectedness of a random graph. *Acta Mathematica Scientia Hungary* 12:261–267.

Erwig, M. 2000. The graph Voronoi diagram with applications. *Networks* 36(3):156–163.

Esipova, N., T. W. Miller, M. D. Zarnecki, J. Elzaurdia, and S. Ponnaiya 2004. Exploring the possibilities of online focus groups. In T. W. Miller and J. Walkowski (eds.), *Qualitative Research Online*, Chapter A, pp. 107–128. Manhattan Beach, Calif.: Research Publishers.

Everitt, B. S., S. Landau, M. Leese, and D. Stahl 2011. *Cluster Analysis* (fifth ed.). New York: Wiley.

Fader, P. S. and B. G. S. Hardie 1996, November. Modeling consumer choice among SKUs. *Journal of Marketing Research* 33:442–452.

Fader, P. S. and B. G. S. Hardie 2002. A note on an integrated model of consumer buying behavior. *European Journal of Operational Research* 139(3):682–687.

Fawcett, T. 2003, January 7. ROC graphs: Notes and practical considerations for researchers. http://www.hpl.hp.com/techreports/2003/HPL-2003-4.pdf.

Feinerer, I. 2014. *Introduction to the tm Package*. Comprehensive R Archive Network. 2014. `http://cran.r-project.org/web/packages/tm/vignettes/tm.pdf`.

Feldman, D. M. 2002a, Winter. The pricing puzzle. *Marketing Research*:14–19.

Feldman, R. 2002b. Text mining. In W. Klösgen and J. M. Żytkow (eds.), *Handbook of Data Mining and Knowledge Discovery*, Chapter 38, pp. 749–757. Oxford: Oxford University Press.

Feldman, R. 2013. Techniques and applications for sentiment analysis. *Communications of the ACM* 56(4):82–89.

Feldman, R. and J. Sanger 2007. *The Text Mining Handbook: Advanced Approaches in Analyzing Unstructured Data*. Cambridge, UK: Cambridge University Press.

Fellows, I. 2014. *wordcloud: Word Clouds*. Comprehensive R Archive Network. 2014. `http://cran.r-project.org/web/packages/wordcloud/wordcloud.pdf`.

Ferrucci, D., E. Brown, J. Chu-Carroll, J. Fan, D. Gondek, A. A. Kalyanpur, A. Lally, J. W. Murdock, E. Nyberg, J. Prager, N. Schlaefer, and C. Welty 2010, Fall. Building Watson: An overview of the DeepQA project. *AI Magazine* 31(3):59–79.

Few, S. 2009. *Now You See It: Simple Visualization Techniques and Quantitative Analysis*. Oakland, Calif.: Analytics Press.

Fienberg, S. E. 2007. *Analysis of Cross-Classified Categorical Data* (second ed.). New York: Springer.

Firth, D. 1991. Generalized linear models. In D. Hinkley and E. Snell (eds.), *Statistical Theory and Modeling: In Honour of Sir David Cox, FRS*, Chapter 3, pp. 55–82. London: Chapman and Hall.

Fisher, R. A. 1970. *Statistical Methods for Research Workers* (fourteenth ed.). Edinburgh: Oliver and Boyd. First edition published in 1925. 264

Fisher, R. A. 1971. *Design of Experiments* (ninth ed.). New York: Macmillan. First edition published in 1935. 264

Fiske, D. W. 1971. *Measuring the Concepts of Personality*. Chicago: Aldine. 297

Flam, F. 2014, September 30. The odds, continually updated. *The New York Times*:D1. Retrieved from the World Wide Web at `http://www.nytimes.com/2014/09/30/science/the-odds-continually-updated.html?_r=0`. 264

Fleisher, C. S. and B. E. Bensoussan 2003. *Strategic and Competitive Analysis: Methods and Techniques for Analyzing Business Competition*. Upper Saddle River, N.J.: Prentice Hall.

Foster, G. 2013, August 26. Behold the power of a free bobblehead doll. *The Wall Street Journal*. Retrieved from the World Wide Web at `http://online.wsj.com/article/SB10001424127887323407104579037182599696524.html`. 372

Fournier, S., S. Dobscha, and D. G. Mick 1998, January–February. Preventing the premature death of relationship marketing. *Harvard Business Review* 76(3):43–51.

Fournier, S. and D. G. Mick 1999, October. Rediscovering satisfaction. *Journal of Marketing* 63(4):5–23.

Fox, J. 2002, January. Robust regression: Appendix to an R and S-PLUS companion to applied regression. Retrieved from the World Wide Web at `http://cran.r-project.org/doc/contrib/Fox-Companion/appendix-robust-regression.pdf`. 273

Fox, J. 2014. *car: Companion to Applied Regression*. Comprehensive R Archive Network. 2014. `http://cran.r-project.org/web/packages/car/car.pdf`.

Fox, J. and S. Weisberg 2011. *An R Companion to Applied Regression* (second ed.). Thousand Oaks, Calif.: Sage.

Franceschet, M. 2011, June. PageRank: Standing on the shoulders of giants. *Communications of the ACM* 54(6):92–101.

Franchi, E. 2014. An agent-based modeling framework for social network simulation. In F. Can, T. Özyer, and F. Polat (eds.), *State of the Art Applications of Social Network Analysis*, Chapter 4, pp. 75–96. New York: Springer.

Frank, R. E., W. F. Massey, and Y. Wind 1972. *Market Segmentation*. Englewood Cliffs, N.J.: Prentice Hall.

Franks, B. 2012. *Taming the Big Data Tidal Wave: Finding Opportunities in Huge Data Streams with Advanced Analytics*. Hoboken, N.J.: Wiley. 259

Franses, P. H. and R. Paap 2001. *Quantitative Models in Marketing Research*. Cambridge: Cambridge University Press.

Freeman, L. C. 2004. *The Development of Social Network Analysis: A Study in the Sociology of Science*. Vancouver, B.C.: Empirical Press.

Freeman, L. C. 2005. Graphic techniques for exploring social network data. In P. J. Carrington, J. Scott, and S. Wasserman (eds.), *Models and Methods in Social Network Analysis*, Chapter 12, pp. 248–269. Cambridge, UK: Cambridge University Press.

Frees, E. W. and T. W. Miller 2004. Sales forecasting with longitudinal data models. *International Journal of Forecasting* 20:99–114.

Friendly, M. 2000. *Visualizing Categorical Data*. Cary, N.C.: SAS Institute.

Fruchterman, T. M. J. and E. M. Reingold 1991. Graph drawing by force-directed placement. *Software—Practice and Experience* 21(11):1129–1164.

Fry, B. 2008. *Visualizing Data: Exploring and Explaining Data with the Processing Environment*. Sebastopol, Calif.: O'Reilly.

Gabriel, K. R. 1971. The biplot graphical display of matrices with application to principal component analysis. *Biometrika* 58:453–467. 96

Garcia-Molina, H., J. D. Ullman, and J. Widom 2009. *Database Systems: The Complete Book* (second ed.). Upper Saddle River, N.J.: Prentice-Hall.

Gasston, P. 2011. *The Book of CSS3: A Developer's Guide to the Future of Web Design*. San Francisco: No Starch Press.

Gasston, P. 2013. *The Modern Web: Multi-Device Web Development with HTML5, CSS3, and JavaScript*. San Francisco: No Starch Press.

Gauntlett, D. (ed.) 2000. *web.studies*. London: Arnold. 349

Geisser, S. 1993. *Predictive Inference: An Introduction*. New York: Chapman & Hall. 259, 264

Gelman, A., J. B. Carlin, H. S. Stern, and D. B. Rubin 1995. *Bayesian Data Analysis*. London: Chapman & Hall. 264

Gelman, A. and J. Hill 2007. *Data Analysis Using Regression and Mulitlevel/Hierarchical Models*. Cambridge, UK: Cambridge University Press. 341

Gelman, A., J. Hill, Y.-S. Su, M. Yajima, and M. G. Pittau 2014. *mi: Missing Data Imputation and Model Checking*. Comprehensive R Archive Network. 2014. `http://cran.r-project.org/web/packages/mi/mi.pdf`.

Ghiselli, E. E. 1964. *Theory of Psychological Measurement*. New York: McGraw-Hill. 297

Gibson, H., J. Faith, and P. Vickers 2013, 07. A survey of two-dimensional graph layout techniques for information visualisation. *Information Visualization* 12(3-4):324–357.

Gilley, O. and R. Pace 1996. On the Harrison and Rubinfeld data. *Journal of Environmental Economics and Management* 31:403–405.

Gnanadesikan, R. 1997. *Methods for Statistical Data Analysis of Multivariate Observations* (second ed.). New York: Wiley.

Goldenberg, A., A. X. Zheng, S. E. Fienberg, and E. M. Airoldi 2009, January. A survey of statistical network models. *Foundations and Trends in Machine Learning* 2(2):129–233.

Goodreau, S. M., M. S. Handcock, D. R. Hunter, C. T. Butts, and M. Morris 2008, 5 8. A statnet tutorial. *Journal of Statistical Software* 24(9):1–26. `http://www.jstatsoft.org/v24/i09`. CODEN JSSOBK. ISSN 1548-7660.

Gower, J. C. and D. J. Hand 1996. *Biplots*. London: Chapman & Hall. 96

Granger, C. W. 1969. Investigating causal relations by econometric models and cross-spectral methods. *Econometrica* 37:424–438.

Granovetter, M. S. 1973. The strength of weak ties. *American Journal of Sociology* 78:1360–1380.

Graybill, F. A. 1961. *Introduction to Linear Statistical Models, Volume 1*. New York: McGraw-Hill.

Graybill, F. A. 2000. *Theory and Application of the Linear Model*. Stamford, Conn.: Cengage Learning.

Greene, W. H. 2012. *Econometric Analysis* (seventh ed.). Upper Saddle River, N.J.: Pearson Prentice Hall. 173

Grieve, T. 2003, October 14. The decline and fall of the Enron empire. *Salon*:165–178. Retrieved from the World Wide Web at `http://www.salon.com/2003/10/14/enron_22/`.

Grolemund, G. and H. Wickham 2011, April 7. Dates and times made easy with lubridate. *Journal of Statistical Software* 40(3):1–25. `http://www.jstatsoft.org/v40/i03`.

Grolemund, G. and H. Wickham 2014. *lubridate: Make Dealing with Dates a Little Easier*. Comprehensive R Archive Network. 2014. `http://cran.r-project.org/web/packages/lubridate/lubridate.pdf`. 228

Gross, J. L., J. Yellen, and P. Zhang (eds.) 2014. *Handbook of Graph Theory* (second ed.). Boca Raton, Fla.: CRC Press/Chapman & Hall.

Groves, R. M. and M. P. Couper (eds.) 1998. *Nonresponse in Household Interview Surveys*. Wiley Series in Probability and Statistics. New York: Wiley.

Groves, R. M., F. J. Fowler, Jr., M. P. Couper, J. M. Lepkowski, E. Singer, and R. Tourangeau 2009. *Survey Methodology* (second ed.). New York: Wiley.

Gubrium, J. F. and J. A. Holstein (eds.) 2002. *Handbook of Interview Research: Context & Method*. Thousand Oaks, Calif.: Sage.

Guilford, J. P. 1954. *Psychometric Methods* (second ed.). New York: McGraw-Hill. First edition published in 1936. 337

Gulliksen, H. 1950. *Theory of Mental Tests*. New York: Wiley. 297

Gunning, D., V. K. Chaudhri, P. Clark, K. Barker, S.-Y. Chaw, M. Greaves, B. Grosof, A. Leung, D. McDonald, S. Mishra, J. Pacheco, B. Porter, A. Spaulding, D. Tecuci, and J. Tien 2010, Fall. Project Halo update—Progress toward digital Aristotle. *AI Magazine* 31(3):33–58.

Gustafsson, A., A. Herrmann, and F. Huber (eds.) 2000. *Conjoint Measurement: Methods and Applications*. New York: Springer-Verlag.

Hagbert, A. and D. Schult 2014. *NetworkX: Python Software for Complex Networks*. NetworkX Development Team. 2014. Retrieved from the World Wide Web at `https://github.com/networkx/`.

Hahsler, M. 2014a. *recommenderlab: A Framework for Developing and Testing Recommendation Algorithms*. Comprehensive R Archive Network. 2014. `http://cran.r-project.org/web/packages/recommenderlab/vignettes/recommenderlab.pdf`.

Hahsler, M. 2014b. *recommenderlab: Lab for Developing and Testing Recommender Algorithms*. Comprehensive R Archive Network. 2014. `http://cran.r-project.org/web/packages/recommenderlab/recommenderlab.pdf`.

Hahsler, M., C. Buchta, B. Grün, and K. Hornik 2014a. *arules: Mining Association Rules and Frequent Itemsets*. Comprehensive R Archive Network. 2014. `http://cran.r-project.org/web/packages/arules/arules.pdf`.

Hahsler, M., C. Buchta, B. Grün, and K. Hornik 2014b. *Introduction to arules: A Computational Environment for Mining Association Rules and Frequent Itemsets*. Comprehensive R Archive Network. 2014. `http://cran.r-project.org/web/packages/arules/vignettes/arules.pdf`.

Hahsler, M., C. Buchta, and K. Hornik 2008. Selective association rule generation. *Computational Statistics* 23:303–315. 153

Hahsler, M. and S. Chelluboina 2014a. *arulesViz: Visualizing Association Rules and Frequent Itemsets*. Comprehensive R Archive Network. 2014. `http://cran.r-project.org/web/packages/arulesViz/arulesViz.pdf`.

Hahsler, M. and S. Chelluboina 2014b. *Visualizing Association Rules: Introduction to the R-extension Package arulesViz*. Comprehensive R Archive Network. 2014. `http://cran.r-project.org/web/packages/arulesViz/vignettes/arulesViz.pdf`.

Hahsler, M., S. Chelluboina, K. Hornik, and C. Buchta 2011. The arules R-package ecosystem: Analyzing interesting patterns from large transaction data sets. *Journal of Machine Learning Research* 12:2021–2025.

Hahsler, M., B. Grün, and K. Hornik 2005, September 29. arules: A computational environment for mining association rules and frequent item sets. *Journal of Statistical Software* 14(15):1–25. `http://www.jstatsoft.org/v14/i15`.

Hahsler, M., K. Hornik, and T. Reutterer 2006. Implications of probabilistic data modeling for mining association rules. In M. Spiliopoulou, R. Kruse, C. Borgelt, A. Nuernberger, and W. Gaul (eds.), *Data and Information Analysis to Knowledge Engineering, Studies in Classification, Data Analysis, and Knowledge Organization*, pp. 598–605. New York: Springer.

Hamilton, J. D. 1994. *Time Series Analysis*. Princeton, N.J.: Princeton University Press.

Hand, D. J. 1997. *Construction and Assessment of Classification Rules*. New York: Wiley.

Handcock, M. S. and K. Gile 2007, April 28. Modeling social networks with sampled or missing data. `http://www.csss.washington.edu/Papers/wp75.pdf`. 311

Handcock, M. S., D. R. Hunter, C. T. Butts, S. M. Goodreau, and M. Morris 2008, May 5. statnet: Software tools for the representation, visualization, analysis and simulation of network data. *Journal of Statistical Software* 24(1):1–11. `http://www.jstatsoft.org/v24/i01`.

Hanneman, R. A. and M. Riddle 2005. *Introduction to Social Network Methods*. Riverside, Calif.: University of California Riverside. `http://www.faculty.ucr.edu/~hanneman/nettext/`.

Hanssens, D. M., L. J. Parsons, and R. L. Schultz 2001. *Market Response Models: Econometric and Time Series Analysis* (second ed.). Boston: Kluwer.

Harrell, Jr., F. E. 2001. *Regression Modeling Strategies: With Applications to Linear Models, Logistic Regression, and Survival Analysis*. New York: Springer.

Harrington, P. 2012. *Machine Learning in Action*. Shelter Island, N.Y.: Manning. 153

Harrison, D. and D. Rubinfeld 1978. Hedonic housing prices and the demand for clean air. *Journal of Environmental Economics and Management* 5:81–102. 358

Hartigan, J. A. and B. Kleiner 1984. A mosaic of television ratings. *The American Statistician* 38(1):32–35.

Hastie, T. 2013. *Package gam: Generalized Additive Models*. Comprehensive R Archive Network. 2013. `http://cran.r-project.org/web/packages/gam/gam.pdf`.

Hastie, T., R. Tibshirani, and J. Friedman 2009. *The Elements of Statistical Learning: Data Mining, Inference, and Prediction* (second ed.). New York: Springer. 153

Hastie, T. J. 1992. Generalized linear models. In J. M. Chambers and T. J. Hastie (eds.), *Statistical Models in S*, Chapter 6, pp. 195–247. Pacific Grove, Calif.: Wadsworth & Brooks/Cole.

Hausser, R. 2001. *Foundations of Computational Linguistics: Human-Computer Communication in Natural Language* (second ed.). New York: Springer-Verlag.

Hayes, B. E. 1998. *Measuring Customer Satisfaction: Survey Design, Use, and Statistical Analysis Methods* (second ed.). Milwaukee: ASQ Quality Press.

Heer, J., M. Bostock, and V. Ogievetsky 2010, May 1. A tour through the visualization zoo: A survey of powerful visualization techniques, from the obvious to the obscure. *acmqueue: Association for Computing Machinery*:1–22. Retrieved from the World Wide Web at `http://queue.acm.org/detail.cfm?id=1805128`.

Heisley, D. D. and S. J. Levy 1991, December. Autodriving: a photoelicitation technique. *Journal of Consumer Research* 18:257–272. 347

Helmer, R. M. and J. K. Johansson 1977. An exposition of the Box-Jenkins transfer function analysis with an application to advertising-sales relationship. *Journal of Marketing Research* 14:227–239.

Hemenway, K. and T. Calishain 2004. *Spidering Hacks: 100 Industrial-Strength Tips & Tools*. Sabastopol, Calif.: O'Reilly. 225

Hensher, D. A. and L. W. Johnson 1981. *Applied Discrete-Choice Modeling*. New York: Wiley.

Hensher, D. A., J. M. Rose, and W. H. Greene 2005. *Applied Choice Analysis: A Primer*. Cambridge: Cambridge University Press.

Herman, A. and T. Swiss (eds.) 2000. *The World Wide Web and Contemporary Cultural Theory*. New York: Routledge. 351

Hessan, D. 2003, November 13. Telephone interview by Dana H. James.

Heyse, J. F. and W. W. S. Wei 1985. Modeling of the advertising-sales relationship through the use of multiple time series techniques. *Journal of Forecasting* 4:165–181. 369

Higgins, R. 2011. *Analysis for Financial Management*. New York: McGraw-Hill/Irwin.

Hinkley, D. V., N. Reid, and E. J. Snell (eds.) 1991. *Statistical Theory and Modeling*. London: Chapman and Hall. 264

Ho, Q. and E. P. Xing 2014. Analyzing time-evolving networks using an evolving cluster mixed membership blockmodel. In E. M. Airoldi, D. Blei, E. A. Erosheva, and S. E. Fienberg (eds.), *Handbook of Mixed Membership Models and Their Applications*, Chapter 23, pp. 489–525. Boca Raton, Fla.: CRC Press.

Hoberman, S. 2014. *Data Modeling for MongoDB: Building Well-Designed and Supportable MongoDB Databases*. Basking Ridge, N.J.: Technics Publications.

Hoerl, A. E. and R. W. Kennard 2000. Ridge regression: biased estimation for nonorthogonal problems. *Technometrics* 42(1):80–86. Reprinted from *Technometrics*, volume 12. 272

Hoffman, P. and Scrapy Developers 2014, June 26. Scrapy documentation release 0.24.0. Supported at `http://scrapy.org/`. Documentation retrieved from the World Wide Web, October 26, 2014, at `https://media.readthedocs.org/pdf/scrapy/0.24/scrapy.pdf`.

Holden, D. 2006. Hierarchical edge bundles: Visualization of adjacency relations in hierarchical data. *IEEE Transactions on Visualization and Computer Graphics* 12(5):741–748.

Holden, K., D. A. Peel, and J. L. Thompson 1990. *Economic Forecasting: An Introduction*. Cambridge, UK: Cambridge University Press.

Holeton, R. (ed.) 1998. *Composing Cyberspace: Identity, Community, and Knowledge in the Electronic Age*. Boston: McGraw-Hill. 351

Holland, P. W., K. B. Laskey, and S. Leinhardt 1983. Stochastic blockmodels: First steps. *Social Networks* 5:109–138.

Hollis, N. 2005, Fall. Branding unmasked: Expose the mysterious consumer purchase process. *Marketing Research* 17(3):24–29.

Holtzman, S. R. 1994. *Digital Mantras: The Languages of Abstract and Virtual Worlds*. Cambridge: MIT Press. 349

Honaker, J., G. King, and M. Blackwell 2014. *Amelia II: A Program for Missing Data*. Comprehensive R Archive Network. 2014. `http://cran.r-project.org/web/packages/Amelia/Amelia.pdf`.

Honomichl, J. J. 1986. *Honomichl on Marketing Research*. Lincolnwood, Ill.: Crain Books.

Hornik, K. 2014a. *RWeka Odds and Ends*. Comprehensive R Archive Network. 2014. `http://cran.r-project.org/web/packages/RWeka/vignettes/RWeka.pdf`.

Hornik, K. 2014b. *RWeka: R/Weka Interface*. Comprehensive R Archive Network. 2014. `http://cran.r-project.org/web/packages/RWeka/RWeka.pdf`.

Hosmer, D. W., S. Lemeshow, and S. May 2013. *Applied Survival Analysis: Regression Modeling of Time to Event Data* (second ed.). New York: Wiley.

Hosmer, D. W., S. Lemeshow, and R. X. Sturdivant 2013. *Applied Logistic Regression* (third ed.). New York: Wiley.

Hothorn, T., A. Zeileis, R. W. Farebrother, C. Cummins, G. Millo, and D. Mitchell 2014. *lmtest: Testing Linear Regression Models*. Comprehensive R Archive Network. 2014. `http://cran.r-project.org/web/packages/lmtest/lmtest.pdf`.

Houston, B., L. Bruzzese, and S. Weinberg 2002. *The Investigative Reporter's Handbook: A Guide to Documents, Databases and Techniques* (fourth ed.). Boston: Bedford/St. Martin's.

Huber, J. and K. Zwerina 1996. The importance of utility balance in efficient choice designs. *Journal of Marketing Research* 33:307–317.

Huberman, B. A. 2001. *The Laws of the Web: Patterns in the Ecology of Information*. Cambridge: MIT Press. 351

Huet, S., A. Bouvier, M.-A. Poursat, and E. Jolivet 2004. *Statistical Tools for Nonlinear Regression: A Practical Guide with S-Plus and R Examples* (second ed.). New York: Springer.

Hughes, A. M. 2000. *Strategic Database Marketing: The Masterplan for Starting and Managing a Profitable Customer-Based Marketing Program* (second ed.). New York: McGraw-Hill.

Hughes-Croucher, T. and M. Wilson 2012. *Node Up and Running*. Sebastopol, Calif.: O'Reilly.

Huisman, M. 2010. Imputation of missing network data: Some simple procedures. *Journal of Social Structure* 10. `http://www.cmu.edu/joss/content/articles/volume10/huisman.pdf`. 311

Hyman, H. H. 1954. *Interviewing in Social Research*. Chicago: University of Chicago Press.

Hyndman, R. J., R. A. Ahmed, G. Athanasopoulos, and H. L. Shang 2011. Optimal combination forecasts for hierarchical time series. *Computational Statistics and Data Analysis* 55:2579–2589.

Hyndman, R. J. and G. Athanasopoulos 2014. *Forecasting: Principles and Practice*. Online: OTexts. `https://www.otexts.org/fpp`.

Hyndman, R. J., G. Athanasopoulos, S. Razbash, D. Schmidt, Z. Zhou, and Y. Khan 2014. *forecast: Forecasting Functions for Time Series and Linear Models*. Comprehensive R Archive Network. 2014. `http://cran.r-project.org/web/packages/forecast/forecast.pdf`.

Hyndman, R. J., A. B. Koehler, J. K. Ord, and R. D. Snyder 2008. *Forecasting with Exponential Smoothing: The State Space Approach*. New York: Springer. 288

Ihaka, R., P. Murrell, K. Hornik, J. C. Fisher, and A. Zeileis 2014. *colorspace: Color Space Manipulation*. Comprehensive R Archive Network. 2014. `http://cran.r-project.org/web/packages/colorspace/colorspace.pdf`.

Indurkhya, N. and F. J. Damerau (eds.) 2010. *Handbook of Natural Language Processing* (second ed.). Boca Raton, Fla.: Chapman and Hall/CRC.

Ingersoll, G. S., T. S. Morton, and A. L. Farris 2013. *Taming Text: How to Find, Organize, and Manipulate It*. Shelter Island, N.Y.: Manning. 275

Inselberg, A. 1985. The plane with parallel coordinates. *The Visual Computer* 1(4):69–91. 166

Izenman, A. J. 2008. *Modern Multivariate Statistical Techniques: Regression, Classification, and Manifold Learning*. New York: Springer.

Jackson, M. O. 2008. *Social and Economic Networks*. Princeton, N.J.: Princeton University Press.

James, D. H. 2002, March 4. This bulletin just in: Online research technique proving invaluable. *Marketing News*:45–46. 346

James, G., D. Witten, T. Hastie, and R. Tibshirani 2013. *An Introduction to Statistical Learning with Applications in R*. New York: Springer.

Janssen, J., M. Hurshman, and N. Kalyaniwalla 2012. Model selection for social networks using graphlets. *Internet Mathematics* 8(4):338–363.

Johnson, R. A. and D. W. Wichern 1998. *Applied Multivariate Statistical Analysis* (fourth ed.). Upper Saddle River, N.J.: Prentice Hall.

Johnson, S. 1997. *Interface Culture*. New York: Basic Books. 351

Jones, S. G. (ed.) 1997. *Virtual Culture: Identity & Communication in Cybersociety*. Thousand Oaks, Calif.: Sage. 351

Jones, S. G. (ed.) 1999. *Doing Internet Research: Critical Issues and Methods for Examining the Net*. Thousand Oaks, Calif.: Sage. 351

Jonscher, C. 1999. *The Evolution of Wired Life: From the Alphabet to the Soul-Catcher Chip—How Information Technologies Change Our World*. New York: Wiley. 351

Joula, P. 2008. *Authorship Attribution*. Hanover, Mass.: Now Publishers. 285

Judge, G. G., W. E. Griffiths, R. C. Hill, H. Lütkepohl, and T.-C. Lee 1985. *The Theory and Practice of Econometrics* (second ed.). New York: Wiley.

Jünger, M. and P. Mutzel (eds.) 2004. *Graph Drawing Software*. New York: Springer.

Jurafsky, D. and J. H. Martin 2009. *Speech and Language Processing: An Introduction to Natural Language Processing, Computational Linguistics, and Speech Recognition* (second ed.). Upper Saddle River, N.J.: Prentice Hall.

Kadushin, C. 2012. *Understanding Social Networks*. New York: Oxford University Press.

Kamada, T. and S. Kawai 1989. An algorithm for drawing general undirected graphs. *Information Processing Letters* 31(1):7–15.

Katz, G. 2011. Rethinking the product development funnel. *Visions* 25(2):24–31.

Kaufman, L. and P. J. Rousseeuw 1990. *Finding Groups in Data: An Introduction to Cluster Analysis*. New York: Wiley.

Keila, P. and D. B. Skillicorn 2005, October. Structure of the Enron email dataset. *Computational and Mathematical Organization Theory* 11(3):183–199.

Kempthorne, O. 1952. *The Design and Analysis of Experiments*. New York: Wiley. Also see Hinkelmann and Kempthorne (1994).

Kennedy, P. 2008. *A Guide to Econometrics* (sixth ed.). New York: Wiley.

Keppel, G. and T. D. Wickens 2004. *Design and Analysis: A Researcher's Handbook* (fourth ed.). Upper Saddle River: N.J.: Pearson.

Kim, J. H. 2005. Investigating the advertising-sales relationship in the Lydia Pinkham data: A bootstrap approach. *Applied Economics* 37:347–354.

Kirk, R. E. 2013. *Experimental Design: Procedures for the Behavioral Sciences* (fourth ed.). Thousand Oaks, Calif.: Sage.

Kleiber, C. and A. Zeileis 2008. *Applied Econometrics with R*. New York: Springer.

Klimmt, B. and Y. Yang 2004. Introducing the Enron corpus. http://ceas.cc/2004/168.pdf.

Kolaczyk, E. D. 2009. *Statistical Analysis of Network Data: Methods and Models*. New York: Springer. 202

Kolaczyk, E. D. and G. Csárdi 2014. *Statistical Analysis of Network Data with R*. New York: Springer.

Koller, M. 2014. *Simulations for Sharpening Wald-type Inference in Robust Regression for Small Samples*. Comprehensive R Archive Network. 2014. http://cran.r-project.org/web/packages/robustbase/vignettes/lmrob_simulation.pdf. 273

Koller, M. and W. A. Stahel 2011. Sharpening Wald-type inference in robust regression for small samples. *Computational Statistics and Data Analysis* 55(8):2504–2515. 273

Kossinets, G. 2006. Effects of missing data in social networks. *Social Networks* 28(3):247–268. 311

Kotler, P. and K. L. Keller 2012. *Marketing Management* (fourteenth ed.). Upper Saddle River, N.J.: Prentice Hall. 51, 131

Kratochvíl, J. (ed.) 1999. *Graph Drawing: 7th International Symposium / Proceedings GD'99*. New York: Springer.

Krippendorff, K. H. 2012. *Content Analysis: An Introduction to Its Methodology* (third ed.). Thousand Oaks, Calif.: Sage. 283

Krishnamurthi, L. 2001. Pricing strategies and tactics. In D. Iacobucci (ed.), *Kellogg on Marketing*, Chapter 12, pp. 279–301. New York: Wiley.

Krug, S. 2014. *Don't Make Me Think: A Common Sense Approach to Web Usability* (third ed.). Upper Saddle River, N.J.: Pearson/New Riders.

Kuhfeld, W. F., R. D. Tobias, and M. Garratt 1994. Efficient experimental design with marketing research applications. *Journal of Marketing Research* 31:545–557.

Kuhn, M. 2014. *caret: Classification and Regression Training*. Comprehensive R Archive Network. 2014. http://cran.r-project.org/web/packages/caret/caret.pdf.

Kuhn, M. and K. Johnson 2013. *Applied Predictive Modeling*. New York: Springer.

Kutner, M. H., C. J. Nachtsheim, J. Neter, and W. Li 2004. *Applied Linear Statistical Models* (fifth ed.). Boston: McGraw-Hill.

Kvale, S. 1994. Ten standard objections to qualitative research interviews. *Journal of Phenomenological Psychology* 25(2):147–173. 351

Kvale, S. 1996. *InterViews: An Introduction to Qualitative Research Interviewing*. Thousand Oaks, Calif.: Sage.

Lacy, L. W. 2005. *Owl: Representing Information Using the Web Ontology Language*. Victoria, B.C.: Trafford.

Lamp, G. 2014. *ggplot for Python*. GitHub. 2014. https://github.com/yhat/ggplot. 276

Landauer, T. K., P. W. Foltz, and D. Laham 1998. An introduction to latent semantic analysis. *Discourse Processes* 25:259–284. Retrieved from the World Wide Web, October 31, 2014, at http://lsa.colorado.edu/papers/dp1.LSAintro.pdf.

Landauer, T. K., D. S. McNamara, S. Dennis, and W. Kintsch (eds.) 2014. *Handbook of Latent Semantic Analysis* (reprint ed.). New York: Psychology Press.

Lang, D. T. 2014a. *Package RCurl: General Network (HTTP/FTP/...) Client Interface for R*. Comprehensive R Archive Network. 2014. `http://cran.r-project.org/web/packages/RCurl/RCurl.pdf`. 228

Lang, D. T. 2014b. *Package XML: Tools for Parsing and Generating XML within R and S-Plus*. Comprehensive R Archive Network. 2014. `http://cran.r-project.org/web/packages/XML/XML.pdf`. 228

Langer, J. 2001. *The Mirrowed Window: Focus Groups from a Moderator's Point of View*. Ithaca, N.Y.: Paramount Market Publishing.

Langer, J. 2002, November 1. Telephone interview by Dana H. James. 346

Langville, A. N. and C. D. Meyer 2006. *Google's Page Rank and Beyond: The Science of Search Engine Rankings*. Princeton, N.J.: Princeton University Press.

Lantz, B. 2013. *Machine Learning with R*. Birmingham, U.K.: Packt Publishing.

Laursen, G. H. N. and J. Thorlund 2010. *Business Analytics for Managers: Taking Business Intelligence Beyond Reporting*. Hoboken, N.J.: Wiley. 259

Law, A. M. 2014. *Simulation Modeling and Analysis* (fifth ed.). New York: McGraw-Hill.

Le, C. T. 1997. *Applied Survival Analysis*. New York: Wiley.

Le, C. T. 1998. *Applied Categorical Data Analysis*. New York: Wiley.

Leber, J. 2013. The immortal life of the Enron e-mails. *MIT Technology Review*. `http://www.technologyreview.com/news/515801/the-immortal-life-of-the-enron-e-mails/`.

Ledoiter, J. 2013. *Data Mining and Business Analytics with R*. New York: Wiley.

Leeflang, P. S. H., D. R. Wittink, M. Wedel, and P. A. Naert 2000. *Building Models for Marketing Decisions*. Boston: Kluwer.

Leetaru, K. 2011. *Data Mining Methods for Content Analysis: An Introduction to the Computational Analysis of Content*. New York: Routledge. 283

Lefkowitz, M. 2014. Twisted documentation. Contributors listed as Twisted Matrix Laboratories at `http://twistedmatrix.com/trac/wiki/TwistedMatrixLaboratories`. Documentation available at `http://twistedmatrix.com/trac/wiki/Documentation`.

Lehmann, D. R. and R. S. Winer 2001. *Analysis for Marketing Planning* (fifth ed.). New York: McGraw-Hill/Irwin.

Lehmann, J., R. Isele, M. Jakob, A. Jentzsch, D. Kontokostas, P. N. Mendes, S. Hellmann, M. Morsey, P. van Kleef, S. Auer, and C. Bizer 2014. DBpedia—A large-scale, multilingual knowledge base extracted from wikipedia. *Semantic Web Journal*:in press. Retrieved from the World Wide Web at `http://svn.aksw.org/papers/2013/SWJ_DBpedia/public.pdf`.

Leisch, F. and B. Gruen 2014. *CRAN Task View: Cluster Analysis & Finite Mixture Models*. Comprehensive R Archive Network. 2014. `http://cran.r-project.org/web/views/Cluster.html`.

Lemke, R. J., M. Leonard, and K. Tlhokwane 2010. Estimating attendance at major league baseball games for the 2007 season. *Journal of Sports Economics* 11(3):316–348. 372

Lenat, D., M. Witbrock, D. Baxter, E. Blackstone, C. Deaton, D. Schneider, J. Scott, and B. Shepard 2010, Fall. Harnessing Cyc to answer clinical researchers' ad hoc queries. *AI Magazine* 31(3):13–32.

Lengauer, T. and R. E. Tarjan 1979. A fast algorithm for finding dominators in a flowgraph. *ACM Transactions on Programming Languages and Systems (TOPLAS)* 1(1):121–141.

Leskovec, J. and C. Faloutsos 2006. Sampling from large graphs. Retrieved from the World Wide Web at `http://cs.stanford.edu/people/jure/pubs/sampling-kdd06.pdf`. 311

Leskovec, J., D. Huttenlocher, and J. Kleinbert 2010a. Predicting positive and negative links in online social networks. Retrieved from the World Wide Web at `http://cs.stanford.edu/people/jure/pubs/triads-chi10.pdf`.

Leskovec, J., D. Huttenlocher, and J. Kleinbert 2010b. Signed networks in social media. Retrieved from the World Wide Web at `http://cs.stanford.edu/people/jure/pubs/triads-chi10.pdf`.

Leskovec, J., K. J. Lang, A. Dasgupta, and M. W. Mahoney 2009. Community structure in large networks: Natural cluster sizes and the absence of large well-defined clusters. *Internet Mathematics* 6(1):29–123. `http://www.technologyreview.com/news/515801/the-immortal-life-of-the-enron-e-mails/`.

Levinson, J. D. 1986. *Fleamarkets*. Berlin: Braus. 351

Levitt, T. 1986. *The Marketing Imagination (New Expanded Edition)*. New York: The Free Press.

Lévy, P. 1997. *Collective Intelligence: Mankind's Emerging World in Cyberspace*. Cambridge, Mass.: Perseus Books. 351

Lévy, P. 2001. *Cyberculture*. Minneapolis: University of Minnesota Press. 351

Levy, P. S. and S. Lemeshow 1999. *Sampling of Populations: Methods and Applications* (third ed.). New York: Wiley.

Lewin-Koh, N. 2014. *Hexagon Binning: an Overview*. Comprehensive R Archive Network. 2014. `http://cran.r-project.org/web/packages/hexbin/vignettes/hexagon_binning.pdf`. 276

Lewis, T. G. 2009. *Network Science: Theory and Applications*. New York: Wiley. 202

Liaw, A. and M. Wiener 2014. *randomForest: Breiman and Cutler's Random Forests for Classification and Regression*. Comprehensive R Archive Network. 2014. `http://cran.r-project.org/web/packages/randomForest/randomForest.pdf`.

Likert, R. 1932. A technique for the measurement of attitudes. *Archives of Psychology* 140: 1–55.

Lilien, G. L., P. Kotler, and K. S. Moorthy 1992. *Marketing Models*. Englewood Cliffs, N.J.: Prentice-Hall.

Lilien, G. L. and A. Rangaswamy 2003. *Marketing Engineering: Computer-Assisted Marketing Analysis and Planning* (second ed.). Upper Saddle River, N.J.: Prentice Hall.

Lindsey, J. K. 1997. *Applying Generalized Linear Models*. New York: Springer.

Little, R. J. A. and D. B. Rubin 1987. *Statistical Analysis with Missing Data*. New York: Wiley. 261

Liu, B. 2010. Sentiment analysis and subjectivity. In N. Indurkhya and F. J. Damerau (eds.), *Handbook of Natural Language Processing* (second ed.)., pp. 627–665. Boca Raton, Fla.: Chapman and Hall/CRC.

Liu, B. 2011. *Web Data Mining: Exploring Hyperlinks, Contents, and Usage Data*. New York: Springer. 275

Liu, B. 2012. *Sentiment Analysis and Opinion Mining*. San Rafael, Calif.: Morgan & Claypool.

Lloyd, C. J. 1999. *Statistical Analysis of Categorical Data*. New York: Wiley.

Lodish, L. M., M. M. Abraham, J. Livelsberger, B. Lubetkin, B. Richardson, and M. E. Stevens 1995. A summary of fifty-five in-market experimental estimates of the long-term effect of advertising. *Marketing Science* 14(3):G133–G140.

Lopez, V., C. Unger, P. Cimiano, and E. Motta 2013. Evaluating question answering over linked data. *Web Semantics* 21:3–13.

Lord, F. M. and M. R. Novick 1968. *Statistical Theories of Mental Test Scores*. Reading, Mass.: Addison-Wesley. 297

Louviere, J. J. 1993. The best-worst or maximum difference measurement model: Applications to behavioral research in marketing. Paper presented at the American Marketing Association Behavioral Research Conference, Phoenix.

Louviere, J. J., D. A. Hensher, and J. D. Swait 2000. *Stated Choice Methods: Analysis and Application*. Cambridge: Cambridge University Press.

Lovelock, C. and J. Wirtz (eds.) 2006. *Services Marketing: People, Technology, Strategy* (sixth ed.). Upper Saddle River, N.J.: Prentice Hall. 77

Luangkesorn, L. 2014. *Simulation Programming with Python*. University of Pittsburgh. 2014. Retrieved from the World Wide Web at `http://users.iems.northwestern.edu/~nelsonb/IEMS435/PythonSim.pdf`. Translation of chapter 4 of Nelson (2014).

Lubanovic, B. 2015. *Introducing Python: Modern Computing in Simple Packages*. Sebastopol, Calif.: O'Reilly.

Luce, D. and J. Tukey 1964. Simultaneous conjoint measurement: A new type of fundamental measurement. *Journal of Mathematical Psychology* 1:1–27.

Luger, G. F. 2008. *Artificial Intelligence: Structures and Strategies for Complex Problem Solving* (sixth ed.). Boston: Addison-Wesley. 275

Lumley, T. 2010. *Complex Surveys: A Guide to Analysis Using R*. New York: Wiley. 261

Lumley, T. 2014. *mitools: Tools for Multiple Imputation of Missing Data*. Comprehensive R Archive Network. 2014. `http://cran.r-project.org/web/packages/mitools/mitools.pdf`.

Lunenfeld, P. 2000. *Snap to Grid: A User's Guide to Digital Arts, Media, and Cultures*. Cambridge: MIT Press. 349

Lyon, D. W. 2000. Pricing research. In C. Chakrapani (ed.), *Marketing Research: State-of-the-Art Perspectives*, Chapter 19, pp. 551–582. Chicago: American Marketing Association.

Lyon, D. W. 2002, Winter. The price is right (or is it)? *Marketing Research*:8–13.

Macal, C. M. and M. J. North 2006. *Introduction to Agent-Based Modeling and Simulation*. `http://www.docstoc.com/docs/36015647/Introduction-to-Agent-based-Modeling-and-Simulation`.

Maechler, M. 2014a. *Package cluster*. Comprehensive R Archive Network. 2014. `http://cran.r-project.org/web/packages/cluster/cluster.pdf`.

Maechler, M. 2014b. *robustbase: Basic Robust Statistics*. Comprehensive R Archive Network. 2014. `http://cran.r-project.org/web/packages/robustbase/robustbase.pdf`. 273

Mahajan, V., E. Muller, and F. M. Bass 1995. Diffusion of new products: empirical generalizations and managerial uses. *Marketing Science* 14(3):G79–G88. 209

Mahajan, V., E. Muller, and Y. Wind 2000. *New Product Diffusion Models*. New York: Springer. 209

Maisel, L. S. and G. Cokins 2014. *Predictive Business Analytics: Forward-Looking Capabilities to Improve Business Performance*. New York: Wiley. 259

Makridakis, S., S. C. Wheelwright, and R. J. Hyndman 2005. *Forecasting Methods and Applications* (third ed.). New York: Wiley.

Malhotra, N. K. 2009. *Marketing Research: An Applied Orientation* (sixth ed.). Upper Saddle River, N.J.: Pearson/Prentice Hall.

Manly, B. F. J. 1992. *The Design and Analysis of Research Studies*. Cambridge: Cambridge University Press.

Manly, B. F. J. 1994. *Multivariate Statistical Methods: A Primer* (second ed.). London: Chapman & Hall.

Mann, C. and F. Stewart 2000. *Internet Communication and Qualitative Research: A Handbook for Researching Online*. London: Sage. 351

Manning, C. D. and H. Schütze 1999. *Foundations of Statistical Natural Language Processing*. Cambridge: MIT Press.

Marder, E. 1997. *The Laws of Choice: Predicting Consumer Behavior*. New York: Free Press.

Maren, A., C. Harston, and R. Pap 1990. *Handbook of Neural Computing Applications*. New York: Academic Press.

Markham, A. N. 1998. *Life Online: Researching Real Experience in Virtual Space*. Walnut Creek, Calif.: AltaMira Press. 351

Maronna, R. A., D. R. Martin, and V. J. Yohai 2006. *Robust Statistics Theory and Methods*. New York: Wiley. 273

Marshall, P. and E. T. Bradlow 2002. A unified approach to conjoint analysis methods. *Journal of the American Statistical Association* 97(459):674–682.

Matloff, N. 2011. *The Art of R Programming*. San Francisco: no starch press.

Maybury, M. T. (ed.) 1997. *Intelligent Multimedia Information Retrieval*. Menlo Park, Calif./ Cambridge: AAAI Press / MIT Press.

McArthur, R. and P. Bruza 2001. The ABCs of online community. In N. Zhong, Y. Yao, J. Liu, and S. Ohsuga (eds.), *Web Intelligence: Research and Development*, pp. 141–147. New York: Springer.

McCallum, Q. E. (ed.) 2013. *Bad Data Handbook*. Sebastopol, Calif.: O'Reilly.

McCullagh, P. and J. A. Nelder 1989. *Generalized Linear Models* (second ed.). New York: Chapman and Hall.

McFadden, D. 2001. Economic choices. *American Economic Review* 91:351–378.

McGrayne, S. B. 2011. *The Theory that Would Not Die: How Bayes' Rule Cracked the Enigma Code, Hunted Down Russian Submarines and Emerged Triumphant from Two Centuries of Controversy*. New Haven, Conn.: Yale University Press. 264

McKellar, J. and A. Fettig 2013. *Twisted Network Programming Essentials: Event-driven Network Programming with Python* (second ed.). Sebastopol, Calif.: O'Reilly.

McLean, B. and P. Elkind 2003. *The Smartest Guys in the Room: The Amazing Rise and Scandalous Fall of Enron*. New York: Penguin.

McPherson, M., L. Smith-Lovin, and J. M. Cook 2001. Birds of a feather: homophily in social networks. *Annual Review of Sociology* 27:415–444.

McTaggart, R. and G. Daroczi 2014. *Package Quandl: Quandl Data Connection*. Comprehensive R Archive Network. 2014. `http://cran.r-project.org/web/packages/Quandl/Quandl.pdf` with additional online documentation available at `https://www.quandl.com/`. 228

Meadow, C. T., B. R. Boyce, and D. H. Kraft 2000. *Text Information Retrieval Systems* (second ed.). San Diego: Academic Press.

Mehrens, W. A. and R. L. Ebel (eds.) 1967. *Principles of Educational and Psychological Measurement: A Book of Selected Readings*. Chicago: Rand McNally.

Merkl, D. 2002. Text mining with self-organizing maps. In W. Klösgen and J. M. Żytkow (eds.), *Handbook of Data Mining and Knowledge Discovery*, Chapter 46.9, pp. 903–910. Oxford: Oxford University Press.

Mertz, D. 2002. *Charming Python: SimPy Simplifies Complex Models*. IBM developerWorks. 2002. Retrieved from the World Wide Web at `http://www.ibm.com/developerworks/linux/library/l-simpy/l-simpy-pdf.pdf`.

Meyer, D., E. Dimitriadou, K. Hornik, A. Weingessel, and F. Leisch 2014. *e1071: Misc Functions of the Department of Statistics (e1071), TU Wien*. Comprehensive R Archive Network. 2014. `http://cran.r-project.org/web/packages/e1071/e1071.pdf`.

Meyer, D., A. Zeileis, and K. Hornik 2006, October 19. The strucplot framework: Visualizing multi-way contingency tables with vcd. *Journal of Statistical Software* 17(3):1–48. `http://www.jstatsoft.org/v17/i03`.

Meyer, D., A. Zeileis, K. Hornik, and M. Friendly 2014. *vcd: Visualizing Categorical Data*. Comprehensive R Archive Network. 2014. `http://cran.r-project.org/web/packages/vcd/vcd.pdf`.

Michalawicz, Z. and D. B. Fogel 2004. *How to Solve It: Modern Heuristics*. New York: Springer. 275

Mikowski, M. S. and J. C. Powell 2014. *Single Page Web Applications JavaScript End-to-End*. Shelter Island, N.Y.: Manning.

Milborrow, S. 2014. *rpart.plot: Plot rpart models. An Enhanced Version of plot.rpart*. Comprehensive R Archive Network. 2014. `http://cran.r-project.org/web/packages/rpart.plot/rpart.plot.pdf`.

Milgram, S. 1967. The small world problem. *Psychology Today* 1(1):60–67. 196

Miller, D. and D. Slater 2000. *The Internet: An Ethnographic Approach*. Oxford: Berg. 351

Miller, J. H. and S. E. Page 2007. *Complex Adaptive Systems: An Introduction to Computational Models of Social Life*. Princeton, N. J.: Princeton University Press.

Miller, J. P. 2000a. *Millennium Intelligence: Understanding and Conducting Competitive Intelligence in the Digital Age*. Medford, N.J.: Information Today, Inc. 225

Miller, T. W. 1999. The Boston splits: Sample size requirements for modern regression. *1999 Proceedings of the Statistical Computing Section of the American Statistical Association*:210–215.

Miller, T. W. 2000b. Marketing research and the information industry. *CASRO Journal 2000*:21–26.

Miller, T. W. 2001, Summer. Can we trust the data of online research? *Marketing Research*: 26–32. Reprinted as "Make the call: Online results are a mixed bag," *Marketing News*, September 24, 2001:20–25. 305

Miller, T. W. 2002. Propensity scoring for multimethod research. *Canadian Journal of Marketing Research* 20(2):46–61.

Miller, T. W. 2005. *Data and Text Mining: A Business Applications Approach*. Upper Saddle River, N.J.: Pearson Prentice Hall.

Miller, T. W. 2008. *Research and Information Services: An Integrated Approach to Business*. Manhattan Beach, Calif.: Research Publishers LLC. ix

Miller, T. W. 2015a. *Modeling Techniques in Predictive Analytics with Python and R: A Guide to Data Science*. Upper Saddle River, N.J.: Pearson Education. `http://www.ftpress.com/miller`.

Miller, T. W. 2015b. *Modeling Techniques in Predictive Analytics: Business Problems and Solutions with R* (revised and expanded ed.). Upper Saddle River, N.J.: Pearson Education. `http://www.ftpress.com/miller`.

Miller, T. W. 2015c. *Web and Network Data Science: Modeling Techniques in Predictive Analytics with Python and R*. Upper Saddle River, N.J.: Pearson Education. `http://www.ftpress.com/miller`. 254, 311

Miller, T. W. and P. R. Dickson 2001. On-line market research. *International Journal of Electronic Commerce* 5(3):139–167.

Miller, T. W., D. Rake, T. Sumimoto, and P. S. Hollman 2001. Reliability and comparability of choice-based measures: Online and paper-and-pencil methods of administration. *2001 Sawtooth Software Conference Proceedings*:123–130. Published by Sawtooth Software, Sequim, WA.

Miller, T. W. and J. Walkowski (eds.) 2004. *Qualitative Research Online*. Manhattan Beach, Calif.: Research Publishers LLC.

Mintel International Group, Ltd. 2001. The cinema & movie theater market: US consumer Intelligence, April 2001. Technical Report, Chicago.

Mitchell, M. 1996. *An Introduction to Genetic Algorithms*. Cambridge: MIT Press. 275

Mitchell, M. 2009. *Complexity: A Guided Tour*. Oxford, U.K.: Oxford University Press. 209

Moreno, J. L. 1934. Who shall survive?: Foundations of sociometry, group psychotherapy, and sociodrama. Reprinted in 1953 (second edition) and in 1978 (third edition) by Beacon House, Inc., Beacon, N.Y. 207

Morgan, D. L. and R. A. Krueger 1998. *The Focus Group Kit*. Thousand Oaks, Calif.: Sage.

Moro, S., P. Cortez, and P. Rita 2014. A data-driven approach to predict the success of bank telemarketing. *Decision Support Systems* 62:22–31.

Moro, S., R. M. S. Laureano, and P. Cortez 2011, October 24–26. Using data mining for bank direct marketing: an application of the CRISP-DM methodology. In P. Novais et al. (ed.), *Proceedings of the European Simulation and Modeling Conference*, ESM'2011, pp. 117–121. Hotel de Guimaraes, Guimaraes, Portugal: EUROSIS. Retrieved from the World Wide Web at `http://repositorium.sdum.uminho.pt/bitstream/1822/14838/1/MoroCortezLaureano_DMApproach4DirectMKT.pdf`.

Mosteller, F. and D. L. Wallace 1984. *Applied Bayesian and Classical Inference: The Case of "The Federalist" Papers* (second ed.). New York: Springer. Earlier edition published in 1964 by Addison-Wesley, Reading, Mass. The previous title was *Inference and Disputed Authorship: The Federalist*.

Moustafa, R. and E. Wegman 2006. Multivariate continuous data—parallel coordinates. In A. Unwin, M. Theus, and H. Hoffman (eds.), *Graphics of Large Databases: Visualizing a Million*, Chapter 7, pp. 143–155. New York: Springer. 166

Müller, K., T. Vignaux, and C. Chui 2014. *SimPy: Discrete Event Simulation for Python*. SimPy Development Team. 2014. Retrieved from the World Wide Web at `https://simpy.readthedocs.org/en/latest/` with code available at `https://bitbucket.org/simpy/simpy/`.

Murphy, K. P. 2012. *Machine Learning: A Probabilistic Perspective*. Cambridge, Mass.: MIT Press. 264, 275

Murrell, P. 2011. *R Graphics* (second ed.). Boca Raton, Fla.: CRC Press.

Nagle, T. T. and J. Hogan 2005. *The Strategy and Tactics of Pricing: A Guide to Growing More Profitably* (fourth ed.). Upper Saddle River, N.J.: Prentice Hall.

Nair, V. G. 2014. *Getting Started with Beautiful Soup: Build Your Own Web Scraper and Learn All About Web Scraping with Beautiful Soup*. Birmingham, UK: PACKT Publishing.

Nash, E. L. 1995. *Direct Marketing: Strategy/Planning/Execution* (third ed.). New York: McGraw-Hill.

Nash, E. L. (ed.) 2000. *The Direct Marketing Handbook* (second ed.). New York: McGraw-Hill.

Neal, W. D. 2000. Market segmentation. In C. Chakrapani (ed.), *Marketing Research: State-of-the-Art Perspectives*, Chapter 1, pp. 375–399. American Marketing Association.

Nelson, B. L. 2013. *Foundations and Methods of Stochastic Simulation: A First Course*. New York: Springer. Supporting materials available from the World Wide Web at `http://users.iems.northwestern.edu/~nelsonb/IEMS435/`.

Nelson, W. B. 2003. *Recurrent Events Data Analysis for Product Repairs, Disease Recurrences, and Other Applications*. Series on Statistics and Applied Probability. Philadelphia and Alexandria, Va.: ASA-SIAM. 77

Newman, M. E. J. 2010. *Networks: An Introduction*. Oxford, UK: Oxford University Press. 202

Nolan, D. and D. T. Lang 2014. *XML and Web Technologies for Data Sciences with R*. New York: Springer.

North, M. J. and C. M. Macal 2007. *Managing Business Complexity: Discovering Strategic Solutions with Agent-Based Modeling and Simulation*. Oxford, U.K.: Oxford University Press. 204

Nunnally, J. C. 1967. *Psychometric Theory*. New York: McGraw-Hill. 297

Nunnally, J. C. and I. H. Bernstein 1994. *Psychometric Theory* (third ed.). New York: McGraw-Hill. 297

Obe, R. and L. Hsu 2012. *PostgreSQL up and Running: A Practical Guide to the Advanced Open Source Database*. Sebastopol, Calif.: O'Reilly.

O'Hagan, A. 2010. *Kendall's Advanced Theory of Statistics: Bayesian Inference*, Volume 2B. New York: Wiley. 264

Orme, B. K. 2013. *Getting Started with Conjoint Analysis: Strategies for Product Design and Pricing Research* (third ed.). Glendale, Calif.: Research Publishers LLC. `http://research-publishers.com/rp/gsca.htm`.

Osborne, J. W. 2013. *Best Practices in Data Cleaning: A Complete Guide to Everything You Need to Do Before and After Collecting Your Data*. Los Angeles: Sage.

Osgood, C. 1962. Studies in the generality of affective meaning systems. *American Psychologist* 17:10–28. 284

Osgood, C., G. Suci, and P. Tannenbaum (eds.) 1957. *The Measurement of Meaning*. Urbana, Ill.: University of Illinois Press. 284

Pal, N. and A. Rangaswamy (eds.) 2003. *The Power of One: Gaining Business Value from Personalization Technologies*. Victoria, BC: Trafford.

Pasquale, F. 2015. *The Black Box Society: The Secret Algorithms that Control Money and Information*. Cambridge, Mass.: Harvard University Press.

Patzer, G. L. 1996. *Experiment-Research Methodology in Marketing: Types and Applications*. Westport, Conn.: Quorum Books.

Pearce, S. M. 1999. *On Collecting (Collecting Cultures)*. New York: Routledge. 351

Penenberg, A. L. and M. Barry 2000. *Spooked: Espionage in Corporate America*. Cambridge, Mass.: Perseus Publishing. 223

Pentland, A. 2014. *Social Physics: How Good Ideas Spread—The Lessons from a New Science*. New York: Penguin.

Peppers, D. and M. Rogers 1993. *The One to One Future: Building Relationships One Customer at a Time*. New York: Doubleday.

Peterson, K. 2004. *The Power of Place: Advanced Customer and Location Analytics for Market Planning*. San Diego: Integras.

Petris, G. 2010, October 13. An R package for dynamic linear models. *Journal of Statistical Software* 36(12):1–16. `http://www.jstatsoft.org/v36/i12`.

Petris, G. and W. Gilks 2014. *dlm: Bayesian and Likelihood Analysis of Dynamic Linear Models*. Comprehensive R Archive Network. 2014. `http://cran.r-project.org/web/packages/dlm/dlm.pdf`.

Petris, G., S. Petrone, and P. Campagnoli 2009. *Dynamic Linear Models with R*. New York: Springer.

Pindyck, R. and D. Rubinfeld 2012. *Microeconomics* (eighth ed.). Upper Saddle River, N.J.: Pearson. 173

Pinheiro, J. C. and D. M. Bates 2009. *Mixed-Effects Models in S and S-PLUS*. New York: Springer-Verlag. 341

Pinker, S. 1994. *The Language Instinct*. New York: W. Morrow and Co.

Pinker, S. 1997. *How the Mind Works*. New York: W.W. Norton & Company.

Pinker, S. 1999. *Words and Rules: The Ingredients of Language*. New York: HarperCollins.

Pinkham, L. E. 1900. *Lydia E. Pinkham's Private Text-book upon Ailments Peculiar to Women*. Lynn, Mass.: Lydia E. Pinkham Medicine Co. 369

Popping, R. 2000. *Computer-Assisted Text Analysis*. Thousand Oaks, Calif.: Sage. 283

Porter, D. (ed.) 1997. *Internet Culture*. New York: Routledge. 351

Porter, M. E. 1980. *Competitive Strategy: Techniques for Analyzing Industries and Competitors*. New York: The Free Press.

Powazak, D. M. 2001. *Design for Community: The Art of Connecting Real People in Virtual Places*. Indianapolis: New Riders.

Powers, S. 2003. *Practical RDF*. Sebastopol, Calif.: O'Reilly.

Preece, J. 2000. *Online Communities: Designing Usability, Supporting Sociability*. New York: Wiley.

Priebe, C. E., J. M. Conroy, D. J. Marchette, and Y. Park 2005, October. Scan statistics on Enron graphs. *Computational and Mathematical Organization Theory* 11(3):229–247.

Provost, F. and T. Fawcett 2014. *Data Science for Business: What You Need to Know About Data Mining and Data-Analytic Thinking*. Sebastopol, Calif.: O'Reilly. 259

Pustejovsky, J. and A. Stubbs 2013. *Natural Language Annotation for Machine Learning*. Sebastopol, Calif.: O'Reilly. 297

Putler, D. S. and R. E. Krider 2012. *Customer and Business Analytics: Applied Data Mining for Business Decision Making Using R*. Boca Raton, Fla: Chapman & Hall/CRC.

Radcliffe-Brown, A. R. 1940. On social structure. *Journal of the Royal Anthropological Society of Great Britain and Ireland* 70:1–12. 207

Rajaraman, A. and J. D. Ullman 2012. *Mining of Massive Datasets*. Cambridge, UK: Cambridge University Press. 153

Rao, V. R. 2014. *Applied Conjoint Analysis*. New York: Springer.

Rasch, D., J. Pilz, R. Verdooren, and A. Gebhardt 2011. *Optimal Experimental Design with R*. Boca Raton, Fla.: CRC Press.

Reingold, E. M. and J. S. Tilford 1981. A fast algorithm for finding dominators in a flowgraph. *IEEE Transactions on Software Engineering* 7:223–228.

Reis, P. N. and H. Smith 1963. The use of chi-square for preference testing in multidimensional problems. *Chemical Engineering Progress Symposium Series* 59(42):39–43.

Reitz, K. 2014a, October 15. Python guide documentation (release 0.0.1). Retrieved from the World Wide Web, October 23, 2014, at `https://media.readthedocs.org/pdf/python-guide/latest/python-guide.pdf`.

Reitz, K. 2014b. Requests: HTTP for humans. Documentation available from the World Wide Web at `http://docs.python-requests.org/en/latest/`.

Rencher, A. C. and G. B. Schaalje 2008. *Linear Models in Statistics* (second ed.). New York: Wiley.

Resig, J. and BearBibeault 2013. *Secrets of the JavaScript Ninja*. Shelter Island, N.Y.: Manning.

Resnick, M. 1998. *Turtles, Termites, and Traffic Jams: Explorations in Massively Parallel Microworlds*. Cambridge, Mass.: MIT Press. 209

Revelle, W. 2014. *psych: Procedures for Psychological, Psychometric, and Personality Research*. Comprehensive R Archive Network. 2014. `http://cran.r-project.org/web/packages/psych/psych.pdf`.

Rheingold, H. 2000. *The Virtual Community: Homesteading on the Electronic Frontier* (revised ed.). Cambridge: The MIT Press. 351

Ricci, F., L. Rokach, B. Shapira, and P. B. Kantor (eds.) 2011. *Recommender Systems Handbook*. New York: Springer.

Richardson, L. 2014. Beautiful soup documentation. Available from the World Wide Web at `http://www.crummy.com/software/BeautifulSoup/bs4/doc/`.

Richardson, M. 1938. Multidimensional psychophysics. *Psychological Bulletin* 35:659–660. 96

Ripley, B. and W. Venables 2015. *Package nnet: Feed-Forward Neural Networks with a Single Hidden Layer, and for Multinomial Log-Linear Models*. Comprehensive R Archive Network. 2015. `http://cran.r-project.org/web/packages/nnet/nnet.pdf`.

Ripley, B., W. Venables, D. M. Bates, K. Hornik, A. Gebhardt, and D. Firth 2015. *Package MASS: Functions and Datasets to Support Venables and Ripley "Modern Applied Statistics with S" (4th edition 2002)*. Cambridge, USA: Comprehensive R Archive Network. 2015. `http://cran.r-project.org/web/packages/MASS/MASS.pdf`.

Ripley, B. D. 1996. *Pattern Recognition and Neural Networks*. Cambridge: Cambridge University Press.

Robert, C. P. 2007. *The Bayesian Choice: From Decision Theoretic Foundations to Computational Implementation* (second ed.). New York: Springer. 264

Robert, C. P. and G. Casella 2009. *Introducing Monte Carlo Methods with R*. New York: Springer. 264

Roberts, C. W. (ed.) 1997. *Text Analysis for the Social Sciences: Methods for Drawing Statistical Inferences from Texts and Transcripts*. Mahwah, N.J.: Lawrence Erlbaum Associates. 283

Robinson, I., J. Webber, and E. Eifrem 2013. *Graph Databases*. Sebastopol, Calif.: O'Reilly.

Rogers, E. M. 2003. *Diffusion of Innovations* (fifth ed.). New York: Free Press. 209

Rogers, H. . J., H. Swaminathan, and R. K. Hambleton 1991. *Fundamentals of Item Response Theory*. Newbury Park, Calif.: Sage.

Rosenbaum, P. R. 1995. *Observational Studies*. New York: Springer. 341

Rosenberg, S. 2009. *Say Everything: How Blogging Began, What It's Becoming, and Why It Matters*. New York: Crown.

Rossant, C. 2014. *IPython Interactive Computing and Visualization Cookbook*. Birmingham, UK: Packt Publishing.

Rossi, P. 2014. *bayesm: Bayesian Inference for Marketing/Micro-econometrics*. Comprehensive R Archive Network. 2014. `http://cran.r-project.org/web/packages/bayesm/bayesm.pdf`.

Rossi, P. E., G. M. Allenby, and R. McCulloch 2005. *Bayesian Statistics and Marketing*. New York: Wiley.

Rounds, J. B., T. W. Miller, and R. V. Dawis 1978. Comparability of multiple rank order and paired comparison methods. *Applied Psychological Measurement* 2(3):415–422.

Rousseeuw, P. J. 1987. Silhouettes: A graphical aid to the interpretation and validation of cluster analysis. *Journal of Computational and Applied Mathematics* 20:53–65. 53

Rubin, D. B. 1987. *Multiple Imputation for Nonresponse in Surveys*. New York: Wiley. 261

Rudder, C. 2014. *Dataclysm: Who We Are (When We Think No One's Looking)*. New York: Crown/Penguin Random House.

Russell, M. A. 2014. *Mining the Social Web: Data Mining Facebook, Twitter, LinkedIn, Google+, GitHub, and More*. Sebastopol, Calif.: O'Reilly. 228

Russell, S. and P. Norvig 2009. *Artificial Intelligence: A Modern Approach* (third ed.). Upper Saddle River, N.J.: Prentice Hall. 275

Rusu, A. 2014. Tree drawing algorithms. In R. Tamassia (ed.), *Handbook of Graph Drawing and Visualization*, Chapter 5, pp. 155–192. Boca Raton, Fla.: CRC Press/Chapman & Hall.

Ryan, J. A. 2014. *Package quantmod: Quantitative Financial Modelling Framework*. Comprehensive R Archive Network. 2014. `http://cran.r-project.org/web/packages/quantmod/quantmod.pdf`. 228

Ryan, J. A. and J. M. Ulrich 2014. *Package xts: eXtensible Time Series*. Comprehensive R Archive Network. 2014. `http://cran.r-project.org/web/packages/xts/xts.pdf`. 228

Ryan, M.-L. (ed.) 1999. *Cyberspace Textuality: Computer Technology and Literary Theory*. Bloomington, Ind.: Indiana University Press. 349

Ryan, T. P. 2008. *Modern Regression Methods* (second ed.). New York: Wiley.

Šalamon, T. 2011. *Design of Agent-Based Models: Developing Computer Simulations for a Better Understanding of Social Processes*. Czech Republic: Tomáš Bruckner, Řepín-Živonín.

Sanjek, R. (ed.) 1990. *Fieldnotes: The Makings of Anthropology*. Ithica, N.Y.: Cornell University Press.

Sarkar, D. 2008. *Lattice: Multivariate Data Visualization with R*. New York: Springer.

Sarkar, D. 2014. *lattice: Lattice Graphics*. Comprehensive R Archive Network. 2014. `http://cran.r-project.org/web/packages/lattice/lattice.pdf`.

Sarkar, D. and F. Andrews 2014. *latticeExtra: Extra Graphical Utilities Based on Lattice*. Comprehensive R Archive Network. 2014. `http://cran.r-project.org/web/packages/latticeExtra/latticeExtra.pdf`.

Savage, D., X. Zhang, X. Yu, P. Chou, and Q. Wang 2014. Anomaly detection in online social networks. *Social Networks* 39:62–70.

Schafer, J. L. 1997. *Analysis of Incomplete Multivariate Data*. London: Chapman & Hall. 261

Schauerhuber, M., A. Zeileis, D. Meyer, and K. Hornik 2008. Benchmarking open-source tree learners in R/RWeka. In C. Preisach, H. Burkhardt, L. Schmidt-Thieme, and R. Decker (eds.), *Data Analysis, Machine Learning, and Applications*, pp. 389–396. New York: Springer.

Schaul, J. 2014. *ComplexNetworkSim Package Documentation*. pypi.python.org. 2014. Retrieved from the World Wide Web at `https://pythonhosted.org/ComplexNetworkSim/` with code available at `https://github.com/jschaul/ComplexNetworkSim`.

Schnettler, S. 2009. A structured overview of 50 years of small-world research. *Social Networks* 31:165–178. 196

Schoenfeld, E. 1998, September 28. The customized, digitized, have-it-your-way economy. *Fortune*:115–124.

Scholder, A. and J. Crandall (eds.) 2001. *Interaction: Artistic Practice in the Network*. New York: Distributed Art Publishers. 349

Scott:, J. 2013. *Social Network Analysis* (third ed.). Thousand Oaks, Calif.: Sage.

Sebastiani, F. 2002. Machine learning in automated text categorization. *ACM Computing Surveys* 34(1):1–47.

Seber, G. A. F. 2000. *Multivariate Observations*. New York: Wiley. Originally published in 1984.

Segaran, T., C. Evans, and J. Taylor 2009. *Programming the Semantic Web*. Sebastopol, Calif.: O'Reilly.

Sen, R. and M. H. Hansen 2003, March. Predicting web user's next access based on log data. *Journal of Computational and Graphical Statistics* 12(1):143–155.

Senkul, P. and S. Salin 2012, March. Improving pattern quality in web usage mining by using semantic information. *Knowledge and Information Systems* 30(3):527–541.

Sermas, R. 2014. *ChoiceModelR: Choice Modeling in R*. Comprehensive R Archive Network. 2014. `http://cran.r-project.org/web/packages/ChoiceModelR/ChoiceModelR.pdf`.

Shakhnarovich, G., T. Darrell, and P. Indyk (eds.) 2006. *Nearest-Neighbor Methods in Learning and Vision: Theory and Practice*. Cambridge, Mass.: MIT Press.

Sharda, R. and D. Delen 2006. Predicting box office success of motion pictures with neural networks. *Expert Systems with Applications* 30:243–254. 284

Sharma, S. 1996. *Applied Multivariate Techniques*. New York: Wiley.

Sherry, Jr., J. F. 1999. A sociocultural analysis of a midwestern American flea market. *Journal of Consumer Research* 17(1):13–30. 351

Sherry, Jr., J. F. and R. V. Kozinets 2001. Qualitative inquiry in marketing and consumer research. In D. Iacobucci (ed.), *Kellogg on Marketing*, Chapter 8, pp. 165–194. New York: Wiley.

Shmueli, G. 2010. To explain or predict? *Statistical Science* 25(3):289–310.

Shroff, G. 2013. *The Intelligent Web: Search, Smart Algorithms, and Big Data*. Oxford, UK: Oxford University Press.

Shrout, P. E. and S. T. Fiske (eds.) 1995. *Personality Research, Methods, and Theory: A Festschrift Honoring Donald W. Fiske*. Hillsdale, N.J.: Lawrence Erlbaum Associates.

Siegel, E. 2013. *Predictive Analytics: The Power to Predict Who Will Click, Buy, Lie, or Die*. Hoboken, N.J.: Wiley. 259

Silver, N. 2012. *The Signal and the Noise: Why So Many Predictions Fail—But Some Don't*. New York: The Penguin Press.

Sing, T., O. Sander, N. Beerenwinkel, and T. Lengauer 2005. ROCR: Visualizing classifier performance in R. *Bioinformatics* 21(20):3940–3941.

Sing, T., O. Sander, N. Beerenwinkel, and T. Lengauer 2015. *Package ROCR: Visualizing the Performance of Scoring Classifiers*. Comprehensive R Archive Network. 2015. `http://cran.r-project.org/web/packages/ROCR/ROCR.pdf`.

Smith, J. A. and J. Moody 2013. Structural effects of network sampling coverage I: Nodes missing at random. *Social Networks* 35:652–668. 311

Smith, M. A. and P. Kollock (eds.) 1999. *Communities in Cyberspace*. New York: Routledge. 351

Snedecor, G. W. and W. G. Cochran 1989. *Statistical Methods* (eighth ed.). Ames, Iowa: Iowa State University Press. First edition published by Snedecor in 1937. 264

Snijders, T. A. B. and R. J. Bosker 2012. *Multilevel Analysis: An Introduction to Basic and Advanced Multilevel Modeling* (second ed.). Thousand Oaks, Calif.: Sage. 261

Socher, R., J. Pennington, E. H. Huang, A. Y. Ng, and C. D. Manning 2011. Semi-Supervised Recursive Autoencoders for Predicting Sentiment Distributions. In *Proceedings of the 2011 Conference on Empirical Methods in Natural Language Processing (EMNLP)*.

Spradley, J. P. and D. W. McCurdy 1972. *The Cultural Experience: Ethnography in Complex Society*. Chicago: Science Research Associates.

Srivastava, A. N. and M. Sahami (eds.) 2009. *Text Mining: Classification, Clustering, and Applications*. Boca Raton, Fla.: CRC Press.

Stage, S. 1979. *Female Complaints: Lydia Pinkham and the Business of Women's Health*. New York: W. W. Norton.

Stein, M. L. and S. F. Paterno 2001. *Talk Straight, Listen Carefully: The Art of Interviewing*. Ames, Iowa: Iowa State University Press.

Sternthal, B. and A. M. Tybout 2001. Segmentation and targeting. In D. Iacobucci (ed.), *Kellogg on Marketing*, Chapter 1, pp. 3–30. New York: Wiley.

Stevens, S. S. 1946, June 7. On the theory of scales of measurement. *Science* 103(2684): 677–680. `http://www.jstor.org/stable/1671815`. 296, 337

Stigler, G. J. 1987. *The Theory of Price* (fourth ed.). New York: Macmillan. 173

Stone, P. J. 1997. Thematic text analysis: New agendas for analyzing text content. In C. W. Roberts (ed.), *Text Analysis for the Social Sciences: Methods for Drawing Statistical Inferences from Texts and Transcripts*, Chapter 2, pp. 35–54. Mahwah, N.J.: Lawrence Erlbaum Associates.

Stone, P. J., D. C. Dunphy, M. S. Smith, and D. M. Ogilvie 1966. *The General Inquirer: A Computer Approach to Content Analysis*. Cambridge: MIT Press.

Stuart, A., K. Ord, and S. Arnold 2010. *Kendall's Advanced Theory of Statistics: Classical Inference and the Linear Model*, Volume 2A. New York: Wiley. 264

Studenmund, A. H. 1992. *Using Econometrics: A Practical Guide* (second ed.). New York: HarperCollins.

Studenmund, A. H. 2010. *Using Econometrics: A Practical Guide* (sixth ed.). Upper Saddle River, N.J.: Addison-Wesley. 373

Suess, E. A. and B. E. Trumbo 2010. *Introduction to Probability Simulation and Gibbs Sampling with R*. New York: Springer. 264

Sullivan, D. 2001. *Document Warehousing and Text Mining: Techniques for Improving Business Operations, Marketing, and Sales*. New York: Wiley.

Supowit, K. J. and E. M. Reingold 1983. The complexity of drawing trees nicely. *Acta Informatica* 18:177–392.

Sutton, C. J. 1980. *Economics and Corporate Strategy*. Cambridge: Cambridge University Press.

Szymanski, C. 2014. *dlmodeler: Generalized Dynamic Linear Modeler*. Comprehensive R Archive Network. 2014. `http://cran.r-project.org/web/packages/dlmodeler/dlmodeler.pdf`.

Taddy, M. 2013a. Measuring political sentiment on Twitter: factor-optimal design for multinomial inverse regression. Retrieved from the World Wide Web at `http://arxiv.org/pdf/1206.3776v5.pdf`.

Taddy, M. 2013b. Multinomial inverse regression for text analysis. Retrieved from the World Wide Web at `http://arxiv.org/pdf/1012.2098v6.pdf`.

Taddy, M. 2014. *textir: Inverse Regression for Text Analysis*. 2014. `http://cran.r-project.org/web/packages/textir/textir.pdf`.

Tamassia, R. (ed.) 2014. *Handbook of Graph Drawing and Visualization*. Boca Raton, Fla.: CRC Press/Chapman & Hall.

Tan, P.-N., M. Steinbach, and V. Kumar 2006. *Introduction to Data Mining*. Boston: Addison-Wesley. 153

Tancer, B. 2014. *Everyone's a Critic: Winning Consumers in a Review-Driven World*. New York: Portfolio/Penguin.

Tang, W., H. He, and X. M. Tu 2012. *Applied Categorical and Count Data Analysis*. Boca Raton, Fla.: Chapman & Hall/CRC.

Tannenbaum, A. S. and D. J. Wetherall 2010. *Computer Networks* (fifth ed.). Upper Saddle River, N.J.: Pearson/Prentice Hall.

Tanner, M. A. 1996. *Tools for Statistical Inference: Methods for the Exploration of Posterior Distributions and Likelihood Functions* (third ed.). New York: Springer. 264

Tayman, J. and L. Pol 1995, Spring. Retail site selection and geographical information systems. *Journal of Applied Business Research* 11(2):46–54.

Therneau, T. 2014. *survival: Survival Analysis*. Comprehensive R Archive Network. 2014. `http://cran.r-project.org/web/packages/survival/survival.pdf`. 174

Therneau, T., B. Atkinson, and B. Ripley 2014. *rpart: Recursive Partitioning*. Comprehensive R Archive Network. 2014. `http://cran.r-project.org/web/packages/rpart/rpart.pdf`.

Therneau, T. and C. Crowson 2014. *Using Time Dependent Covariates and Time Dependent Coefficients in the Cox Model*. Comprehensive R Archive Network. 2014. `http://cran.r-project.org/web/packages/survival/vignettes/timedep.pdf`.

Therneau, T. M. and P. M. Grambsch 2000. *Modeling Survival Data: Extending the Cox Model*. New York: Springer.

Thiele, J. C. 2014. *Package RNetLogo: Provides an Interface to the Agent-Based Modelling Platform NetLogo*. Comprehensive R Archive Network. 2014. User manual: `http://cran.r-project.org/web/packages/RNetLogo/RNetLogo.pdf`. Additional documentation available at `http://rnetlogo.r-forge.r-project.org/`.

Thompson, T. (ed.) 2000. *Writing about Business* (second ed.). New York: Columbia University Press.

T.Hothorn, F. Leisch, A. Zeileis, and K. Hornik 2005, September. The design and analysis of benchmark experiments. *Journal of Computational and Graphical Statistics* 14(3):675–699.

Thurman, W. N. and M. E. Fisher 1988. Chickens, eggs, and causality, or which came first? *American Journal of Agricultural Economics* 70(2):237–238. 290

Thurstone, L. L. 1927. A law of comparative judgment. *Psychological Review* 34:273–286.

Tibshirani, R. 1996. Regression shrinkage and selection via the lasso. *Journal of the Royal Statistical Society, Series B* 58:267–288. 272

Tidwell, J. 2011. *Designing Interfaces: Patterns for Effective Interactive Design* (second ed.). Sebastopol, Calif.: O'Reilly.

Toivonen, R., L. Kovanen, J.-P. O. Mikko Kivelä, J. SaramŁki, and K. Kaski 2009. A comparative study of social network models: Network evolution models and nodal attribute models. *Social Networks* 31:240–254.

Torgerson, W. S. 1958. *Theory and Methods of Scaling*. New York: Wiley. 337

Train, K. and Y. Croissant 2014. *Kenneth Train's exercises using the mlogit Package for R*. Comprehensive R Archive Network. 2014. `http://cran.r-project.org/web/packages/mlogit/vignettes/Exercises.pdf`. 174

Train, K. E. 1985. *Qualitative Choice Analysis*. Cambridge, Mass.: MIT Press.

Train, K. E. 2003. *Discrete Choice Methods with Simulation*. Cambridge: Cambridge University Press.

Travers, J. and S. Milgram 1969, December. Experimental study of small world problem. *Sociometryy* 32(4):425–443. 196

Trybula, W. J. 1999. Text mining. In M. E. Williams (ed.), *Annual Review of Information Science and Technology*, Volume 34, Chapter 7, pp. 385–420. Medford, N.J.: Information Today, Inc.

Tsay, R. S. 2013. *An Introduction to Analysis of Financial Data with R*. New York: Wiley.

Tsvetovat, M. and A. Kouznetsov 2011. *Social Network Analysis for Startups: Finding Connections on the Social Web*. Sebastopol, Calif.: O'Reilly.

Tufte, E. R. 1990. *Envisioning Information*. Cheshire, Conn.: Graphics Press.

Tufte, E. R. 1997. *Visual Explanations: Images and Quantities, Evidence and Narrative*. Cheshire, Conn.: Graphics Press.

Tufte, E. R. 2004. *The Visual Display of Quantitative Information* (second ed.). Cheshire, Conn.: Graphics Press.

Tufte, E. R. 2006. *Beautiful Evidence*. Cheshire, Conn.: Graphics Press.

Tukey, J. W. 1977. *Exploratory Data Analysis*. Reading, Mass.: Addison-Wesley.

Tukey, J. W. and F. Mosteller 1977. *Data Analysis and Regression: A Second Course in Statistics*. Reading, Mass.: Addison-Wesley.

Tull, D. S. and D. I. Hawkins 1993. *Marketing Research: Measurement and Method* (sixth ed.). New York: Macmillan.

Turney, P. D. 2002, July 8–10. Thumbs up or thumbs down? Semantic orientation applied to unsupervised classification of reviews. *Proceedings of the 40th Annual Meeting of the Association for Computational Linguistics (ACL '02)*:417–424. Available from the National Research Council Canada publications archive. 284

Uddin, S., J. Hamra, and L. Hossain 2013. Exploring communication networks to understand organizational crisis using exponential random graph models. *Computational and Mathematical Organization Theory* 19:25–41.

Uddin, S., A. Khan, L. Hossain, M. Piraveenan, and S. Carlsson 2014. A topological framework to explore longitudinal social networks. *Computational and Mathematical Organization Theory*:21 pages. Manuscript published online by Springer Science+Business Media, New York.

Underhill, P. 1999. *Why We Buy: The Science of Shopping*. New York: Simon & Schuster.

Underhill, P. 2004. *Call of the Mall*. New York: Simon & Schuster.

Unwin, A., M. Theus, and H. Hofmann (eds.) 2006. *Graphics of Large Datasets: Visualizing a Million*. New York: Springer. 276

Urban, G. L. and J. R. Hauser 1993. *Design and Marketing of New Products* (second ed.). Upper Saddle River, N.J.: Prentice Hall.

Valente, T. W. 1995. *Network Models of the Diffusion of Innovations*. Cresskill, N.J.: Hampton Press.

Vapnik, V. N. 1998. *Statistical Learning Theory*. New York: Wiley. 78

Vapnik, V. N. 2000. *The Nature of Statistical Learning Theory* (second ed.). New York: Springer. 78

Varian, H. R. 2005. *Intermediate Microeconomics: A Modern Approach* (seventh ed.). New York: Norton. 173

Velleman, P. F. and L. Wilkinson 1993, February. Nominal, ordinal, interval, and ratio typologies are misleading. *The American Statistician* 47(1):65–72.

Venables, W. N. and B. D. Ripley 2002. *Modern Applied Statistics with S* (fourth ed.). New York: Springer-Verlag. Champions of S, S-Plus, and R call this *the mustard book*.

Vine, D. 2000. *Internet Business Intelligence: How to Build a Big Company System on a Small Company Budget*. Medford, N.J.: Information Today, Inc. 225

Walther, J. B. 1996. Computer-mediated communication: impersonal, interpersonal, and hyperpersonal interaction. *Communication Research* 23(1):3–41. 351

Wanderschneider, M. 2013. *Learning Node.js: A Hands-On Guide to Building Web Applications in JavaScript*. Upper Saddle River, N.J.: Pearson Education/Addison-Wesley.

Wang, X., H. Tao, Z. Xie, and D. Yi 2013. Mining social networks using wave propogation. *Computational and Mathematical Organization Theory* 19:569–579.

Washburn, R. C. 1931. *The Life and Times of Lydia E. Pinkham*. New York: G. P. Putnam's Sons.

Wasserman, L. 2010. *All of Statistics: A Concise Course in Statistical Inference*. New York: Springer. 264

Wasserman, S. and C. Anderson 1987. Stochastic *a posteriori* blockmodels: Construction and assessment. *Social Networks* 9:1–36.

Wasserman, S. and K. Faust 1994. *Social Network Analysis: Methods and Applications*. Cambridge, UK: Cambridge University Press. Chapter 4: Graphs and Matrices contributed by D. Iacobucci.

Wasserman, S. and D. Iacobucci 1986. Statistical analysis of discrete relational data. *British Journal of Mathematical and Statistical Psychology* 39:41–64.

Wasserman, S. and P. Pattison 1996, September. Logit models and logistic regression for social networks: I An introduction to markov graphs and $p^*$. *Psychometrika* 61(3):401–425.

Wassertheil-Smoller, S. 1990. *Biostatistics and Epidemiology: A Primer for Health Professionals*. New York: Springer.

Watts, D. J. 1999. *Small Worlds: The Dynamics of Networks between Order and Randomness*. Princeton, N.J.: Princeton University Press. 196

Watts, D. J. 2003. *Six Degrees: The Science of a Connected Age*. New York: W.W. Norton.

Watts, D. J. and S. H. Strogatz 1998. Collective dynamics of 'small-world' networks. *Nature* 393(6684):440–442. 194

Webb, E. J., D. T. Campbell, R. D. Schwartz, and L. Sechrest 1966. *Unobtrusive Measures: Nonreactive Research in the Social Sciences*. Chicago: Rand McNally.

Wedel, M. and W. Kamakura 2000. *Market Segmentation: Conceptual and Methodological Foundations* (second ed.). Boston: Kluwer.

Wegman, E. J. 1990. Hyperdimensional data analysis using parallel coordinates. *Journal of the American Statistical Association* 85:664–675. 166

Weibel, P. and T. Druckrey (eds.) 2001. *net_condition: art and global media*. Cambridge: MIT Press. 349

Weiss, S. M., N. Indurkhay, and T. Zhang 2010. *Fundamentals of Predictive Text Mining*. New York: Springer.

Weiss, S. M., N. Indurkhya, and T. Zhang 2010. *Fundamentals of Predictive Text Mining*. New York: Springer.

Wellner, A. S. 2003, March. The new science of focus groups. *American Demographics*:29–33. 346

Werry, C. and M. Mowbray (eds.) 2001. *Online Communities: Commerce, Community Action, and the Virtual University*. Upper Saddle River, N.J.: Prentice-Hall.

West, M. D. (ed.) 2001. *Theory, Method, and Practice in Computer Content Analysis*. Westport, Conn.: Ablex. 283

White, T. 2011. *Hadoop: The Definitive Guide* (second ed.). Sebastopol, Calif.: O'Reilly.

Wickham, H. 2015. *Advanced R*. Boca Raton, Fla.: Chapman & Hall/CRC.

Wickham, H. and W. Chang 2014. *ggplot2: An Implementation of the Grammar of Graphics*. Comprehensive R Archive Network. 2014. `http://cran.r-project.org/web/packages/ggplot2/ggplot2.pdf`. 228, 276

Wikipedia 2014a. 2011 Virginia earthquake—Wikipedia, the free encyclopedia. Retrieved from the World Wide Web, September 18, 2014, at url = `http://en.wikipedia.org/w/index.php?title=2011_Virginia_earthquake&oldid=625563344`.

Wikipedia 2014b. History of Wikipedia—Wikipedia, the free encyclopedia. Retrieved from the World Wide Web, September 23, 2014, at `http://en.wikipedia.org/wiki/History_of_Wikipedia`.

Wilensky, U. 1999. Netlogo. NetLogo 5.1.0 User Manual and documentation available from the Center for Connected Learning and Computer-Based Modeling, Northwestern University, Evanston, IL at `http://ccl.northwestern.edu/netlogo/`.

Wilkinson, L. 2005. *The Grammar of Graphics* (second ed.). New York: Springer.

Winer, B. J., D. R. Brown, and K. M. Michels 1991. *Statistical Principles in Experimental Design* (third ed.). New York: McGraw-Hill.

Winer, R. S. 1979. An analysis of the time-varying effects of advertising: The case of Lydia Pinkham. *Journal of Business* 52:563–576.

Winston, W. L. 2014. *Marketing Analytics: Data-Driven Techniques with Microsoft Excel*. New York: Wiley.

Witten, I. H., E. Frank, and M. A. Hall 2011. *Data Mining: Practical Machine Learning Tools and Techniques*. Burlington, Mass.: Morgan Kaufmann. 153

Witten, I. H., A. Moffat, and T. C. Bell 1999. *Managing Gigabytes: Compressing and Indexing Documents and Images* (second ed.). San Francisco: Morgan Kaufmann.

Wolcott, H. F. 1994. *Transforming Qualitative Data: Description, Analysis, and Interpretation*. Thousand Oaks, Calif.: Sage.

Wolcott, H. F. 1999. *Ethnography: A Way of Seeing*. Walnut Creek, Calif.: AltaMira.

Wolcott, H. F. 2001. *The Art of Fieldwork*. Walnut Creek, Calif.: AltaMira. 349

Wood, D., M. Zaidman, and L. Ruth 2014. *Linked Data: Structured Data on the Web*. Shelter Island, N.Y.: Manning.

Worsley, J. C. and J. D. Drake 2002. *Practical PostgreSQL*. Sebastopol, Calif.: O'Reilly.

Wrigley, N. (ed.) 1988. *Store Choice, Store Location, and Market Analysis*. London, UK: Routledge.

Wunderman, L. 1996. *Being Direct: Making Advertising Pay*. New York: Random House.

Yau, N. 2011. *Visualize This: The FlowingData Guide to Design, Visualization, and Statistics*. New York: Wiley.

Yau, N. 2013. *Data Points: Visualization That Means Something*. New York: Wiley.

Young, G. and A. Householder 1938. Discussion of set of points in terms of their mutual distances. *Psychometrika* 3:19–22. 96

Zabin, J., G. Brebach, and P. Kotler 2004. *Precision Marketing: The New Rules for Attracting, Retaining and Leveraging Profitable Customers*. New York: Wiley.

Zeileis, A., G. Grothendieck, and J. A. Ryan 2014. *Package zoo: S3 Infrastructure for Regular and Irregular Time Series (Z's ordered observations)*. Comprehensive R Archive Network. 2014. `http://cran.r-project.org/web/packages/zoo/zoo.pdf`. 228

Zeileis, A., K. Hornik, and P. Murrell 2009, July. Escaping RGBland: Selecting colors for statistical graphics. *Computational Statistics and Data Analysis* 53(9):3259–3270.

Zeileis, A., K. Hornik, and P. Murrell 2014. *HCL-Based Color Palettes in R*. Comprehensive R Archive Network. 2014. `http://cran.r-project.org/web/packages/colorspace/vignettes/hcl-colors.pdf`.

Zeithaml, V. A., M. J. Bitner, and D. D. Gremler 2005. *Services Marketing* (fourth ed.). New York: McGraw-Hill/Irwin. 77

Zivot, E. and J. Wang 2003. *Modeling Financial Time Series with S-PLUS*. Seattle: Insightful Corporation.

Zubcsek, P. P., I. Chowdhury, and Z. Katona 2014. Information communities: The network structure of communication. *Social Networks* 38:50–62.

Zwerina, K. 1997. *Discrete Choice Experiments in Marketing: Use of Priors in Efficient Choice Designs and Their Application to Individual Preference Measurement*. New York: Physica-Verlag.

# Index

## D



## E

## F



## G



## H



## I



## K



## L



## M

## N



## O



## P



## Q



## R

## S